Microwave Sensing and Imaging

Microwave Sensing and Imaging

Editors

Andrea Randazzo
Cristina Ponti
Alessandro Fedeli

MDPI • Basel • Beijing • Wuhan • Barcelona • Belgrade • Manchester • Tokyo • Cluj • Tianjin

Editors
Andrea Randazzo
University of Genoa
Italy

Cristina Ponti
"Roma Tre" University
Italy

Alessandro Fedeli
University of Genoa
Italy

Editorial Office
MDPI
St. Alban-Anlage 66
4052 Basel, Switzerland

This is a reprint of articles from the Special Issue published online in the open access journal *Sensors* (ISSN 1424-8220) (available at: https://www.mdpi.com/journal/sensors/special_issues/microwaveimaging).

For citation purposes, cite each article independently as indicated on the article page online and as indicated below:

LastName, A.A.; LastName, B.B.; LastName, C.C. Article Title. *Journal Name* **Year**, *Volume Number*, Page Range.

ISBN 978-3-0365-4203-4 (Hbk)
ISBN 978-3-0365-4204-1 (PDF)

Contents

About the Editors

Andrea Randazzo

Andrea Randazzo (Full Professor) received a laurea degree in telecommunication engineering and Ph.D. degree in information and communication technologies from the University of Genoa, Italy, in 2001 and 2006, respectively. He is currently a Full Professor of electromagnetic fields at the Department of Electrical, Electronic, Telecommunication Engineering, and Naval Architecture, University of Genoa. He co-authored the book Microwave Imaging Methods and Applications (Artech House, 2018) and about 300 articles published in journals and conference proceedings. His primary research interests are in the field of microwave imaging, inverse-scattering techniques, radar processing, numerical methods for electromagnetic scattering and propagation, electrical tomography, and smart antennas. He is a Senior Member of the IEEE and member of the Italian Society of Electromagnetism, Inter-University Research Centre for Interactions Between Electromagnetic Fields and Biological Systems (ICEMB), and National Inter-University Consortium for Telecommunications (CNIT).

Cristina Ponti

Cristina Ponti (Assistant Professor) received a B.Sc. and M.Sc. (cum laude) in Electronic Engineering, at "Sapienza" University of Rome, in 2004 and 2006, respectively. In 2010 she received her PhD degree in Applied Electronics at "Roma Tre" University of Rome. At present, she is a tenure-track Assistant Professor in Electromagnetic Fields at the Department of Industrial, Electronic and Mechanical Engineering of "Roma Tre" University. Her main research interests are in electromagnetic theory, numerical methods, scattering by buried objects, ground-penetrating radar, through-the-wall radar, Electromagnetic Band-Gap (EBG) structures, antennas for microwave and millimeter applications, and nuclear fusion. She is a member of the IEEE, IEEE Antennas and Propagation, Microwave Theory and Technique, and Women in Engineering Societies, Italian Society of Electromagnetism, National Inter-University Consortium for Telecommunications (CNIT), and European Geosciences Union.

Alessandro Fedeli

Alessandro Fedeli (Assistant Professor) received B.Sc. and M.Sc. degrees in Electronic Engineering from the University of Genoa, Genoa, Italy, in 2011 and 2013, respectively, where he received his Ph.D. degree in Science and Technology for Electronic and Telecommunications Engineering in 2017. He is currently an Assistant Professor of electromagnetic fields at the Department of Electrical, Electronics, Telecommunication Engineering and Naval Architecture (DITEN) of the University of Genoa, and he teaches courses on guided propagation, electromagnetic imaging and remote sensing. His main research interests include computational methods to solve forward and inverse electromagnetic scattering problems, microwave imaging, and processing techniques applied to Ground Penetrating Radar (GPR) systems. He is a member of the IEEE, of the IEEE Antennas and Propagation Society, the Italian Society of Electromagnetism, the Inter-University Research Centre for Interactions Between Electromagnetic Fields and Biological Systems (ICEMB), the European Geosciences Union, and the Italian Georadar Association.

Preface to "Microwave Sensing and Imaging"

Microwave sensing and imaging is acquiring an ever-growing importance in several applicative fields, such as non-destructive evaluations in industry and civil engineering, subsurface prospection, security, and biomedical imaging. Microwave techniques can be, in principle, used to retrieve information on some physical parameters of the inspected targets (dielectric properties, shape, etc.) by using safe electromagnetic radiations and cost-effective systems.

Although great technological advances have been attained in recent years, there is still a great deal of scientific research activity in this field, with the aim of further improving imaging systems and techniques. More efficient and reliable measurement systems are continuously being designed and validated on a case-by-case basis to address specific scenarios. Moreover, great attention has been paid to the development of effective data-processing algorithms, which are able to solve the underlying electromagnetic inverse scattering problem (which is generally nonlinear and ill-posed) in order to retrieve the required information on the inspected targets from the measured scattered-field samples. Finally, efficient forward solvers, which are fundamental for modeling the electromagnetic interactions between the interrogating fields and the targets, are proposed for both the development of the inversion approaches and for the definition/validation of the imaging setups.

This book, which reprints a Special Issue of the journal Sensors, provides some recent insights into microwave sensing and imaging systems and techniques, with reference to the topics outlined above.

Andrea Randazzo, Cristina Ponti, and Alessandro Fedeli
Editors

Article

A Multi-Frequency Tomographic Inverse Scattering Using Beam Basis Functions

Ram Tuvi

John A. and Katherine G. Jackson School of Geosciences, Institute for Geophysics, The University of Texas at Austin, Austin, TX 78758, USA; ram@ig.utexas.edu

Abstract: We present an overview of a beam-based approach to ultra-wide band (UWB) tomographic inverse scattering, where beam-waves are used for local data-processing and local imaging, as an alternative to the conventional plane-wave and Green's function approaches. Specifically, the method utilizes a phase–space set of iso-diffracting beam-waves that emerge from a discrete set of points and directions in the source domain. It is shown that with a proper choice of parameters, this set constitutes *a frame* (an overcomplete generalization of a basis), termed "beam frame", over the entire propagation domain. An important feature of these beam frames is that they need to be calculated once and then used for all frequencies, hence the method can be implemented either in the multi-frequency domain (FD), or directly in the time domain (TD). The algorithm consists of two phases: in the processing phase, the scattering data is transformed to the beam domain using windowed phase–space transformations, while in the imaging phase, the beams are backpropagated to the target domain to form the image. The beam-domain data is not only localized and compressed, but it is also physically related to the local Radon transform (RT) of the scatterer via a local Snell's reflection of the beam-waves. This expresses the imaging as an inverse local RT that can be applied to any local domain of interest (DoI). In previous publications, the emphasis has been set on TD data processing using a special class of localized space–time beam-waves (wave-packets). The goal of the present paper is to present the imaging scheme in the UWB FD, utilizing simpler Fourier-based data-processing tools in the space and time domains.

Keywords: inverse scattering; imaging; wave propagation; beam summation methods

Citation: Tuvi, R. A Multi-Frequency Tomographic Inverse Scattering Using Beam Basis Functions. *Sensors* **2022**, *22*, 1684. https://doi.org/10.3390/s22041684

Academic Editors: Andrea Randazzo, Cristina Ponti and Alessandro Fedeli

Received: 2 December 2021
Accepted: 3 February 2022
Published: 21 February 2022

Publisher's Note: MDPI stays neutral with regard to jurisdictional claims in published maps and institutional affiliations.

1. Introduction

Inverse scattering deals with determining the shape and the composition of an unknown object from measurements of the scattering field data due to a known illumination. This area has a wide range of medical, geophysical, oceanographical, industrial, etc., applications, using electromagnetic, acoustic, elastic, or seismic waves [1–5]. Inverse scattering problems are, in general, non-linear and highly ill-posed, hence accurate solutions typically require iterative schemes and are limited to rather small configurations in the order of wavelengths. For large domains, practical algorithms rely on linearized weak scattering formulations using the Born, Rytov, Physical optics, or other single scattering approximations [2,5] which linearize the relation between the target and the field and provide the basis for diffraction tomography (DT) reconstruction [6].

Inverse scattering requires diversity and relies on the wave data in hand. Depending on the application, it may involve multiple excitation frequencies (or short-pulse response) and/or several illumination directions. With the overall complexity of the problem, full utilization of the wave data is essential to formulate an efficient, robust, and accurate algorithm.

Beam summation (BS) methods are when the wave field is expressed as a superposition of collimated beam waves. Here, we use the generic term "beam" for both the FD formulations where the propagators are Gaussian beams, and for the TD formulations where the propagators are pulsed beams. This provides the proper physical basis for such robust DT reconstruction. Like the plane-wave (PW) spectrum approach, the BS provides a complete spectral basis which is required for DT reconstruction, yet unlike plane-waves, the beam waves provide spatial resolution and they can easily and efficiently be propagated in inhomogeneous media along ray trajectories. Unlike rays, on the other hand, they provide a uniform spectral basis and are insensitive to geometrical optics transition regions. Thus, beams provide a way to convert the wave problem to a ray-based skeleton.

BS methods can be classified into two classes (see review in [7,8]). The first class addresses radiation from localized (point) sources by expressing the field as an angular superposition of collimated beams that emerge radially from the source. This formulation has been derived asymptotically [9,10], but later on it was formulated as an exact spectral identity using complex source beams [11,12] and was also extended to the time domain (TD) [12]. Consequently, it has been used in various applications of propagation, scattering, and inverse scattering.

The other class addresses radiation from extended (aperture) sources, where the field is expressed as a sum of beam propagators that emerge from a discrete phase–space lattice of points and directions in the aperture. These formulations utilize local window (e.g., Gaussian) functions to transform the data to the beam domain, and then propagate the data using beams. These formulations are utilized in the present work where we analyze the data on the measurement aperture and then backpropagate it to the scatterer domain. Early implementation of this approach was based on a Gabor series expansion of the planar field [13–16] and therefore suffered from two major drawbacks: (a) The Gabor expansion coefficients (the beam amplitudes) are notoriously non-local and unstable, and (b) the beam lattice is frequency-dependent, hence a new lattice should be calculated at each frequency. Both difficulties have been mitigated in the ultra-wide band phase–space beam summation (UWB-PS-BS) method [17,18] which utilizes the overcomplete windowed Fourier transform (WFT) frames. In linear algebra, a frame of an inner product space is a generalization of a basis of a vector space to overcomplete sets. In signal processing, frames provide a redundant, stable way of representing a signal, instead of the Gabor series. The formulation is structured upon a *frequency independent* lattice of beams that emerge from a discrete phase–space lattice of points and directions in the aperture, and utilizes iso-diffracting Gaussian beams (ID-GB) whose propagation parameters are *frequency independent*. These properties make this representation efficient for wideband applications and also allow an extension of the theory to the time-domain (TD) [19].

A major step forward has been the proof in [20] that these phase–space sets of beams constitute a frame not only in the aperture domains but actually everywhere in the propagation domain. This implies that these beam basis functions may be used to expand not only the sources and the field, but also any function of space and in particular the medium inhomogeneities, a property that is being used in our beam-based inverse scattering theory. The theory has been proved originally in the frequency domain (FD) [20] and then extended to the TD in [21].

As noted above, it has been recognized long ago that BS provides the proper physical basis for a robust DT reconstruction. Examples for point-sources configurations can be found in [22–30]. For configuration where the data is measured over a wide aperture (see Figure 1a), it is more suitable to use the extended sources approach noted above: see [31–37] and [38–44] for medium reconstruction over a homogenous or inhomogeneous background, respectively. Without going into detailed comparison, the main difference between the methods is in the data representation phase (see [42] for a detailed comparison).

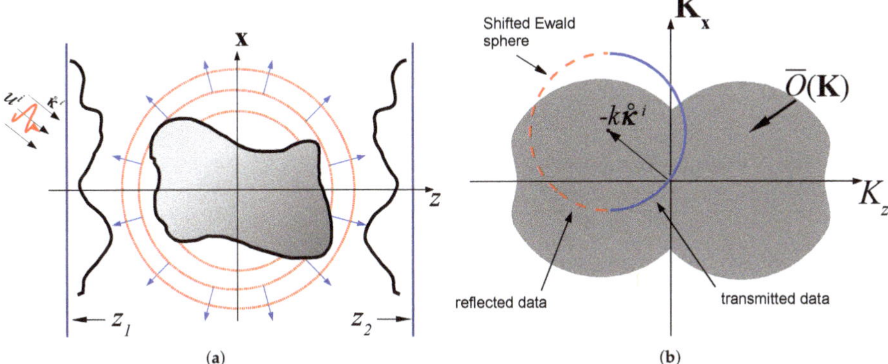

Figure 1. Diffraction tomography and the K-space diffraction tomography identity in the spatial and spectral domains. (**a**) The physical configuration. The unknown object $O(\mathbf{r})$ is located between two measurement planes. At $z = z_1$ we have an array of sources/receivers, while at $z = z_2$ we have an array of receivers. The object is illuminated by the plane-wave (black arrows) of Equation (3) propagating in the direction $\overset{\circ}{\kappa}{}^i$. In red we plot pulsed plane-wave illumination of Equation (3). The scattered field is measured on the z_j planes. (**b**) The DT identity of (8). The plane-wave spectrum of the scattered field $\hat{u}_j^s(\boldsymbol{\xi})$ is mapped to values of $\bar{O}(\mathbf{K})$ over a shifted Ewald sphere $\mathbf{K} = k(\overset{\circ}{\kappa}_j - \overset{\circ}{\kappa}{}^i)$. The data measured on the $j = 1, 2$ planes are illustrated by red dashed and blue solid-line hemispheres, respectively.

In this work we review the beam-based local inverse scattering theory derived in [36,37], which is based on the frame-based UWB-PS-BS theory discussed above. As noted there, the theory is structured on a frequency-independent phase–space sets of beams that constitute frames everywhere in the propagation domain. This beam frame formulation enables the expansion of both the medium inhomogeneities and the scattering data with the same set of beam-basis functions, thus enabling a direct inversion over the beam domain. In previous publications, the emphasis has been set on TD data processing using special localized space–time beam-waves (wave-packets). This requires somewhat sophisticated mathematics and processing tools. In the present paper, on the other hand, we utilize a simpler FD Fourier-based data-processing approach followed by an integration over the relevant frequency band. The paper makes extensive references to equations or figures in [36,37]. Therefore, to simplify the presentation, we refer to them by the prefixes I and II, respectively.

The advantages of our beam-based DT over the conventional DT approach are:

a. Data localization: The phase–space processing of the scattered data extract the local direction of arrival. The phase space representation of the data is therefore localized along well defined trajectories corresponding to the local direction and time of arrival from the relevant regions in the target domain.

b. Under the Born approximation, the beam-wave scattering mechanism by the medium inhomogeneities is described by local Snell's reflections from the local stratification, which is related to the local Radon transform (LRT) of the medium inhomogeneities (Section 5 in [36]).

c. As follows from the discussion in items *a* and *b*, the phase space data is directly related to the LRT of the medium inhomogeneities about a given region.

d. The beam-based imaging enables local imaging within a given domain of interest (DoI) by considering only the data that correspond to beams that pass through or near the DoI. This not only reduces the problem complexity, but also reduces the noise level, since data and noise arriving from other regions are a priori filtered out.

e.   The beam-based imaging enables backpropagation and imaging over a non-homogeneous background.

The presentation below starts in Section 2 with a review of the main concepts in DT. We proceed in Section 3 by reviewing elements of the beam representation, and in particular of the UWB-PS-BS and the BF concept. The beam-based DT is presented in Section 4, where, as discussed above, we emphasize the multi-frequency data processing as opposed to the more complicated TD processing used in [36,37]. The presentation ends in Section 5 with a practical description of the algorithm, including the choice of the various parameters and numerical examples.

2. UWB Diffraction Tomography in the Spectral Plane-Wave Domain: A Review

This section reviews the conventional plane-wave DT algorithms. Referring to the configuration in Figure 1a, we consider the two alternative schemes: The *angular diversity* scheme where the data is measured for several illumination directions $\mathring{\kappa}^i$ at a given frequency (Section 2.3), and the *frequency diversity* scheme where the data is measured over a wide frequency band for a single illumination direction $\mathring{\kappa}^i$ (Section 2.4). The frequency diversity scheme may also be performed using a short-pulse illumination and calculated directly in the TD [45] (see also I– Section 3-B in [36]).

2.1. Problem Description—Physical Configuration

The physical configuration is illustrated in Figure 1a, where the object is located between two measurement planes, at $z = z_1 < 0$ and at $z = z_2 > 0$. We assume a 3D coordinate frame $\mathbf{r} = (\mathbf{x}, z)$ where the z-coordinate is normal to the measurement planes, and $\mathbf{x} = (x_1; x_2)$ are the transversal coordinates. The data is collected over a wide frequency band $\Omega \in [\omega_{\min}, \omega_{\max}]$. The theory is presented here in the FD, but we also discuss the TD formulation for completeness and clearer interpretation. Field constituents in these domains are related via the temporal Fourier transform

$$\hat{u}(\omega) = \int dt\, u(t)\, e^{i\omega t}, \tag{1}$$

where FD constituents are tagged by an over-hat $\hat{\,}$.

The unknown object is embedded in a uniform background wavespeed v_0 and assumed to be lossless and non-dispersive. It is described by the unknown wavespeed $v(\mathbf{r})$, and we define the "object function"

$$O(\mathbf{r}) = (v_0/v(\mathbf{r}))^2 - 1 \tag{2}$$

(see Equation (7)).

The scattering data may be collected as a function of frequency using time-harmonic plane-wave excitation, or directly in the TD utilizing short-pulse plane-wave. These excitations are given by

$$\hat{u}^i(\mathbf{r}, \omega) = \hat{F}(\omega) e^{ik\mathring{\kappa}^i \cdot \mathbf{r}}, \quad u^i(\mathbf{r}, t) = F(t - v_0^{-1}\mathring{\kappa}^i \cdot \mathbf{r}), \tag{3}$$

where $\hat{F}(\omega)$ is the source spectrum and $k = \omega/v_0$ is the wavenumber. The incident wave propagates in the direction

$$\mathring{\kappa}^i = (\mathring{\xi}^i, \zeta^i) = \sin\theta^i \cos\phi^i \mathring{x}_1 + \sin\theta^i \sin\phi^i \mathring{x}_2 + \cos\theta^i \mathring{z}, \tag{4}$$

with (θ^i, ϕ^i) being the polar angles with respect to the z axis, and over-circles denote unit vectors.

The scattered fields measured over the z_j planes, $j = 1, 2$ are denoted as $\hat{u}_j^s(\mathbf{x}, \omega)$ (see Figure 1a). The PW spectral representation of $\hat{u}_j^s(\mathbf{r})$ is defined via

$$\hat{u}_j^s(\boldsymbol{\xi}) = e^{\pm ik\zeta z_j} \int d^2x \, \hat{u}_j^s(\mathbf{x}) e^{-ik\boldsymbol{\xi}\cdot\mathbf{x}}, \quad \zeta = \sqrt{1 - \boldsymbol{\xi}\cdot\boldsymbol{\xi}}, \quad \mathrm{Im}\,\zeta \geq 0. \tag{5}$$

where lower and upper signs correspond to $j = 1$ and 2, respectfully. We added the $e^{\pm ik\zeta z_j}$ phase term in (5) in order to normalize the spectral PWs to the $z = 0$ plane instead of $z = z_j$ planes.

Note that we use here the frequency normalized spectral coordinates $\boldsymbol{\xi} = \mathbf{k}_x/k$ which are related to the PW direction via $\boldsymbol{\xi} = (\xi_1, \xi_2) = \sin\theta(\cos\phi, \sin\phi)$, where (θ, ϕ) are the conventional spherical angles with respect to the z-axis so that the scattered PWs propagate in the unit vector direction

$$\mathring{\boldsymbol{\kappa}}_j = (\boldsymbol{\xi}, \mp\zeta) = (\sin\theta\cos\phi, \sin\theta\sin\phi, \cos\theta), \quad j = 1, 2. \tag{6}$$

The spectral ranges $|\boldsymbol{\xi}| < 1$ and $|\boldsymbol{\xi}| > 1$ define the *propagation spectrum* and *evanescent spectrum*, respectively. Typically, DT formulations are restricted only to the propagation spectrum data (see discussion after Equation (9)).

2.2. The DT Identity

According to the weak scattering (first Born) approximation of the Lippmann–Schwinger integral equation, the scattered field can be expressed as [5]

$$\hat{u}^s(\mathbf{r}) = k^2 \int_V d^3r' \hat{u}^i(\mathbf{r}') O(\mathbf{r}') \hat{G}(\mathbf{r}, \mathbf{r}'), \tag{7}$$

where $\hat{G} = \frac{e^{ik|\mathbf{r}-\mathbf{r}'|}}{4\pi|\mathbf{r}-\mathbf{r}'|}$ is the 3D Green's function in the uniform background. This approximation is valid if $O(\mathbf{r}) \ll 1$ and in addition $kLO_{\max} < 1$, where L is the spatial support of O and $O_{\max}$ is its maximal value.

Inserting (7) into (5) and using the spectral representation of $\hat{G}$, we obtain (I-7),

$$\hat{u}_j^s(\boldsymbol{\xi}) \simeq \frac{k}{-2i\zeta} \bar{O}(\mathbf{K})\Big|_{\mathbf{K}=k(\mathring{\boldsymbol{\kappa}}_j - \mathring{\boldsymbol{\kappa}}^i)} \qquad |\boldsymbol{\xi}| < 1, \tag{8}$$

where $\mathring{\boldsymbol{\kappa}}_j$ and $\mathring{\boldsymbol{\kappa}}^i$ are given by (6) and (4), and

$$\bar{O}(\mathbf{K}) = \int d^3r \, O(\mathbf{r}) e^{-i\mathbf{K}\cdot\mathbf{r}}, \qquad \mathbf{K} = (K_1, K_2, K_z), \tag{9}$$

is the K-space distribution of $O(\mathbf{r})$. Equation (8) is referred to as the *DT identity*. It relates the scattering data in the $\mathring{\boldsymbol{\kappa}}_j$ directions to values of $\bar{O}(\mathbf{K})$ at the points $\mathbf{K} = k(\mathring{\boldsymbol{\kappa}}_j - \mathring{\boldsymbol{\kappa}}^i)$. As illustrated in Figure 1b, these points define a K-space sphere with radius k that is centered at $\mathbf{K} = -k\mathring{\boldsymbol{\kappa}}^i$, which is referred to as the *shifted Ewald sphere*. Note from (6) that the left and right hemispheres (plotted as red and blue, respectively) correspond to data from the z_j measurement plane with $j = 1, 2$, respectively.

The DT identity above applies only to the propagation spectrum $|\boldsymbol{\xi}| < 1$. Adding the evanescent spectrum may improve the resolution. However, the contribution of the evanescent spectrum is exponentially weak and hence has a low signal to noise ratio. In addition, backpropagating this data to form the image amplifies the noise level exponentially. For these reasons, the evanescent spectrum contribution is usually neglected except for near field imaging schemes.

In view of the DT identity, one may obtain a full $K-$space coverage of the object function by measuring the scattering response for several illumination directions or several frequencies [2,5]. These alternative schemes are reviewed in the following sections.

2.3. Object Reconstruction via Angular Diversity

The angular diversity approach is illustrated in Figure 2a. Changing the illumination directions $\overset{\circ}{\kappa}{}^i$ while keeping the operational frequency k constant changes the centers of the shifted Ewald sphere and provides a different coverage of the K space. Aggregating the response for several illumination directions recovers $\bar{O}(\mathbf{K})$. Note that for lossless (real) objects, $\bar{O}(\mathbf{K}) = \bar{O}^*(-\mathbf{K})$, so that only half of the K-space needs to be recovered.

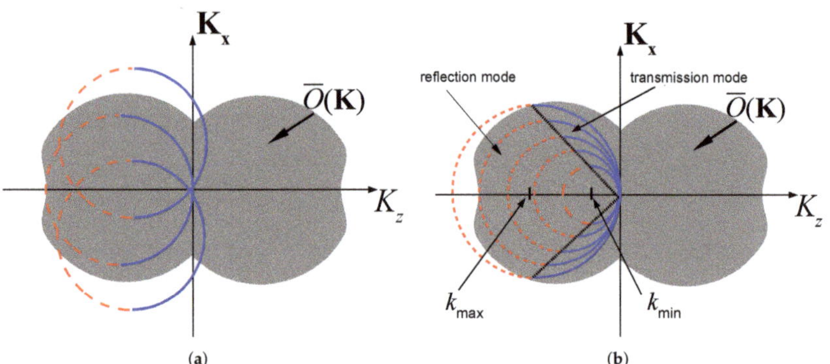

Figure 2. K-space reconstruction. (**a**) Angular diversity reconstruction: Changing the direction of illumination $\overset{\circ}{\kappa}{}^i$ and measuring the transmitted data only provides a coverage of the $K-$space within a sphere of radius $\sqrt{2}k$ [5]. (**b**) Frequency diversity reconstruction: Changing the excitation frequency for a single illumination direction $\overset{\circ}{\kappa}{}^i$ provides coverage of the $K-$ space as indicated. The figure is plotted for $\overset{\circ}{\kappa}{}^i = \overset{\circ}{z}$.

As one observes from Figure 2a, the transmitted data on z_2 recovers the K-space distribution of the object in a sphere of radius $\sqrt{2}k$ about the origin. Thus, the object may be recovered using only transmitted data, as long as k is chosen to be large enough to provide full coverage of $\bar{O}(\mathbf{K})$. Note that in the limit of $k \to \infty$, the transmitted data hemispheres in Figure 2a reduce to planar surfaces normal to $\overset{\circ}{\kappa}{}^i$ that pass through the origin, thus providing the K-space representation of conventional X-ray tomography [46].

One option to reconstruct $O(\mathbf{r})$ is to recover $\bar{O}(\mathbf{K})$ and then apply the inverse Fourier transform of (9). This approach requires interpolation of the data from the shifted Ewald spheres to a Cartesian K-domain grid [47], and therefore requires a large number of illuminations.

The "filtered backpropagation" reconstruction algorithm [6,47] overcomes this difficulty by circumventing the need to recover $\bar{O}(\mathbf{K})$ and operating, instead, directly on the scattering data. In this approach, the scattering data is multiplied by a spectral filter and then back-propagated to the object domain. This reconstruction approach is analogous to the X-ray tomography filtered backprojection algorithm of [48], where the filtered data is back-projected along straight lines.

2.4. Object Reconstruction via Frequency Diversity (UWB Tomography)

In the frequency diversity approach, the data is collected over a wide frequency spectrum $\Omega \in [\omega_{\min}, \omega_{\max}]$ for a fixed illumination direction. This can be done either in a frequency by frequency approach or by using a short-pulse illumination. As noted earlier, in this paper we emphasize the multi-frequency approach. The readers are referred to I–Section 3-B in [36] for the TD formulation, which has an important cogent physical interpretation, but is not utilized here.

As illustrated in Figure 2b for illumination along the positive z axis, changing the illumination frequency changes the radius of the shifted Ewald sphere. One observes that the reflected data recovers the K-space distribution of O inside a $\pi/4$ cone with an axis along the negative K_z axis and base radii between $k_{\min}$ and $k_{\max}$, while the transmitted data recovers the complementary K-space part. As noted before, only half of the K-space is needed to recover the real function O. Otherwise, several illumination directions are required. More illuminations also add robustness.

As follows from Figure 2b, the reflection data on the z_1 plane mainly recovers the object variations along the z axis, while the transmitted data on the z_2 plane recovers the transversal variation (see also Snell's law interpretation in I-Section 3-B in [36]. Thus, for quasi-stratified media with weak transversal variations, it may be sufficient to measure only the reflected data on the z_1 plane, but may not be sufficient for objects with a large transversal variations. Another limitation is the missing data for $|K_z| < 2k_{\min}$ (see Figure 2b), while the missing data for $|K_z| > 2k_{\max}$ can be measured by using higher frequencies. As follows from Figure 2b, several illumination directions may increase the transversal resolution and also add data at small $|\mathbf{K}|$. Note also that the maximal axial resolution for the case of normal incidence is $\delta_z = \pi/k_{\max}$.

The object can be reconstructed using an inverse transform of $\bar{O}(\mathbf{K})$. However, for the same reasons discussed in Section 2.3, a filtered backpropagation approach is preferable. Backpropagation can be calculated in several alternative ways. For simplicity, we present here the spectral integration approach. Given the scattering data $\hat{u}^s(\mathbf{x}, \omega)$ over the z_j planes, the backpropagated fields into the $z > z_1$ and $z < z_2$ regions are given by (see (5))

$$\hat{u}_j^b(\mathbf{r}) = \left(\frac{k}{2\pi}\right)^2 \int_{|\boldsymbol{\xi}|<1} d^2\xi \, \hat{u}_j^s(\boldsymbol{\xi}) e^{ik(\boldsymbol{\xi}\cdot\mathbf{x}\mp\zeta z)}, \tag{10}$$

where we restrict the integration to the visible spectrum $|\boldsymbol{\xi}| < 1$.

The "imaging field," or the "filtered backpropagated field" corresponding to the data on the $j = 1, 2$ plane is given by (II-2)

$$\hat{I}_j(\mathbf{r}, \omega) = v_0^{-1} k^{-2} \overset{\circ}{\boldsymbol{\kappa}}{}^i \cdot \nabla[e^{-ik\overset{\circ}{\boldsymbol{\kappa}}{}^i\cdot\mathbf{r}} \hat{u}_j^b(\mathbf{r}, \omega)]. \tag{11}$$

The corresponding "partial images" are obtained by summing over the relevant frequency band (II-3)

$$\breve{O}(\mathbf{r}) = 2\mathrm{Re}\,\frac{1}{\pi}\int_0^\infty d\omega\, \hat{I}_j(\mathbf{r}, \omega). \tag{12}$$

If the data is given on both planes, then the "complete image" is given by

$$\breve{O}(\mathbf{r}) = \breve{O}_1(\mathbf{r}) + \breve{O}_2(\mathbf{r}). \tag{13}$$

The features of $O(\mathbf{r})$ that are described by $\breve{O}_j$ have been discussed above in connection with Figure 2b. As noted there, in many situations it is sufficient to recover only $\breve{O}_1$.

The derivation of the filtered backpropagation imaging algorithm in (10)–(12) is done by inserting the Born approximated data of (8) into (10).

3. Mathematical Background on the UWB-PS-BS

3.1. The Windowed Fourier Transform (WFT) Frame Representation of the Field

As noted in the introduction above, the phase–space beam summation representation is based on the theory of WFT frame expansion of the field. Following [17], the theory is presented here in the context of radiation into the half-space $z > 0$ in a 3D coordinate space $\mathbf{r} = (\mathbf{x}, z)$, $\mathbf{x} = (x_1, x_2)$, due to a time harmonic field $\hat{u}_0(\mathbf{x}, \omega)$ defined over the plane $z = 0$. The WFT frame set $\{\hat{\psi}_\mu(\mathbf{x}, \omega)\}$ is defined by (Equation (22)] [17])

$$\hat{\psi}_\mu(\mathbf{x}, \omega) = \hat{\psi}(\mathbf{x} - \mathbf{m}\bar{x}) e^{ikn\bar{\boldsymbol{\xi}}\cdot(\mathbf{x}-\mathbf{m}\bar{x})}, \tag{14}$$

with $\hat{\psi}$ being a localized window function (typically a Gaussian, see more details below) and $\mu = (\mathbf{m}, \mathbf{n})$ being a 4-index. The frame elements are localized about the spatial ($\mathbf{x}$) and spectral ($\boldsymbol{\xi}$) phase space lattice

$$(\mathbf{x_m}, \boldsymbol{\xi_n}) = (\mathbf{m}\bar{x}, \mathbf{n}\bar{\xi}) = (m_1\bar{x}, m_2\bar{x}; n_1\bar{\xi}, n_2\bar{\xi}), \tag{15}$$

where $(\bar{x}, \bar{\xi})$ defines the lattice unit cell. As will be shown, the points $(\mathbf{x_m}, \boldsymbol{\xi_n})$ define the beams' initiation points and propagation directions (see Equation (21) below). To constitute a frame, the set above needs to fully cover the phase space, i.e., the unit cell area should be less than 2π, implying that

$$k\bar{x}\bar{\xi} = 2\pi\nu, \tag{16}$$

with $\nu < 1$ being the overcompleteness parameter and the limit $\nu = 1$ define the critically complete limit. As will be shown in Section 3.2 below, the frame over completeness provides a local and stable representation of the field (Equation (13) [17]), with it also being used to derive an UWB representation of the field (see Equations (Equations (33)–(35) [17]))).

The WFT frame can be used to expand $\hat{u}_0(\mathbf{x})$ in the form

$$\hat{u}_0(\mathbf{x}) = \sum_\mu \hat{a}_\mu \, \hat{\psi}_\mu(\mathbf{x}). \tag{17}$$

In view of the overcompleteness, the coefficients set $\{\hat{a}_\mu\}$ is not unique. A particularly convenient set with a minimum ℓ_2 norm is obtained by using the dual frame $\{\hat{\varphi}_\mu(\mathbf{x})\}$ which has the same structure as $\{\hat{\psi}_\mu\}$ in (15) except that the mother window $\hat{\psi}(\mathbf{x})$ is replaced by the dual mother window $\hat{\varphi}(\mathbf{x})$. The resulting canonical coefficient set is given by (Equation (23) [17]).

$$\hat{a}_\mu = \langle \hat{u}_0(\mathbf{x}), \hat{\varphi}_\mu(\mathbf{x}) \rangle \tag{18}$$

where $\langle f, g \rangle = \int fg^*$ is the conventional $\mathbb{L}_2$ inner product in the transverse coordinate $\mathbf{x}$. The canonical coefficients $\hat{a}_\mu$ in (18) are readily identified as the local spectrum of $\hat{u}_0(\mathbf{x})$ windowed with respect to $\hat{\varphi}_\mu$ about the phase–space points $(\mathbf{x_m}, \boldsymbol{\xi_n})$.

Generally, $\hat{\varphi}$ should be calculated numerically, for a given $\hat{\psi}$ and lattice $(\bar{x}, \bar{\xi})$. However, if the lattice is sufficiently overcomplete, ($\nu \lesssim 1/3$) $\hat{\varphi} \propto \hat{\psi}$ can be approximated by (Equation (11) [17])

$$\hat{\varphi}(\mathbf{x}) \approx \nu^2 \hat{\psi}(\mathbf{x}) / \|\psi\|^2. \tag{19}$$

There are mainly two reasons to prefer the use of this highly overcomplete parameter regime, even though it implies a larger number of terms in the phase–space expansion (17): (*i*)—as follows from (19), in this parameter regime $\hat{\varphi}$ is localized both spatially and spectrally, so that the expansion (17) comprises local and stable coefficients. (*ii*)—$\hat{\varphi}$ is given analytically via (19) and does not have to be to calculated numerically. Reason (*ii*) is critical for UWB problems where $\hat{\varphi}$ needs to be found for each ω.

The radiated field in $z > 0$ is obtained now by replacing $\hat{\psi}_\mu(\mathbf{x})$ in Equation (17) by beam propagators (Equation (24) [17])

$$\hat{u}(\mathbf{r}) = \sum_\mu \hat{a}_\mu \, \hat{\Psi}_\mu^+(\mathbf{r}), \tag{20}$$

$$\hat{\Psi}_\mu^+(\mathbf{r}) = \left(\frac{k}{2\pi}\right)^2 \int d^2\xi \, \hat{\tilde{\psi}}_\mu(\boldsymbol{\xi}) e^{ik(\boldsymbol{\xi}\cdot\mathbf{x}+\zeta z)}. \tag{21}$$

$\hat{\Psi}_\mu^+(\mathbf{r})$ are identified as the fields that are radiated forward into $z > 0$ by $\hat{\psi}_\mu(\mathbf{x})$. In (21), $\hat{\tilde{\psi}}_\mu(\boldsymbol{\xi}) = \hat{\tilde{\psi}}(\boldsymbol{\xi} - \boldsymbol{\xi_n}) e^{-ik\boldsymbol{\xi}\cdot\mathbf{x_m}}$ is the PW spectrum (5) of $\hat{\psi}_\mu(\mathbf{x})$, with $\hat{\tilde{\psi}}(\boldsymbol{\xi})$ being the spectrum of the "mother window" $\hat{\psi}(\mathbf{x})$. If $\hat{\psi}(\mathbf{x})$ is wide on a wavelength scale, then $\hat{\Psi}_\mu^+(\mathbf{r})$ behave like collimated beams, emerging from the points $\mathbf{x_m}$ over the $z = 0$ plane and directions $\mathring{\kappa}_\mathbf{n} = (\boldsymbol{\xi_n}, \zeta_\mathbf{n}) = (\sin\theta_\mathbf{n}\cos\phi_\mathbf{n}, \sin\theta_\mathbf{n}\sin\phi_\mathbf{n}, \cos\theta_\mathbf{n})$ with $\zeta_\mathbf{n} = \sqrt{1 - |\boldsymbol{\xi_n}|^2}$.

3.2. UWB Considerations

In general, the applications in hand require UWB excitations. Following [17], we use the following frequency-scaling of the WFT frame set that renders the theory amenable for UWB field representations:

(1) Frequency independent beam skeleton: As implied from Equation (16) above, the beam lattice should be recalculated for each frequency. For efficient UWB representations, it is required to have the same beam lattice $(\mathbf{x_m}, \boldsymbol{\zeta_n})$ over the entire frequency band. In view of (16), this requirement implies (Equation (10) [21])

$$\nu(\omega) = \nu_{max} \frac{\omega}{\omega_{max}}, \qquad \omega \in [\omega_{min}, \omega_{max}], \tag{22}$$

with ν_{max} being the value of ν at ω_{max}, so that $\nu < \nu_{max}$ for all $\omega < \omega_{max}$. Typically, we use $\nu_{max} = 1/3$ (see discussion in (25) below).

(2) Iso-diffracting propagators: We use iso-diffracting (ID) Gaussian windows which are scaled with frequency in the form (Equation (27) [17])

$$\hat{\psi}_{ID}(x) = e^{-k|x|^2/2b}, \quad k > 0, \tag{23}$$

where $b > 0$ is a frequency independent parameter. Inserting (23) into (21) and evaluating the integral one finds that the resulting propagators are ID-GBs, with b being the collimation distance. The ID designation of these Gaussian beams is due to the fact that their collimation distance b is frequency independent. This property implies that the beam propagation parameters are frequency independent even in inhomogeneous medium. Furthermore, when transformed into the TD, they give rise to ID-Pulsed beams (ID-PB) which are space time wave-packets that maintain their wave-packet structure even through propagation in inhomogeneous medium [49]. Explicit expressions for the corresponding phase–space beam propagators of (26) in free space are given in (Equations (28)–(29)) [17].

Typically b is chosen by the molder and depends on the application (see discussion in the numerical example below), but also should satisfy the condition $k_{min} b \gg 1$, which implies that the beams are highly collimated over the entire frequency band.

(3) Snug frame: The frame is optimal (or snug) when the frame elements are matched to the phase–space lattice $(\bar{x}, \bar{\xi})$ (in the sense that they should provide a balanced phase–space coverage). This requirement implies the relation $b = \bar{x}/\bar{\xi}$ [17]. Combining this condition with (16) one obtains (Equation (A2) [21]),

$$(\bar{x}, \bar{\xi}) = \sqrt{\frac{2\pi \nu_{max}}{k_{max}}} \left(b^{1/2}, b^{-1/2}\right). \tag{24}$$

(4) Simple expression for the dual frame function: In view of (19) we have for $\nu_{max} = 1/3$ (Equation (A3) [21]),

$$\hat{\varphi}_{ID}(\mathbf{x}) \simeq \frac{\nu_{max}^2}{\pi b k_{max}^2} k^3 \, \hat{\psi}_{ID}(\mathbf{x}), \quad \omega < \omega_{max}. \tag{25}$$

Over this regime $\hat{\varphi}$ is spatially and spectrally localized, and leads to a stable and localized expansion coefficients [17].

The properties above yield an efficient multi-frequency representation where the phase–space lattice and propagation parameters should be calculated only once for all frequencies in the band. These advantages also allow a simple transformation of the beam representation to the TD [19,21].

3.3. The Beam Frame Theorem

Following (21), we define the set of forward and backward propagators $\{\hat{\Psi}^{\pm}_{\mu}(\mathbf{r})\}$ (compare Equation (21))

$$\hat{\Psi}^{\pm}_{\mu}(\mathbf{r}) = \left(\frac{k}{2\pi}\right)^2 \int_{|\boldsymbol{\xi}|<1} d^2\boldsymbol{\xi}\,\hat{\psi}_{\mu}(\boldsymbol{\xi})e^{ik(\boldsymbol{\xi}\cdot\mathbf{x}\pm\zeta z)}, \quad |\boldsymbol{\xi}_{\mathbf{n}}| \leq \zeta_0 < 1. \tag{26}$$

where the parameter ζ_0 is typically chosen close to 1. Note that this subset includes only "propagating beams" whose spectrum, which is localized around $\boldsymbol{\xi}_{\mathbf{n}}$, is located in the propagating spectrum range $|\boldsymbol{\xi}| < 1$. We denote this subset by the index μ_P. Inserting the ID Gaussian windows of (23) into (26) and evaluating the integrals asymptotically one readily identifies $\hat{\Psi}^{\pm}_{\mu}$ as forward and backward ID-GB that propagate from $z = \mp\infty$ to $\pm\infty$ in the directions $\mathring{\boldsymbol{\kappa}}^{\pm}_{\mathbf{n}} = (\boldsymbol{\xi}_{\mathbf{n}}, \pm\zeta_{\mathbf{n}}) = (\sin\theta_{\mathbf{n}}\cos\phi_{\mathbf{n}}, \sin\theta_{\mathbf{n}}\sin\phi_{\mathbf{n}}, \cos\theta_{\mathbf{n}})$ (see Figure 3), while for $z = 0$ they converge to the PS lattice of Section 3.1 as illustrated in Figure 3.

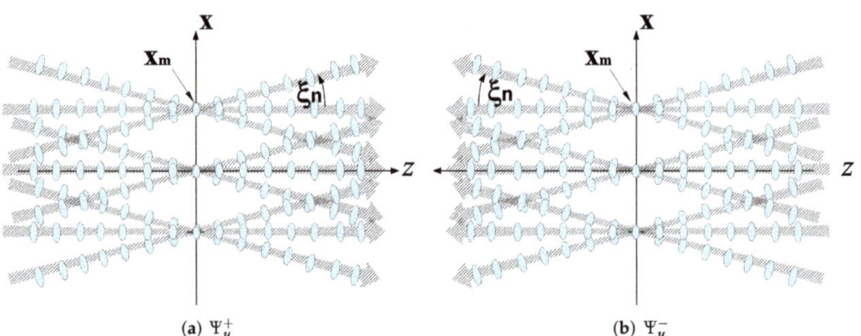

Figure 3. The forward/backward propagating beam frames $\Psi^{\pm}_{\mu}$. (a) The forward and (b) the backward beam frames $\Psi^{\pm}_{\mu}$. The BFs are illustrated by the hatched arrows. The ellipses correspond to the pulsed-beam-frames that are used in the TD formulations and are not considered here (see [21]).

As has been established by **the beam frame theorem** in [20], the beam-sets $\left\{\hat{\Psi}^{\pm}_{\mu}(\mathbf{r})\right\}_{\mu_P}$ constitute frames (hence referred to as "beam frames" (BF)) at any $z = const.$ plane over the Hilbert space $\mathbb{H}_P$ of functions whose spectrum is bounded in the propagation domain $|\boldsymbol{\xi}| < \zeta_0$, with the set $\{\hat{\Phi}^{\pm}_{\mu}(\mathbf{r})\}$ being the dual frames. The propagators $\hat{\Phi}^{\pm}_{\mu}$ have the same form as $\hat{\Psi}^{\pm}_{\mu}$ in (26), except that $\hat{\psi}_{\mu}$ are now replaced by $\hat{\phi}_{\mu}$. Note that in view of (25), $\hat{\Phi}^{\pm}_{\mu}$ are proportional to $\hat{\Psi}^{\pm}_{\mu}$.

It follows from the beam frame theorem that any function over $\mathbb{H}_P$ may be expanded by the BF. This observation is very important in the context of inverse scattering since it implies that both the scattered field and the medium are expanded on the same basis.

An important special case of the above is when the BF are used to expand forward or backward propagating wave-fields $\hat{u}^{\pm}(\mathbf{r})$. In view of the theorem, u^+ may be expanded using $\hat{\Psi}^{+}_{\mu}$, and u^- may be expanded using $\hat{\Psi}^{-}_{\mu}$, but the physically meaningful choice is to expand $u^{\pm}$ using $\hat{\Psi}^{\pm}_{\mu}$, respectively, viz (Equation (32) [20])

$$\hat{u}^{\pm}(\mathbf{r}) = \sum_{\mu\in\mu_P} \hat{A}^{\pm}_{\mu}\hat{\Psi}^{\pm}_{\mu}(\mathbf{r}), \tag{27}$$

where the summation includes only "μ_P propagating" frame-beams with no evanescent spectrum. As has been established by **the expansion coefficient invariance theorem** in [20], $\hat{A}^{\pm}_{\mu}$ may be calculated by projecting $\hat{u}^{\pm}(\mathbf{r})$ on the dual frame $\hat{\Phi}^{\pm}_{\mu}(\mathbf{r})$ over any $z = z'$ plane, giving the same result, i.e., (Equation (33) [20])

$$\hat{A}^{\pm}_{\mu} = \left\langle \hat{u}^{\pm}(\mathbf{r}), \hat{\Phi}^{\pm}_{\mu}(\mathbf{r})\right\rangle\Big|_{z'} = \left\langle \hat{u}^{\pm}(\mathbf{x}), \hat{\phi}_{\mu}(\mathbf{x})\right\rangle\Big|_{z=0} \tag{28}$$

where the last expression describes the canonical WFT coefficients of (18) evaluated over the $z = 0$ plane.

Finally we note that in [21], the BF theorem has been extended to the TD using ID-PB propagators.

4. UWB Beam-Based Diffraction Tomography: Multi-Frequency Formulation

The beam frame concept provides a framework to formulate the beam-based inverse scattering algorithm. Using the BFs, we may use the same set of beam basis functions to expand both the scattering data and the medium (actually, the sources that are induced due to the medium heterogeneities). As illustrated in Figure 4, the inverse problem is thereby described by the local interaction between the beam amplitudes and the unknown object. As noted in the introduction, optimal localization is obtained in the time-domain formulation, using localized space–time wave-packets. This, however, requires somewhat sophisticated processing tools [21]. In the present section we present only the multi-frequency formulation that utilizes conventional FD data-processing tools followed by integration over all the frequencies. The readers are referred to [36,37] for the TD interpretation.

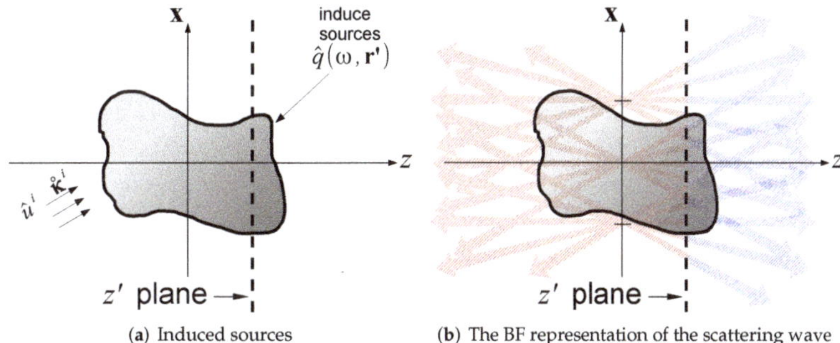

(a) Induced sources

(b) The BF representation of the scattering wave

Figure 4. The scattering mechanism within the propagating frame formulation. The incident field that propagates through the medium (see subplot (**a**)) gives rise to induced sources. At each $z = const.$ plane, these sources are expanded by the forward/backward propagating BFs, giving rise to the forward/backward scattered fields depicted in subplot (**b**) in blue and red, respectively.

4.1. The Inversion Algorithm

Given the scattering data over the z_j planes, the BF representation of the scattering fields into the $z \lessgtr z_j$ half spaces are given by (see (27))

$$\hat{u}_j^s(\mathbf{r}, \omega) = \sum_{\mu \in \mu_P} \hat{A}_\mu^j(\omega) \hat{\Psi}_\mu^\mp(\mathbf{r}, \omega), \quad z \lessgtr z_j, \tag{29}$$

where, as before, upper and lower signs correspond to $j = 1$ and $j = 2$, respectively. The expansion coefficients calculated via (28),

$$\hat{A}_\mu^j(\omega) = \left\langle \hat{u}_j^s(\mathbf{x}, \omega), \hat{\Phi}_\mu^\mp(\mathbf{r}, \omega)\big|_{z_j} \right\rangle. \tag{30}$$

Following the discussion after (7), these coefficients extract the local PW spectrum of the scattering data. Note that, as was done in the PW spectrum of Equation (5), the scattering WFT operation normalizes the scattering on the z_j planes to their phase centers on the $z = 0$ plane. The coefficients in (30) are referred to as the beam-domain data.

The backpropagated fields $\hat{u}_j^b(\mathbf{r}, \omega)$ are obtained by extending (29) as is to $z \gtrless z_j$ (see (II-9)). The "imaging fields" are then calculated by inserting (29) into Equation (11). In view of the local structure of the $\hat{\Psi}_\mu^\mp$ propagators, we obtain the explicit expression (II-11)

$$\hat{I}_j(\mathbf{r}) \simeq \frac{2}{i\omega} e^{-ik\overset{\circ}{\kappa}{}^i \mathbf{r}} \sum_\mu \hat{A}_\mu^j(\omega) \cos^2\left(\frac{\gamma_\mathbf{n}^\mp}{2}\right) \hat{\Psi}_\mu^\mp(\mathbf{r}, \omega), \tag{31}$$

where $\gamma_\mathbf{n}^\mp$ represents the angle between the illumination direction $-\overset{\circ}{\kappa}{}^i$ and the beam direction $\overset{\circ}{\kappa}{}_\mu^j$ (which actually depends only on $\mathbf{n}$. Finally, the reconstructed object is calculated via (12) and (13). For full details, the reader should refer to Appendices II-A,B.

4.2. Interpretation within the Born Approximation

In order to gain insight into the structure of the beam-domain data, we insert the Born approximation of the scattered field in (7) into (30). The resulting relation between $O(\mathbf{r})$ and the beam data is given by (I-21)

$$\hat{A}_\mu^j(\omega) \simeq \langle O(\mathbf{r}), \hat{\Lambda}_\mu^j(\mathbf{r}, \omega) \rangle_V, \quad \hat{\Lambda}_\mu^j(\mathbf{r}, \omega) = k/2i \cos\theta_\mathbf{n} e^{-ik\overset{\circ}{\kappa}{}^i \mathbf{r}} \hat{\Phi}_\mu^\mp(\mathbf{r}, \omega), \tag{32}$$

where the integration covers the entire scatterer domain. Thus, within the Born approximation, the data is described as projections of $O(\mathbf{r})$ on the beam axis, using the projection kernels $\hat{\Lambda}_\mu^j(\mathbf{r}, \omega)$. As shown in I-Section 5-B in [36], this projection extracts the local stratification of O along the beam axis at an angle $\gamma_\mathbf{n}^\mp$ defined in (31). This implies that the scattering amplitudes $\hat{A}_\mu^j$ are obtained from Snell's reflections due the stratification components in $O(\mathbf{r})$ along the μ beam axis, so that strong amplitudes are obtained only for those μ (locations and directions) that correspond to the stratification of $O(\mathbf{r})$ along the μ beam axis. Note that (32) is the local generalization of (7), where the BF basis is used instead of the conventional Green's function that radiates in all directions.

Further localization along the beam axis is provided by using the TD formulation in (I-27)-(I-32). However, as noted earlier, the TD formulation is not presented here since our goal in this paper is to present the pragmatic and practical formulation in the FD where all the operations are based on Fourier-type integrals. The readers are referred to [36,37] for more details on the TD formulations.

To further explore the FD beam data representation, we consider the spectral representation of $\hat{A}_\mu^j$. Substituting (5) into (30) and changing the order of integrations, $\hat{A}_\mu^j$ can be expressed as (I-22)

$$\hat{A}_\mu^j \simeq \left(\frac{k}{2\pi}\right)^2 \int d^2\xi \frac{ik}{2\zeta} \hat{\phi}_\mu^*(\boldsymbol{\xi}) \bar{O}(\mathbf{K}) \Big|_{\mathbf{K}=k(\overset{\circ}{\kappa}{}_j - \overset{\circ}{\kappa}{}^i)}. \tag{33}$$

The expression is the local alternative to the plane-wave-based DT identity in (8). It is recognized as $\overset{\circ}{\kappa}{}_\mathbf{n}^\pm$ samples of the value of $\bar{O}(\mathbf{K})$ over the shifted Ewald sphere defined in (9). The spectral width of these samples is that of $\hat{\phi}_\mu(\boldsymbol{\xi})$.

5. Numerical Examples

In this section, we demonstrate the beam-based DT algorithm through a numerical example. We begin with a step-by-step summary of the algorithm, including guidelines for choosing the parameters.

5.1. A Step by Step Summary of the Algorithm

Phase I—the experimental setup

1. We consider a realistic case where the object is illuminated by an array of M independent point transducers over the z_1 plane, as illustrated in Figure 5a. We illustrate here only the reflected data on the z_1 plane, since in many applications (e.g., geophysics)

2. The data is measured by exciting the sources one at a time by short pulse $F(t)$ that spans the desired frequency band $\Omega \in [\omega_{min}, \omega_{max}]$ as needed to obtain the desired $K-$space coverage. The result is an $M \times M$ data matrix $\mathbf{U}^s_{pq}(t)$ describing the response at the p receiver due to an excitation by the q source.

3. The data is sampled at the proper Nyquist rate and then converted to the FD via FFT, giving rise to the data matrix $\hat{\mathbf{U}}^s_{pq}(\omega)$. Before the calculation, the time-series are padded by zeros to avoid aliasing of the final image when it is generated by integration over all the frequencies. Note that in some applications, $\hat{\mathbf{U}}^s_{pq}(\omega)$ may be measure directly in the FD.

4. The response to time-harmonic PW excitations at different angles is synthesized from $\hat{\mathbf{U}}^s_{pq}(\omega)$ by q-stacking the array data with proper phase terms. The result provides the PW data to the phase–space beam-based processing.

   One may also calculate the time-harmonic PW spectrum of the scattered field via p-stacking with proper phase terms. The result is an $M \times M$ data matrix $\hat{\mathbf{U}}^s_{p'q'}(\omega)$ describing the p' spectral PW due to an excitation by the q' incident PW. As noted earlier, before we do the stacking, the array dimensions should be zero- padded to avoid aliasing of the final image when the images are generated by spectral integrations.

5. As illustrated in Figure 5a, the spectral information that can be covered by the array is determined by the size of the array and by the target range R. The size of the array should be chosen to provide sufficient spectral coverage of the target. In general, R should be within the Fresnel zone of the array, i.e., $R \ll (D\zeta^i)^2/\lambda$. Note also that due to the finite size of the array, one should avoid the array transition zone illustrated in Figure 5a.

6. The array elements inner spacing should, in general, satisfy the Nyquist sampling rate $kd = \pi$. However, since the target range satisfies $kR \gg 1$, it is only required that the phase difference between adjacent elements will be small, yielding a sparser array with $d < \sqrt{\frac{\pi}{8k_{max}}R}$.

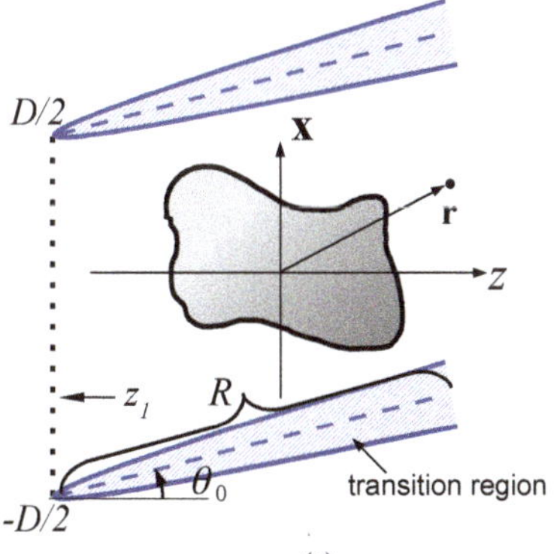

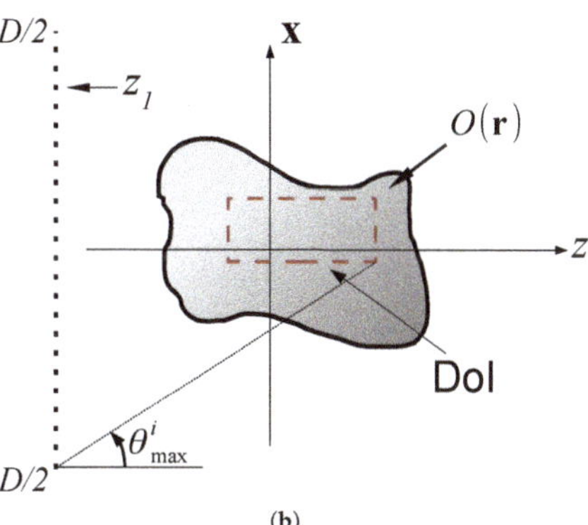

Figure 5. Guidelines for choosing the experimental setup. (**a**) The size of the array should be wide enough to provide a plane-wave illumination at the target range R. The scan angle is limited in order to be sufficiently far for the end-point diffraction zone. (**b**) For local imaging, only the part of the object inside the DOI should satisfy the conditions above.

Phase II—Constructing the phase space lattice

The next step is to set up the phase–space lattice and choose the expansion parameters

1. **Choosing the beam parameter** b: As discussed in Section 3.2, the ID Gaussian windows in (23) are fully determined by the parameter b. The considerations of choosing b were widely studied in [21,36] for the application of local inverse scattering. This parameter balances between the beam collimation length and the beamwidth. We choose b to be on the order of the DoI domain so that the beams are collimated throughout the DoI while being small enough for transversal resolution.

2. **The phase–space lattice:** The guidelines for constructing the UWB beam lattice are discussed in Section 3.2. As discussed there, we choose $\nu_{max} = 1/3$ which balances between stable expansion frame and moderate over-completeness (relatively small number of elements). The optimal values for $(\bar{x}, \bar{\xi})$ are given by (24).

3. **The phase–space propagators:** The frame elements $\{\hat{\Psi}_\mu^\pm(\mathbf{r}), \hat{\Phi}_\mu^\pm(\mathbf{r})\}$ are calculated via (26). For the Gaussian windows (23), the results are ID-GB propagators (see explicit expressions in (Appendices A,C [21]).

4. **Limited physical data:** We need to consider only beams whose initiation point $\mathbf{x_m}$ are supported by the array size. The maximal value of $\xi_\mathbf{n}$ is determined by the scan angle. If, for example, the scan angle is $60°$, then $|\xi_\mathbf{n}| < \frac{\sqrt{3}}{2}$.

Phase III—Calculating the beam data

1. **Calculating the expansion coefficients:** The coefficients $\hat{A}_\mu^j$ are calculated via (30).

2. **Filtering out low amplitude data:** As discussed in Section 4.2, the beam data is related to the local Snell's reflections of the beams by the local stratification in $O(\mathbf{r})$, which in turns are determined by the LRT of $O(\mathbf{r})$. We therefore threshold low amplitude beams at a level of 40 dB.

Phase IV—Local reconstruction via beam backpropagation

1. **Beam backpropagation within the DoI:** Next, following Section 4.1, we backpropagate the beams whose amplitudes $\hat{A}_\mu^j$ are larger than the threshold set above. We consider only the beams passing through the DoI (see Figure 6) or no further than 3 beam-widths away from the DoI (this distance is consistent with an effective threshold of 40 dB).

2. **The image:** The imaging fields are calculated via (31), and finally the image is calculated via Equations (12)–(13).

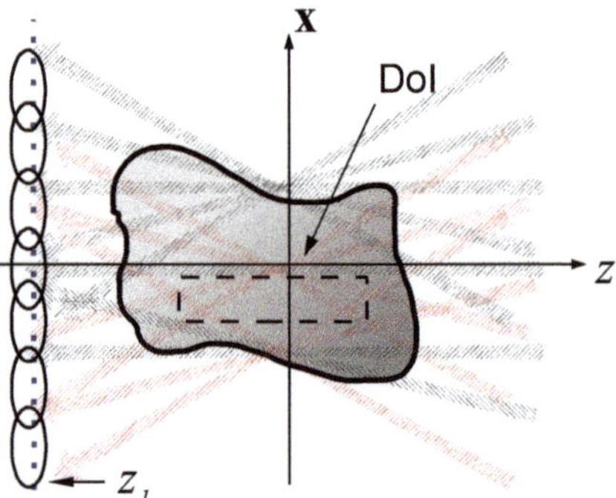

Figure 6. An overview of the local inversion algorithm. The beam expansion of the scattered field is plotted as gray arrows. Only those covering the array are plotted. The scattering data is then transformed to beam amplitudes by stacking the receivers data via (30), as schematized by the black ellipses. Only beams with high amplitude are considered and backpropagated via (31). The image in the DoI (black rectangle) is obtained by aggregating the contribution of beams that pass inside the DoI (red beams).

5.2. Example A: A Smooth and Quasi Stratified Medium. UWB Reflection Mode Data

The medium is plotted in Figure 7a in a 2D coordinate frame $\mathbf{r} = (x, z)$, with the DoI being the 20×20 black rectangle. For simplicity we normalize all space-units such that the background wave-speed is $v_0 = 1$. Note that the contrast is relatively large with values of $O_{\max} = \pm 0.44$. Note that this example is one of those treated in [37] (see Figure 6, but here the processing is done in the multi-frequency domain as outlined above.

The medium is dominated by stratification along the z direction, hence its K-space distribution is localized near the K_z axis (see discussion below). Referring to the discussion in Section 2.4, it can be recovered using UWB reflection data on the z_1 plane. We therefore use illumination by a z-propagating time-harmonic PW with frequencies in the band $\Omega = [0.1, 1]$. The exact data is generated using method of moments (MoM) simulations. We record only the reflected data over an array of receivers located at $z = -150$ with $|x| < 250$ with inter-element spacing $d = 1.15\,\pi$ (larger than the Nyquist distance).

We set $b = 50$, such that the beams are collimated inside the DoI, while maintaining $k_{\min} b \gg 1$ as required for collimation after (23). Using also $k_{\max} = 1$ and $\nu_{\max} = 1/3$ we obtain from (24) $(\bar{x}, \bar{\xi}) = (9.71, 0.194)$. The resulting BF propagators $\{\hat{\Psi}_\mu^\pm(\mathbf{r}), \hat{\Phi}_\mu^\pm(\mathbf{r})\}$ are calculated via (Equations (C1)–(C5) [21]).

Next we calculate the beam-domain data $\hat{A}_\mu^j$ via (30). The reconstructed object inside the DoI is found using the reflected data imaging field $\hat{I}_1(\omega)$ of (31) where we consider only backpropagated beams whose μ axis passes inside the DoI, and then integrating over all frequencies as in (12). The reconstructed medium is illustrated in Figure 7b. As can be seen, the reconstructed object matches well with the object inside the DoI. To better quantify the image results, in Figure 8 we plot cross-sectional cuts of the object at $x = 0$ and at $x \pm 6$.

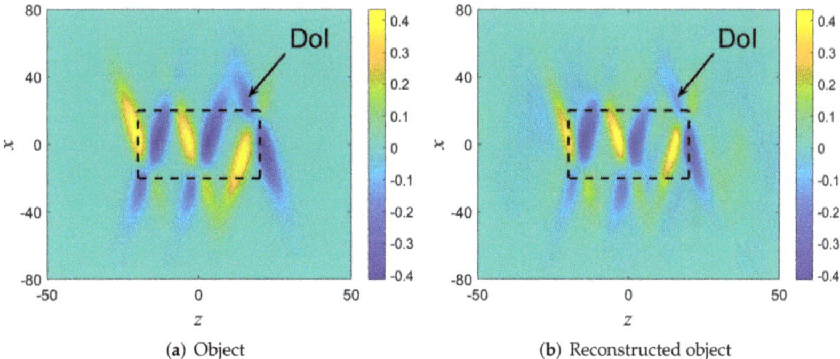

(a) Object

(b) Reconstructed object

Figure 7. Example A: (**a**) The original (**a**) and the reconstructed (**b**) object functions. The DoI is illustrated by the black rectangle.

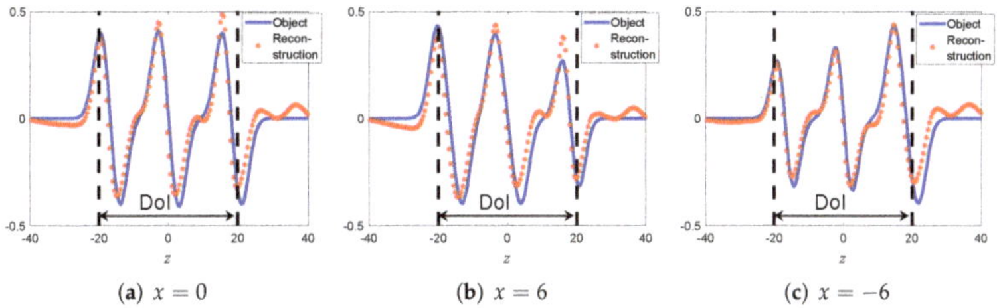

(**a**) $x = 0$

(**b**) $x = 6$

(**c**) $x = -6$

Figure 8. Example A: Comparison of the results along cross sectional cuts parallel to the z-axis.

The sources of error are readily seen in the $K-$space distribution of the original and reconstructed media in Figure 9. Note that the imaging algorithm has recovered most of the object's $K-$space, except for the region around $|\mathbf{K}| \approx 0$. As discussed in Section 2.4, this missing data is due to the low frequency cutoff $k_{min} = 0.1$ in the data. The main drawback of the "UWB reflection mode inversion" schemes is the missing transversal spectrum components and the $|\mathbf{K}| \rightarrow 0$ components, (which are small in this example). In general, one may try to recover this data by using transition mode data (z_2) but in many applications this data is not available. Alternatively, one may use several illumination directions which are synthesized from the array data via the method outlined in Phase I of Section 5. These additional illuminations are also used to reduce the reconstruction noise, as explored in II-Section 6 in [37]. The readers are referred to other examples in [36,37,42–44,50].

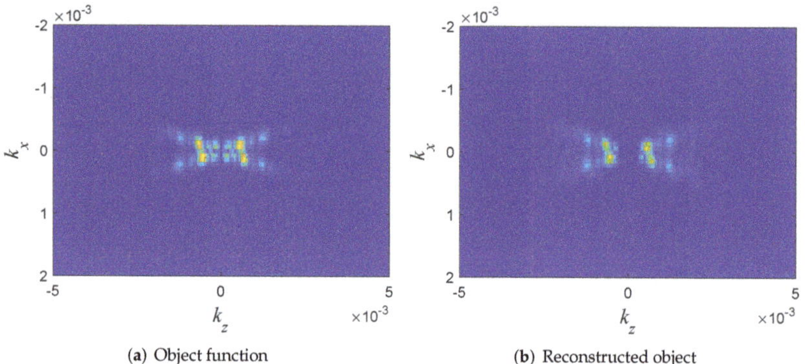

(**a**) Object function   (**b**) Reconstructed object

Figure 9. Example A: Comparison of the $K-$ space distributions of the original (**a**) and reconstructed (**b**) objects.

5.3. *Example B: An Object with Sharp Boundaries. Reflection and Transmission Data*

The object shown in Figure 10a has sharp boundaries, strong transversal K components, and a non-zero average (i.e., $\bar{O}(|\mathbf{K}| = 0) \neq 0$). As before, we consider a $2D$ problem with $\mathbf{r} = (x, z)$ and normalize the space-units such that the background wave-speed is $v_0 = 1$.

The source array is located on the $z_1 = -150$ plane over $|x| < 250$, with inter-element spacing $d = 1.15\pi$. Using this array, we may synthesize PW illumination over a spectral range of $\pm 60°$ (see discussion in Section 5, Phase I(1–4)). The frequency band is $\Omega = [0.1, 1]$. We consider both the reflection and transmission data over similar receiver arrays at $z_1 = -150$ and $z_2 = 150$. The exact scattered data is calculated via the MoM.

For the beam processing we use $b = 20$, such that the beams are collimated inside the DoI, while maintaining $k_{min}b \gg 1$ as required for collimation after (23). Using also $k_{max} = 1$ and $v_{max} = 1/3$ we obtain from (24) $(\bar{x}, \bar{\zeta}) = (6.47, 0.32)$. The resulting BF propagators $\{\hat{\Psi}_{\mu}^{\pm}(\mathbf{r}), \hat{\Phi}_{\mu}^{\pm}(\mathbf{r})\}$ are calculated via (Equations (C1)–(C5) [21]).

Figure 10b depicts the reconstructed objects in the front (left) using a single PW illumination at $\theta_i = 0$ and reflection data at the z_1 plane. As expected, the reflection data provides good longitudinal resolution but poor transversal resolution (see Figure 2b). Note also that the value of the reconstructed object function is approximately one half of the true value due to the missing data at $|\mathbf{K}| = 0$. As expected, the reconstruction of the object outside the DoI is poor.

In Figure 10c we improved the resolution by using several illumination directions (as one may expect by considering Figure 2a,b), yet the reconstruction still suffers from poor transversal resolution and low value of the reconstructed object. These problems are mitigated in Figure 10d where we used both the reflection and transmission data as in (13). Further improvement can be made via iterative schemes [36,37,42–44,50].

Finally, in Figure 11 we demonstrate local imaging within different DoIs. Figure 11a depicts the reconstruction of a cylinder at the front (left-top) using both reflection and transmission data due to illumination at $\theta_i = 0$. The reconstruction is a bit poorer than in Figure 10d where we used several illumination directions. As expected, the reconstruction outside this DoI is poor.

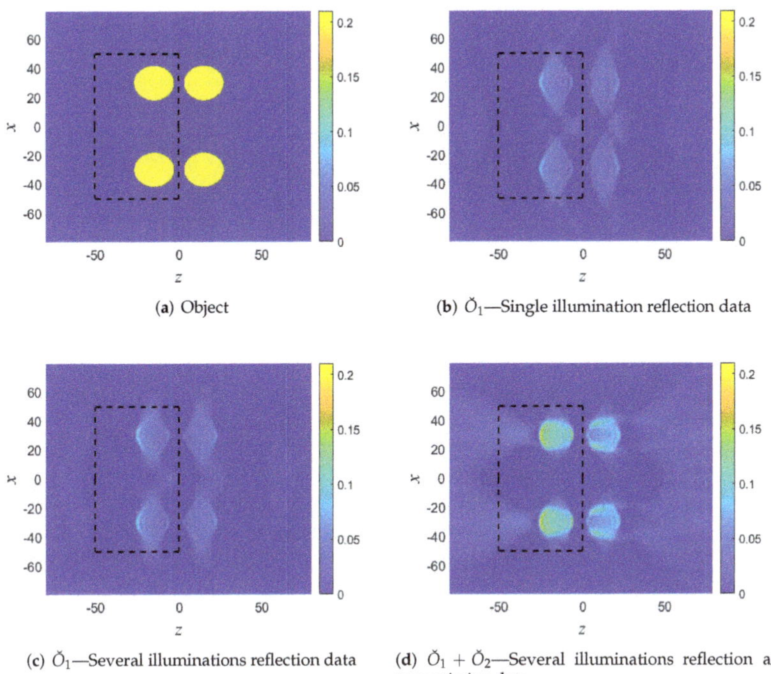

(a) Object

(b) $\breve{O}_1$—Single illumination reflection data

(c) $\breve{O}_1$—Several illuminations reflection data

(d) $\breve{O}_1 + \breve{O}_2$—Several illuminations reflection and transmission data

Figure 10. Numerical example B. (**a**) The object function. (**b**) Reconstruction using single illumination and reflection data. (**c**) Reconstruction using several illuminations and reflection data. (**d**) Reconstruction using several illuminations and reflection and transmission data. The DoI is illustrated in a black rectangle.

Figure 11b depicts the reconstruction of a cylinder at the back (right-top). Since this cylinder is poorly illuminated by normal incidence, we use here reflection and transmission data from several illumination directions $\theta^i = \mp 30^\circ, \mp 40^\circ$. The reconstruction inside the DoI is much better than in Figure 10. As before, the reconstruction outside this DoI is poor.

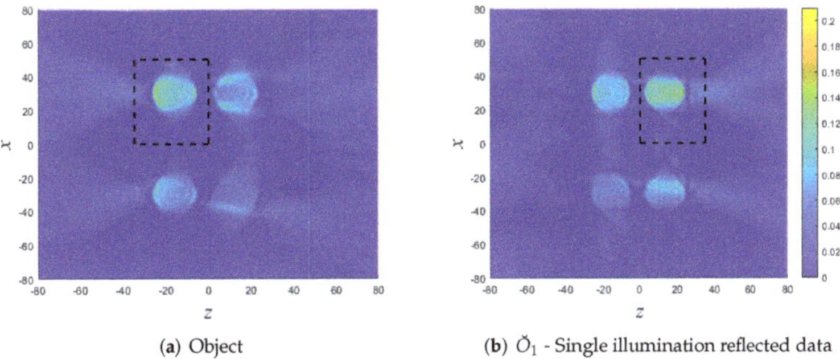

(a) Object

(b) $\breve{O}_1$ - Single illumination reflected data

Figure 11. Numerical example B: Local reconstruction in different DoI's (black rectangles). (**a**) Reconstruction of a cylinder at the front using both reflection and transmission data due to illumination at $\theta_i = 0$. (**b**) Reconstruction of a cylinder in the back using both reflection and transmission data from several illumination directions.

6. Discussion and Conclusions

In this paper, we reviewed the local diffraction–tomography inversion scheme introduced originally in [36,37]. The method is based on a local transformation of the scattering data and local reconstruction using beam backpropagation. It is structured on the concept of beam-frames (BFs). The BFs are a phase–space set of beam-waves that constitute local basis functions (frames) over the propagation domain. As such, they provide an alternative local basis for the global PW or Green's function radiation integrals. We use the BFs to formulate a local inversion algorithm as an alternative to the conventional approaches. In this and other publications, we demonstrated and explored the advantages of the local algorithm over the conventional PW and Green's function DT algorithms:

1. Local imaging within a given domain of interest (DoI).
2. Reduced complexity since it accounts only for the beam basis-functions that cover the DoI.
3. Reduced noise level since data and noise arriving from other regions are a priori filtered out.
4. Backpropagation and imaging over a non-homogeneous background.

In previous publications [36,37] we have emphasized TD data processing, where the beam waves are localized space–time wave-packets. This requires somewhat sophisticated mathematics to construct the wave-packets and use them as signal processing tools. The main advantage of the TD approach is the data localization and interpretation. In the present paper, on the other hand, we utilized FD processing followed by an integration over the relevant frequency band. The motivation has been to provide the reader with more straightforward Fourier-based data-processing tools. We also provide the processing tools and closed-form expressions for the local imaging formula, as well as step-by-step guidelines for choosing the various scheme parameters. The method provides an efficient UWB formulation where one has to calculate the beam lattice and propagators only once and then use them for all the frequencies.

Funding: This work is partially supported by the Israeli Science Foundation (ISF) under Grant Grant 1111/19.

Institutional Review Board Statement: Not applicable.

Informed Consent Statement: Not applicable.

Data Availability Statement: Data associated with this research are available and can be obtained by contacting the author.

Acknowledgments: The author would like to express his sincere appreciation to E. Heyman for helpful discussions pertaining to the preparation of this manuscript.

Conflicts of Interest: The author declare no conflict of interest.

References

1. Claerbout, J.F. Fundamentals of Geophysical Data Processing, Citeseer. McGraw-Hill Inc: New York, NY, USA, 1976; Volume 274.
2. Langenberg, K.J. Applied inverse problems for acoustic, electromagnetic and elastic wave scattering. In *Basic Methods of Tomography and Inverse Problems*; Sabatier, P.C., Ed.; Malvern Physics Series; Adam Hilger: Bristol, UK, 1987.
3. Colton, D.; Kress, R. *Inverse Acoustic and Electromagnetic Scattering Theory*; Springer: Berlin/Heidelberg, Germany, 1998; Volume 93.
4. Cakoni, F.; Colton, D. *A Qualitative Approach to Inverse Scattering Theory*; Springer: New York, NY, USA 2014; Volume 188.
5. Devaney, A.J. *Mathematical Foundations of Imaging, Tomography and Wavefield Inversion*; Cambridge University Press: Cambridge, UK, 2012.
6. Devaney, A.J. Diffraction tomography. In *Inverse Methods in Electromagnetic Imaging, Part 2*; Boerner, W.M., Ed.; D. Reidel Publ. Comp.: Dordrecht, The Netherlands, 1985; pp. 1107–1135.
7. Heyman, E.; Felsen, L.B. Gaussian beam and pulsed beam dynamics: Complex-source and complex spectrum formulations within and beyond paraxial asymptotics. *JOSAA* **2001**, *18*, 1588–1611. [CrossRef]
8. Heyman, E. Pulsed beam solution for propagation and scattering problems. In *Scattering and Inverse Scattering in Pure and Applied Science*; Pike, R., Sabatier, P., Eds.; Academic Press: London, UK, 2002; Chapter 1.5.4, pp. 295–315.
9. Popov, M.M. A new method of computation of wave fields using Gaussian beams. *Wave Motion* **1982**, *4*, 85–97. [CrossRef]

10. Červený, V. Expansion of a plane wave into Gaussian beams. *Stud. Geophys. Geod.* **1982**, *26*, 120–131. [CrossRef]
11. Norris, A.N. Complex point-source representation of real sources and the Gaussian beam summation method. *J. Opt. Soc. A.* **1986**, *3*, 205–210. [CrossRef]
12. Heyman, E. Complex source pulsed beam expansion of transient radiation. *Wave Motion* **1989**, *11*, 337–349. [CrossRef]
13. Bastiaans, M.J. The expansion of an optical signal into a discrete set of Gaussian beams. *Optik* **1980**, *57*, 95–102.
14. Einziger, P.D.; Raz, S.; Shapira, M. Gabor representation and aperture theory. *JOSAA* **1986**, *3*, 508–522. [CrossRef]
15. Steinberg, B.Z.; Heyman, E.; Felsen, L.B. Phase space beam summation for time-harmonic radiation from large apertures. *JOSAA* **1991**, *8*, 41–59. [CrossRef]
16. Maciel, J.J.; Felsen, L.B. Gaussian beam analysis of propagation from an extended plane aperture distribution through dielectric layers, Part I—Plane layer. *IEEE Trans. Antennas Propag.* **1990**, *38*, 1107–1617.
17. Shlivinski, A.; Heyman, E.; Boag, A.; Letrou, C. A phase-space beam summation formulation for ultra wideband radiation. *IEEE Trans. Antennas Propag.* **2004**, *52*, 2042–2056. [CrossRef]
18. Shlivinski, A.; Heyman, E.; Boag, A. A phase-space beam summation formulation for ultrawide-band radiation. II: A multiband scheme. *IEEE Trans. Antennas Propag.* **2005**, *53*, 948–957. [CrossRef]
19. Shlivinski, A.; Heyman, E.; Boag, A. A pulsed beam summation formulation for short pulse radiation based on windowed radon transform (WRT) frames. *IEEE Trans. Antennas Propag.* **2005**, *53*, 3030–3048. [CrossRef]
20. Leibovich, M.; Heyman, E. Beam summation theory for waves in fluctuating media. part I: The beam frame and the beam-domain scattering matrix. *IEEE Trans. Antennas Propag.* **2017**, *65*, 5431–5442. [CrossRef]
21. Tuvi, R.; Heyman, E.; Melamed, T. Beam frame representation for ultrawideband radiation from volume source distributions: Frequency-domain and time-domain formulations. *IEEE Trans. Antennas Propag.* **2019**, *67*, 1010–1024. [CrossRef]
22. Hill, N.R. Gaussian beam migration. *Geophysics* **1990**, *55*, 1416–1428. [CrossRef]
23. Hill, N.R. Prestack Gaussian-beam depth migration. *Geophysics* **2001**, *66*, 1240–1250. [CrossRef]
24. Nowack, R.L.; Sen, M.K.; Paul, S.L. Gassuian beam migration for sparse common-shot and common-receiver data. In Proceedings of the 2003 SEG Annual Meeting, Dallas, TX, USA, 26–31 October 2003; Society of Exploration Geophysicists: Tulsa, OK, USA, 2003.
25. Bleistein, N. Mathematics of modeling, migration and inversion with Gassuian beams. In Proceedings of the 70th EAGE Conference and Exhibition-Workshops and Fieldtrips, Rome, Italy, 9–12 June 2008; European Association of Geoscientists & Engineers: Houten, The Netherlands, 2008; p. 41.
26. Nowack, R.L. Dynamically focused Gaussian beams for seismic imaging. *Int. J. Geophys.* **2011**, *2011*, 316581. [CrossRef]
27. Gray, S.H. Gaussian beam migration of common-shot records. *Geophysics* **2005**, *70*, S71–S77. [CrossRef]
28. Gray, S.H.; Bleistein, N. True-amplitude Gaussian-beam migration. *Geophysics* **2009**, *74*, 11. [CrossRef]
29. Popov, M.M.; Semtchenok, N.M.; Popov, P.M.; Verde, A.R. Depth migration by the Gaussian beam summation method. *Geophysics* **2010**, *75*, 1MA-Z39. [CrossRef]
30. Zacek, K. Gaussian packet pre-stack depth migration. In *SEG Technical Program Expanded Abstracts*; Society of Exploration Geophysicists: Tulsa, OK, USA, 2004; pp. 957–960.
31. Melamed, T.; Heyman, E.; Felsen, L.B. Local spectral analysis of short-pulse-excited scattering from weakly inhomogenous media: Part I–forward scattering. *IEEE Trans. Antennas Propag.* **1999**, *47*, 1208–1217. [CrossRef]
32. Melamed, T.; Heyman, E.; Felsen, L.B. Local spectral analysis of short-pulse-excited scattering from weakly inhomogeneous media: Part II–inverse scattering. *IEEE Trans. Antennas Propag.* **1999**, *47*, 1218–1227. [CrossRef]
33. Shlivinski, A.; Heyman, E.; Langenberg, K.J. Migration based imaging using the UWB beam summation algorithm. In Proceedings of the XXVII General Assembly of the Int. Union of Radio Science (URSI), New Delhi, India, 12–17 September 2005.
34. Heilpern, T.; Shlivinski, A.; Heyman, E. Back-propagation and correlation imaging using phase-space Gaussian-beams processing. *IEEE Trans. Antennas Propag.* **2013**, *61*, 5676–5688. [CrossRef]
35. Heilpern, T.; Heyman, E. MUSIC imaging using phase-space Gaussian-beams processing. *IEEE Trans. Antennas Propag.* **2014**, *62*, 1270–1281. [CrossRef]
36. Tuvi, R.; Heyman, E.; Melamed, T. UWB beam-based local diffraction tomography. part I: Phase-space processing and physical interpretation. *IEEE Trans. Antennas Propag.* **2020**, *68*, 7144–7157. [CrossRef]
37. Tuvi, R.; Heyman, E.; Melamed, T. UWB beam-based local diffraction tomography. part II: The inverse problem. *IEEE Trans. Antennas Propag.* **2020**, *68*, 7158–7169. [CrossRef]
38. Wu, R.S.; Wangand, Y.; Gao, J. Beamlet migration based on local perturbation theory. In Proceedings of the 2000 SEG Annual Meeting, Calgary, AB, Canada, 6–11 August 2000; Society of Exploration Geophysicists: Tulsa, OK, USA, 2000.
39. Wu, R.S.; Chen, L. Beamlet migration using gabor-daubechies frame propagator. In Proceedings of the 63rd EAGE Conference & Exhibition, Amsterdam, The Netherlands, 11–15 June 2001; Netherlands European Association of Geoscientists & Engineers: Houten, The Netherlands, 2001.
40. Wu, R.S.; Wang, Y.; Luo, M. Beamlet migration using local cosine basis. *Geophysics* **2008**, *73*, S207–S217. [CrossRef]
41. Melamed, T. Phase-space Green's functions for modeling time-harmonic scattering from smooth inhomogeneous objects. *J. Math. Phys.* **2004**, *46*, 2232–2246. [CrossRef]
42. Tuvi, R.; Zhao, Z.; Sen, M.K. Multi frequency beam-based migration in inhomogeneous media using windowed Fourier transform frames. *Geophys. J. Int.* **2020**, *8*, 1086–1099. [CrossRef]

43. Tuvi, R.; Zhao, Z.; Sen, M.K. A compressed data approach for image-domain least-squares migration. *Geophysics* **2021**, *86*, 1–37. [CrossRef]
44. Tuvi, R.; Zhao, Z.; Sen, M.K. A fast least-squares migration with ultra-wide-band phase-space beam summation method. *J. Phys. Commun.* **2021**, *5*, 105013. [CrossRef]
45. Melamed, T.; Ehrlich, Y.; Heyman, E. Short-pulse inversion of inhomogeneous media: A time-domain diffraction tomography. *Inverse Probl.* **1996**, *12*, 977. [CrossRef]
46. Dean, S.R. *The Radon Transform and Some of Its Applications*; Wiley—Interscience: New York, NY, USA, 1983.
47. Devaney, A.J. A filtered backpropagation algorithm for diffraction tomography. *Ultrason. Imaging* **1982**, *4*, 336–350. [CrossRef] [PubMed]
48. Kak, A.C. Computerized tomography with X-ray, emission, and ultrasound sources. *Proc. IEEE* **1979**, *67*, 1245–1272. [CrossRef]
49. Heyman, E. Pulsed beam propagation in inhomogeneous medium. *IEEE Trans. Antennas Propag.* **1994**, *42*, 311–319. [CrossRef]
50. Tuvi, R.; Zhao, Z.; Sen, M.K. A time domain seismic imaging method with a sparse pulsed-beams data. *IEEE Trans. Geosci. Remote Sens.* **2021**, 1. [CrossRef]

 *sensors*

Article

Three-Dimensional Microwave Imaging: Fast and Accurate Computations with Block Resolution Algorithms

Corentin Friedrich [1,2], Sébastien Bourguignon [1,*], Jérôme Idier [1] and Yves Goussard [2]

1   Laboratoire des Sciences du Numérique de Nantes, École Centrale de Nantes, 1 rue de la Noë, 44321 Nantes, France; Corentin.Friedrich@gmail.com (C.F.); Jerome.Idier@ls2n.fr (J.I.)
2   École Polytechnique de Montréal, C.P. 6079, Succursale Centre-Ville, Montréal, QC H3C 3A7, Canada; yves.goussard@polymtl.ca
*   Correspondence: Sebastien.Bourguignon@ec-nantes.fr

Received: 1 October 2020; Accepted: 29 October 2020; Published: 4 November 2020

Abstract: This paper considers the microwave imaging reconstruction problem, based on additive penalization and gradient-based optimization. Each evaluation of the cost function and of its gradient requires the resolution of as many high-dimensional linear systems as the number of incident fields, which represents a large amount of computations. Since all such systems involve the same matrix, we propose a block inversion strategy, based on the block-biconjugate gradient stabilized (BiCGStab) algorithm, with efficient implementations specific to the microwave imaging context. Numerical experiments performed on synthetic data and on real measurements show that savings in computing time can reach a factor of two compared to the standard, sequential, BiCGStab implementation. Improvements brought by the block approach are even more important for the most difficult reconstruction problems, that is, with high-frequency illuminations and/or highly contrasted objects. The proposed reconstruction strategy is shown to achieve satisfactory estimates for objects of the Fresnel database, even on the most contrasted ones.

Keywords: microwave imaging; inverse scattering; numerical optimization; block system inversion

---

1. Introduction

Microwave imaging aims at estimating the dielectric properties (permittivity and conductivity) of objects illuminated by incident electromagnetic fields [1]. This technique has been used in a wide range of applications such as biomedical imaging [2,3], geophysics [4,5] and nondestructive testing [6,7]. Quantitative inverse scattering methods estimate the complex permittivity map inside the object from a set of measured scattered electromagnetic fields around it. When the object size is comparable to the wavelength, diffraction occurs and the forward scattering model, which defines the scattered field as a function of the dielectric properties at any point of the object, becomes non-linear. The resulting reconstruction problem is ill-posed [8,9], and its resolution becomes particularly difficult with high-frequency illuminations and/or for highly contrasted objects—that is, if the permittivity inside the object strongly differs from that of the background—where the Born approximation is not valid anymore.

Imaging methods usually rely on solving an optimization problem, whose complexity is strongly impacted by the non-linearity of scattering. In order to circumvent this difficulty, several works introduced additional variables, corresponding to the total fields inside the object, which are jointly estimated together with the contrast function (i.e., the relative complex permittivity). Within this formulation, the domain equation, which links the total fields to the contrast function, is not exactly solved, but the resulting bilinear model allows the low-cost computation of iterates in the optimization procedure. The Modified Gradient Method [10] and Contrast Source Inversion methods [11,12] fall within this category, which achieve satisfactory compromise between computation time and reconstruction quality in cases where diffraction remains limited [13]. Although some works addressed 3-D reconstruction based on this formulation [7,14–16], most of them were limited to 2-D problems, where the number of unknowns remained relatively small [13,17–24].

A second class of methods, which dominates in the recent 3-D microwave imaging literature [25–35]. directly incorporates the domain equation into the problem formulation. The price to pay for using such an exact form of the domain equation is a much more complex evaluation of the cost function and of its gradient : both require a numerical solver for computing the forward problem, based on the resolution of very large linear systems, especially in the 3-D case. The related computational effort then represents the major cost of the reconstruction procedure. Therefore, developing efficient strategies for such system resolutions remains one of the current challenges for improving the efficiency of microwave imaging techniques [33,36,37]. In order to reduce the size of the problem, some works focused on the reconstruction of strongly constrained, parameterized, objects. Efficient numerical methods were developed with global convergence properties, e.g., based on evolutionary algorithms [38] or Markov Chain Monte-Carlo methods, coupled with surrogate models [39]. In the particular case of point-like scatterers, the two-step procedure in [40] efficiently accounts for the sparsity of the object.

In this paper, we focus on methods that aim to estimate any permittivity distribution (that is, without any strong prior assumption on the inspected object), based on the exact problem formulation. Previous works based on multiplicative [33] or additive [13], edge-preserving or sparsity-enhancing [41], regularization fall within this category. All these methods rely on repeated computations of forward scattering problems. One generally resorts to iterative system solving algorithms, which are stopped as soon as an acceptable residual error is obtained. Several such algorithms have been used in microwave imaging, e.g., the conjugate gradient (CG) algorithm [42,43], the biconjugate gradient (BiCG) algorithm [44], the quasi-minimal residual (QMR) algorithm [45,46] or the biconjugate gradient stabilized (BiCGStab) algorithm [33,36,37,47]. BiCGStab is now one of the most popular algorithms in this field and it was shown to converge faster than CG, BiCG and QMR [47].

In microwave imaging, as many forward scattering models as the number of illuminating sources must be computed. Such computations are usually performed sequentially or via massively parallel procedures [33,48]. Note that all such linear systems involve the same system matrix, which only depends on the object dielectric properties and on the discretization scheme. A classical solution relies on the preliminary LU-decomposition of the system matrix [4]. However, this possibility becomes prohibitive in 3-D imaging [49]. Block iterative algorithms specifically address the resolution of such linear systems with multiple right-hand sides (RHSs), by jointly solving all systems. Block versions of several standard iterative methods have been proposed in the applied mathematics literature, including the block-conjugate (and bi-conjugate) gradient algorithms [50], the block-QMR algorithm [51] and the block-BiCGStab algorithm [52]. Such block algorithms were shown to achieve better performance than their sequential counterparts in generic contexts. Block-QMR has been used in microwave imaging [49]. The block-BiCGStab algorithm proposed in [52] was shown to outperform block-QMR. Its standard implementation, however, is rather unsatisfactory for microwave imaging forward problems, where the system matrices are much larger, and where there are many more RHS terms which are highly correlated [53].

Here, we propose a reconstruction procedure based on additive penalization, where the evaluation of both the cost function and its gradient are performed with a dedicated block-BiCGStab implementation.

Specific tunings of the block-BiCGStab algorithm are discussed, and the overall efficiency of the inversion method is evaluated as a function of the problem difficulty (illuminating frequency and contrast). We show in particular that improvements are even greater as the problem complexity increases. We provide reconstruction results obtained on the Fresnel database [54], where the proposed implementation allows the accurate reconstruction of the most difficult objects.

In Section 2, the 3-D forward scattering model is built using an integral formulation. A discretization procedure based on the method of moments is used, and our inversion formulation based on additive regularization is detailed. In Section 3, we propose efficient adaptations of block-BiCGStab for forward scattering computations, and for their implementation within the reconstruction procedure. In Section 4, algorithmic performance is evaluated with simulated data, as a function of the object complexity, with different illuminating frequencies and contrast levels. In Section 5, our method is applied to real measurements extracted from the Fresnel database [54], where reconstruction quality and computation times are discussed. Concluding comments are given in Section 6.

2. Formulation of 3-D Forward and Inverse Scattering Problems

2.1. Forward Scattering Problem

The forward scattering model defines the total electric field in a given volume V containing the scattering object, as a function of the spatial distribution of the object permittivity. We adopt a standard approach based on the integral formulation of the domain equation [1], followed by a discretization procedure based on the method of moments [55]—see for example [14,36,37,43,44] for similar choices. We consider a background medium with constant complex permittivity ϵ_b, containing the object under study with complex permittivity $\epsilon(r)$ where r denotes the vector of spatial coordinates. The object is illuminated by an incident wave with angular frequency ω. In the sequel, the time dependency $\exp(-j\omega t)$ of all quantities is omitted. The electric field integral equation gives an implicit relation between the unknown total electric field at any point inside the domain under study and the object permittivity distribution [1]:

$$\vec{E}^{\text{tot}}(r) = \vec{E}^{\text{inc}}(r) + \left(k_b^2 + \nabla\nabla\cdot \right) \int_{r' \in V} g(r - r')\chi(r')\vec{E}^{\text{tot}}(r')dr', \quad \forall r \in V, \tag{1}$$

where $\vec{E}^{\text{tot}}(r)$ and $\vec{E}^{\text{inc}}(r)$ respectively denote the total and the known incident 3-D electric field vectors at point $r \in V$, μ_0 is the vacuum permeability, $k_b = \sqrt{\omega^2 \mu_0 \epsilon_b}$ is the wavenumber in the background medium, $g(r) = \exp(jk_b\|r\|)/4\pi\|r\|$ is the Green function and $\chi(r) = (\epsilon(r) - \epsilon_b)/\epsilon_b$ is the contrast function. The term $\nabla\nabla\cdot$ represents the gradient of the divergence with respect to r.

Using the discretization procedure based on the method of moments (see [14] for the 3-D case), the volume V is divided into N cubic voxels and the integral Equation (1) turns into:

$$e^{\text{tot}} = e^{\text{inc}} + \mathbf{G}_D \mathbf{X} e^{\text{tot}}. \tag{2}$$

The size-$3N$ vectors e^{inc} and e^{tot} respectively contain the three spatial components of the 3-D incident and total electric field vectors, discretized on the N voxels that are sorted according to the lexicographic order. The size-N vector x corresponds to the contrast function discretized on the N voxels, put in vector form, and $\mathbf{X}$ is the $3N \times 3N$ diagonal matrix whose diagonal contains three replicas of x. The $3N \times 3N$ matrix $\mathbf{G}_D$ corresponds to the discretization of the 3×3 Green tensor $\left(k_b^2 \mathcal{I}_3 + \nabla\nabla \right) g(r - r')$ on the N voxels, where $\mathcal{I}_3$ denotes the 3×3 identity tensor (see for example [53] for details). From (2), the discretized total field is the solution to the following linear system:

$$\mathbf{L}_x e^{\text{tot}} = e^{\text{inc}}, \text{ with } \mathbf{L}_x = \mathbf{I}_{3N} - \mathbf{G}_D \mathbf{X}, \tag{3}$$

where $\mathbf{I}_{3N}$ denotes the $3N \times 3N$ identity matrix.

In microwave imaging, the object is successively illuminated with a set of N_S sources at different spatial locations, creating incident fields $e_i^{\text{inc}}, i = 1, \ldots, N_S$. Then, computing the corresponding total fields e_i^{tot} amounts to solving N_S linear systems $\mathbf{L}_x e_i^{\text{tot}} = e_i^{\text{inc}}$, that is:

$$\mathbf{L}_x \mathbf{E}^{\text{tot}} = \mathbf{E}^{\text{inc}}, \text{ with } \mathbf{E}^{\text{tot}} = \left[e_1^{\text{tot}}, \ldots, e_{N_S}^{\text{tot}} \right] \text{ and } \mathbf{E}^{\text{inc}} = \left[e_1^{\text{inc}}, \ldots, e_{N_S}^{\text{inc}} \right], \tag{4}$$

where all linear systems involve the same system matrix $\mathbf{L}_x$. Such a property is the key point for our block resolution strategy.

2.2. Observation Model

The scattered field $\vec{E}^{\text{scat}}$ at any point r outside V is defined by the volume integral observation equation:

$$\vec{E}^{\text{scat}}(r) = \left(k_b^2 + \nabla\nabla \cdot \right) \int_{r' \in V} g(r - r') \chi(r') \vec{E}^{\text{tot}}(r') dr'. \tag{5}$$

Using the same discretization procedure as in Section 2.1, considering a set of N_M measurement points, it becomes:

$$e^{\text{scat}} = \mathbf{G}_O \mathbf{X} e^{\text{tot}}, \tag{6}$$

where the size-N_M vector e^{scat} collects the measurements of the scattered fields at the receiver locations, and the $N_M \times 3N$ matrix $\mathbf{G}_O$ corresponds to the discretization of the Green tensor $\left(k_b^2 \mathcal{I}_3 + \nabla\nabla \right) g(r - r')$, for r' at the N voxels in V and r at the measurement points [56]. Combining both domain (3) and observation (6) equations finally yields the observation model:

$$e_i^{\text{scat}} = \mathbf{G}_O \mathbf{X} \mathbf{L}_x^{-1} e_i^{\text{inc}} = \mathbf{G}_O \mathbf{X} (\mathbf{I}_{3N} - \mathbf{G}_D \mathbf{X})^{-1} e_i^{\text{inc}}, \text{ for any incident field } e_i^{\text{inc}}. \tag{7}$$

2.3. Inverse Problem Formulation

The microwave imaging inverse problem consists of estimating the contrast function from the measured scattered fields at the receivers. This problem is ill-posed [8,9], notably because the relation between the contrast function and the data, obtained from the two coupled Equations (1) and (5) in the continuous formulation (Equations (2) and (6) in the discrete case), is highly nonlinear.

Here, we adopt a classical approach to inverse problems [13,57], which consists of minimizing the misfit between the data $\tilde{e}_i^{\text{scat}}$ and the simulated scattered fields e_i^{scat}, penalized by an additive regularization term. More precisely, we define the reconstructed contrast function $\hat{x}$ by:

$$\hat{x} = \arg\min_x \mathcal{F}(x), \text{ with } \mathcal{F}(x) = \frac{1}{2} \sum_{i=1}^{N_S} \left\| \tilde{e}_i^{\text{scat}} - \mathbf{G}_O \mathbf{X} \mathbf{L}_x^{-1} e_i^{\text{inc}} \right\|^2 + \mathcal{R}(x), \tag{8}$$

where the regularization term operates on differences of the contrast function at neighboring voxels:

$$\mathcal{R}(x) = \gamma \sum_{i,j,\, i \sim j} \varphi(x_i - x_j) \text{ with } \gamma > 0, \tag{9}$$

where notation $i \sim j$ selects indices of neighbor voxels and φ is the "$\ell_2 \ell_1$" function $\varphi(u) = \sqrt{\delta^2 + u^2}$, where δ is a small positive constant ensuring the differentiability of the cost function, which was set to

10^{-2} in all our experiments. Such a penalization favors edge-preserving solutions and leads to a smooth cost function $\mathcal{F}$. Therefore, optimization can be addressed with gradient-based iterative algorithms. The penalization weight γ trades off between fidelity to data and regularization, which is necessary since the problem is ill-posed [8,9]. Its value may depend on the noise level and on the problem size.

Note that alternate regularization strategies, such as multiplicative regularization [33,58], have also been proposed. Here, we chose additive regularization because it was previously shown to be particularly efficient in the case of highly contrasted objects [13]. However, the techniques presented in this paper may also be applied to reconstruction methods based on multiplicative regularization, since both the corresponding objective function and its gradient present similar algebraic characteristics.

2.4. Optimization and Computational Issues

We propose to solve the optimization problem (9) with the L-BFGS algorithm, which is an acknowledged method for solving high-dimensional convex optimization problems ([59], see for example [13] in application to microwave imaging). Each iteration of L-BFGS consists of two steps: computing a descent direction, based on the computation of the gradient at the current point, and finding a convenient step size, for which several evaluations of the cost function may be necessary. The gradient of $\mathcal{F}$ reads:

$$\nabla \mathcal{F}(x) = -\mathbf{Q} \sum_{i=1}^{N_S} \mathbf{J}_i^\dagger \left(\widetilde{e}_i^{\text{scat}} - \mathbf{G}_O \mathbf{X} \mathbf{L}_x^{-1} e_i^{\text{inc}} \right) + \nabla \mathcal{R}(x) \tag{10}$$

where $\mathbf{Q} = [\mathbf{I}_N, \mathbf{I}_N, \mathbf{I}_N]$, the symbol † denotes the conjugate transpose and $\mathbf{J}_i$ is the Jacobian matrix associated to the i-th data misfit term in (8). It can easily be shown that:

$$\mathbf{J}_i = \mathbf{G}_O \left(\mathbf{L}_x^T \right)^{-1} \text{diag} \left\{ \mathbf{L}_x^{-1} e_i^{\text{inc}} \right\}, \tag{11}$$

where $\text{diag}\{u\}$ stands for the diagonal matrix with diagonal u, such that the gradient reads:

$$\nabla \mathcal{F}(x) = -\mathbf{Q} \sum_{i=1}^{N_S} \text{diag} \left\{ \mathbf{L}_x^{-1} e_i^{\text{inc}} \right\}^\dagger \overline{\mathbf{L}_x}^{-1} \mathbf{G}_O^\dagger \left(\widetilde{e}_i^{\text{scat}} - \mathbf{G}_O \mathbf{X} \mathbf{L}_x^{-1} e_i^{\text{inc}} \right) + \nabla \mathcal{R}(x), \tag{12}$$

where $\overline{\mathbf{L}_x}$ denotes the conjugate of $\mathbf{L}_x$.

Inspection of (8) and (12) reveals that the bulk of the computational effort required for evaluating $\mathcal{F}$ and $\nabla \mathcal{F}$ lies in the evaluation of matrix-vector products involving matrices $\mathbf{X}$ and $\mathbf{G}_O$, and in the resolution of linear systems with matrices $\mathbf{L}_x$ and $\overline{\mathbf{L}_x}$. Note that $\mathbf{G}_O$ is an $N_M \times 3N$ matrix and this manageable size makes it possible to explicitly store it for direct computations; similarly, $\mathbf{X}$ is a $3N \times 3N$ diagonal matrix, which results in small memory and computational requirements. Indeed, the major difficulty lies in the resolution of the N_S systems of size $3N \times 3N$ involving matrix $\mathbf{L}_x$ (computation of $\mathbf{L}_x^{-1} e_i^{\text{inc}}$), and in the resolution of another N_S similar systems involving matrix $\overline{\mathbf{L}_x}$ (computation of $\overline{\mathbf{L}_x}^{-1} v_i$, with $v_i = \mathbf{G}_O^\dagger \left(\widetilde{e}_i^{\text{scat}} - \mathbf{G}_O \mathbf{X} \mathbf{L}_x^{-1} e_i^{\text{inc}} \right)$). Note that, in the frequent case where the number of sources exceeds the number of receivers, by virtue of the Lorentz reciprocity theorem (see for example [60]), one can equivalently switch the roles of sources and receivers, then reducing the number of systems to be solved.

In Section 3, we propose a dedicated block resolution strategy in order to jointly solve such multiple systems. This iterative method requires the computation of high-dimensional matrix-vector products involving matrix $\mathbf{L}_x$ (resp. $\overline{\mathbf{L}_x}$), that is, involving the $3N \times 3N$ matrix $\mathbf{G}_D$ (resp. $\overline{\mathbf{G}_D}$). Note that these products are convolution products, which can be efficiently computed with 3-D Fast Fourier Transform (FFT) algorithms [36,47].

3. Simultaneous Resolution of Multiple Forward Scattering Problems

We now address the joint resolution of the multiple forward scattering problems (4) by the block-BiCGStab algorithm proposed in [52]. We present the algorithm and we focus on implementation issues for solving one set of multiple scattering problems.

3.1. Description of the Block-BiCGStab Algorithm

The pseudo-code implementing the block-BiCGStab algorithm for the joint resolution of (4) is given in Algorithm 1. The notation $\langle \mathbf{T}, \mathbf{S} \rangle_F$ represents the Frobenius product of two matrices. For comparison purposes, the sequential implementation with the BiCGStab algorithm, commonly used in microwave imaging, is recalled in Algorithm 2.

Algorithm 1 Joint resolution of N_S linear systems $\mathbf{L}_x \mathbf{E}^{\text{tot}} = \mathbf{E}^{\text{inc}}$ with block-BiCGStab.

1: For an initial matrix guess $\mathbf{E}^{\text{tot}(0)}$,

    set $\mathbf{R}^{(0)} = \mathbf{E}^{\text{inc}} - \mathbf{L}_x \mathbf{E}^{\text{tot}(0)}$, $\mathbf{P}^{(0)} = \mathbf{R}^{(0)}$

2: Choose an arbitrary $3N \times N_S$ matrix $\widetilde{\mathbf{R}}_0$

3: $k = 0$

4: **repeat**

5:    $\mathbf{V} = \mathbf{L}_x \mathbf{P}^{(k)}$

6:    solve $(\widetilde{\mathbf{R}}_0^{\dagger} \mathbf{V})\mathbf{A} = \widetilde{\mathbf{R}}_0^{\dagger} \mathbf{R}^{(k)}$

7:    $\mathbf{S} = \mathbf{R}^{(k)} - \mathbf{V}\mathbf{A}$

8:    $\mathbf{T} = \mathbf{L}_x \mathbf{S}$

9:    $\omega = \langle \mathbf{T}, \mathbf{S} \rangle_F / \langle \mathbf{T}, \mathbf{T} \rangle_F$

10:   $\mathbf{E}^{\text{tot}(k+1)} = \mathbf{E}^{\text{tot}(k)} + \mathbf{P}^{(k)}\mathbf{A} + \omega \mathbf{S}$

11:   $\mathbf{R}^{(k+1)} = \mathbf{S} - \omega \mathbf{T}$

12:   solve $(\widetilde{\mathbf{R}}_0^{\dagger} \mathbf{V})\mathbf{B} = -\widetilde{\mathbf{R}}_0^{\dagger} \mathbf{T}$

13:   $\mathbf{P}^{(k+1)} = \mathbf{R}^{(k+1)} + (\mathbf{P}^{(k)} - \omega \mathbf{V})\mathbf{B}$

14:   $k \leftarrow k + 1$

15: **until** $\forall i, \|r_i^{(k)}\| / \|e_i^{\text{inc}}\| <$ tolerance

---

Algorithm 2 Sequential resolution of N_S linear systems $\mathbf{L}_x e_i^{\text{tot}} = e_i^{\text{inc}}$, $i = 1, \ldots, N_S$, with BiCGStab.

0: **for** $i = 1, \ldots, N_S$ **do**

1:    For an initial guess $e_i^{\text{tot}(0)}$,

    set $r^{(0)} = e_i^{\text{inc}} - \mathbf{L}_x e_i^{\text{tot}(0)}$, $p^{(0)} = r^{(0)}$

2:    Choose an arbitrary vector $\tilde{r}_0$

3:    $k = 0$

4:    **repeat**

5:      $v = \mathbf{L}_x p^{(k)}$

6:      $\alpha = (\tilde{r}_0^{\dagger} r^{(k)}) / (\tilde{r}_0^{\dagger} v)$

7:      $s = r^{(k)} - \alpha v$

8:      $t = \mathbf{L}_x s$

9:      $\omega = t^{\dagger} s / t^{\dagger} t$

10:     $e_i^{\text{tot}(k+1)} = e_i^{\text{tot}(k)} + \alpha p^{(k)} + \omega s$

11:     $r^{(k+1)} = s - \omega t$

12:     $\beta = (\alpha \tilde{r}_0^{\dagger} r^{(k+1)}) / (\omega \tilde{r}_0^{\dagger} r^{(k)})$

13:     $p^{(k+1)} = r^{(k+1)} + \beta(p^{(k)} - \omega v)$

14:     $k \leftarrow k + 1$

15:    **until** $\|r^{(k)}\| / \|e_i^{\text{inc}}\| <$ tolerance,

16: **end for**

In both algorithms, the most time-consuming operations concern the computation of matrix-vector products of the form $\mathbf{L}_x p = p - \mathbf{G}_D \mathbf{X} p$ in steps 5 and 8. Note that the same number ($2N_S$) of such products are performed by one iteration of block-BiCGStab and N_S iterations of BiCGStab. In the remaining steps, a set of N_S dot products and scalar-vector products (one per system inversion) in BiCGStab are replaced by one N_S-dimensional matrix-vector product in block-BiCGStab. Similarly, a set of N_S scalar divisions are replaced by one $N_S \times N_S$ system inversion (steps 6 and 12). Therefore, such block operations generate a slight extra cost in block-BiCGStab compared to BiCGStab. Because of the small size of the involved systems, this remains negligible compared to the computation time required by multiplications with $\mathbf{L}_x$. Most of all, block operations introduce some coupling in the auxiliary variables and, subsequently, in the directions that are computed in order to update the solution at step 10. The rationale behind block versions is that coupling the descent directions yields more efficient update steps, and therefore results in a decrease in the total number of iterations and computing time.

3.2. Efficient Implementation for Solving Multiple Forward Problems

In its original description [52], the block-BiCGStab algorithm was evaluated on problems involving only a small number (5 to 10) of RHS vectors which were drawn independently, in moderate dimensions (up to 2500 × 2500). In microwave imaging, the number of RHS vectors—the number of measurements—is much higher, and vectors are highly correlated, since they correspond to the different incident illuminating fields, which are spatially close to each other. Moreover, matrices involved in 3-D problems are usually much bigger than those used in [52].

Numerical experiments, whose salient results are reported in Section 4.2, revealed that the conditioning of matrix $\widetilde{\mathbf{R}}_0^\dagger \mathbf{V}$, which enters steps 6 and 12 of Algorithm 1, has a critical impact on the behavior and convergence of the block-BiCGStab algorithm. Inspection of steps 1 and 5 clearly indicates that the conditioning of $\mathbf{V}$ strongly depends on that of $\mathbf{R}^{(0)}$. With the standard initialization $\mathbf{E}^{\text{tot}(0)} = \mathbf{E}^{\text{inc}}$, when two sources i and j are close, the corresponding incident fields e_i^{inc} and e_j^{inc} which make up the corresponding columns of $\mathbf{E}^{\text{tot}(0)}$, become highly correlated; hence, $\mathbf{R}^{(0)}$ becomes poorly conditioned and $\widetilde{\mathbf{R}}_0^\dagger \mathbf{V}$ is near-singular, which may lead to numerical errors propagating in the computations of matrices $\mathbf{A}$ and $\mathbf{B}$ (steps 6 and 12). This pathological behavior does not occur in the simulated experiments reported in [52], since only statistically independent RHS vectors are considered.

Here, we propose to add a small random quantity to $\mathbf{E}^{\text{tot}(0)}$ in the initialization step 1. This reduces the correlation between columns of $\mathbf{E}^{\text{tot}(0)}$, and thus improves the conditioning of matrix $\widetilde{\mathbf{R}}_0^\dagger \mathbf{V}$. Empirically, we have found that an initial perturbation with a signal-to-noise ratio ranging from 40 to 100 dB yielded roughly the same favorable effect, while a higher level of perturbation would slow down the convergence, and a smaller one would amount to no perturbation. In all experiments in this paper including such noisy initialization, the signal-to-noise ratio was set to 50 dB. The results reported in Section 4.2 show that the algorithm behaves in a satisfactory manner with such an initialization. Regarding the choice of matrix $\widetilde{\mathbf{R}}_0$, various options were studied; the best results were obtained by using the choice proposed in [52], that is, $\widetilde{\mathbf{R}}_0 = \mathbf{R}^{(0)} = \mathbf{E}^{\text{inc}} - \mathbf{L}_x \mathbf{E}^{\text{tot}(0)}$.

3.3. Implementation within the Reconstruction Procedure

We now consider the use of the block-BiCGStab algorithm for solving the two multiple linear systems involved in the evaluation of the cost function (8) and of its gradient (12) at each iteration of the optimization procedure. The first system reads $\mathcal{S}^{(t)} : \mathbf{E}^{\text{tot}(t)} = \mathbf{L}_{x^{(t)}}^{-1} \mathbf{E}^{\text{inc}}$, and the second one is of the form $\overline{\mathcal{S}}^{(t)} : \mathbf{V}^{(t)} = \overline{\mathbf{L}}_{x^{(t)}}^{-1} \mathbf{U}^{(t)}$, where notation $^{(t)}$ indexes the iteration number. According to the expression of the gradient (12), the RHS term $\mathbf{U}^{(t)}$ in $\overline{\mathcal{S}}^{(t)}$ depends on the solution of $\mathcal{S}^{(t)}$, so the two systems can only be solved sequentially and not jointly. In the following, details are given considering $\mathcal{S}^{(t)}$, but similar arguments hold for $\overline{\mathcal{S}}^{(t)}$.

As discussed in Section 3.2, initialization of block-BiCGStab plays an important role in the algorithm efficiency. Our implementation is based on the fact that two consecutive iterates of the optimization algorithm are close to each other, especially when the reconstruction algorithm is near to convergence. Therefore, the solution $\mathbf{E}^{\text{tot}(t+1)}$ should be close to that obtained at the former iteration $\mathbf{E}^{\text{tot}(t)}$, and $\mathbf{E}^{\text{tot}(t)}$ could be used as an initialization of block-BiCGStab for solving $\mathcal{S}^{(t+1)}$. In particular, $\mathbf{E}^{\text{tot}(t)}$ should be much closer to $\mathbf{E}^{\text{tot}(t+1)}$ than the matrix formed by the incident fields $\mathbf{E}^{\text{inc}}$, that was used for initializing block-BiCGStab for the resolution of a single set of forward problems in Section 3.2. We noticed, however, that implementing this strategy led to poor convergence rates, because the total fields $\mathbf{E}^{\text{tot}(t)}$ are highly correlated with each other, which generated conditioning issues similar to that explained in Section 3.2. Therefore, we propose to initialize iteration $t + 1$ using a random perturbation added to $\mathbf{E}^{\text{tot}(t)}$.

4. Performance of Block-BiCGStab on Synthetic Data

The results presented in this section were obtained on synthetic data representing a typical experimental setup. First, we illustrate the impact of the implementation choices discussed in Section 3.2. Then, we investigate the performance of the proposed block-BiCGStab implementation for computing multiple forward scattering problems as a function of the problem difficulty, by varying the object contrast and the illuminating frequency. Finally, the computational efficiency of block-BiCGStab is evaluated on the full reconstruction process.

4.1. Description of the Numerical Example and Implementation Details

The object used in our experiments was composed of two nested cubes embedded in air. This example was taken from [33] and the contrast was increased by a factor of two in order to make the reconstruction problem more challenging. The object domain V was a cube with a 30-cm edge size, which was discretized onto a $30 \times 30 \times 30$ regular Cartesian grid (1 cm^3 voxels). It is usually acknowledged that the discretization grid must have at least four voxel sides per wavelength for accurate estimation of the scattered field. Here, the highest illuminating frequency is 3 GHz (see Section 4.3.1), that is, $\lambda = 10$ cm, so that the voxel side is $\lambda/10$. The external part of the object was a cube with a 20-cm edge size, the contrast of which was set to $\chi_1 = 0.6 + j0.8$. The internal part was a smaller cube with a 10-cm edge size and contrast $\chi_2 = 1.2 + j0.4$. The object was illuminated by $N_S = 160$ sources (vertical electric dipoles) that were regularly spaced around the object on five circles with a 60 cm diameter, at heights $z = -20$ cm, -10 cm, 0, 10 cm and 20 cm. Measurements were simulated by considering 160 receivers at the same locations as the sources. The object and the acquisition setup are represented in Figure 1.

Data were generated according to model (7) using a grid twice as fine as the one used in our numerical experiments ($60 \times 60 \times 60$ voxels) for more accurate computations, and circular Gaussian white noise was added in order to obtain a 20-dB SNR. Only the z component of the scattered field was used in our tests. Then, the reconstruction problem had $30^3 = 27{,}000$ complex unknowns and $160^2 = 25{,}600$ complex-valued data points.

All computations were performed with an Intel Core i7-5960X (eight cores) clocked at 3 GHz using Matlab, with multithreaded operations. For BiCGStab, a marching-on-in-source procedure [37] was used when it improved convergence. This procedure allows for better initialization of the solutions by using an interpolation based on the total fields previously computed for the nearest sources. The tolerance on the relative residuals imposed for convergence of both BiCGStab and Block-BiCGStab (line 15 in Algorithms 1 and 2) was set to 10^{-6}. Therefore, the solutions obtained by the two algorithms are the same up to such numerical precision.

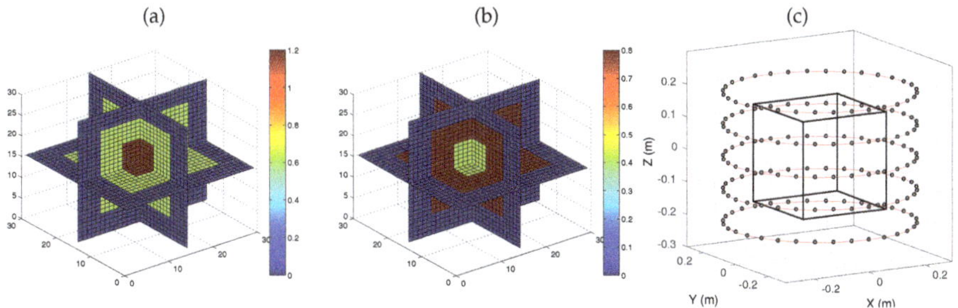

Figure 1. Real (**a**) and imaginary (**b**) parts of the contrast function of the simulated object (two nested cubes in air), with the corresponding discretization grids, and acquisition setup for the simulated object (**c**): the black cube represents the unknown volume V in which the object is placed. The gray circles represent the sources positions around the volume. The receivers are located at the same positions as the sources.

4.2. Impact of Initialization

The solution to the 3D forward problem corresponding to the experimental setup presented in Section 4.1 was computed with both BiCGStab and block-BiCGStab algorithms. First, the solutions were initialized to the incident fields ($e_i^{\mathrm{tot}(0)} = e_i^{\mathrm{inc}}$). Evolution of the iterated relative residuals $\|e_i^{\mathrm{inc}} - \mathbf{L}_x e_i^{\mathrm{tot}(k)}\| / \|e_i^{\mathrm{inc}}\|$ is shown in Figure 2a,b. BiCGStab converges in between 18 and 21 iterations for all problems, whereas block-BiCGStab residuals first slowly decrease, and then increase, leading to divergence.

Then, as described in Section 3.2, noise with 50 dB SNR was added to the initial solution of block-BiCGStab. The corresponding iterates are shown in Figure 2c—BiCGStab iterates, which are not represented, behave similarly to those represented in Figure 2a. On the contrary, block-BiCGStab iterates now converge much faster: in this example, block-BiCGStab converges in 40% less iterations than BiCGStab, corresponding to a similar saving in terms of computation time.

In all remaining experiments in the paper, the block-BiCGStab algorithm was initialized by adding such a noise to the incident fields.

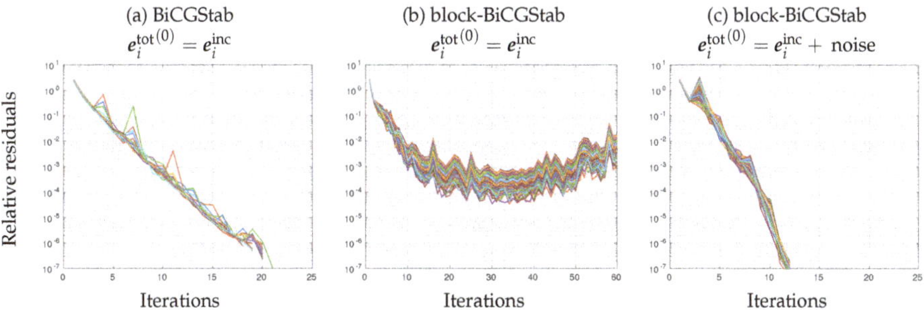

Figure 2. Resolution of 160 forward problems with biconjugate gradient stabilized (BiCGStab) (**a**) and with block-BiCGStab (**b,c**): evolution of the residuals for the 160 iterates. (**a,b**): initialization is set to the incident fields. (**c**) initialization of block-BiCGStab is perturbed with 50 dB circular Gaussian noise. Note the different abscissa scale on panel (**b**).

4.3. Performance for Different Forward Scattering Configurations

We now compare BiCGStab and block-BiCGStab for several configurations of the 3-D forward problem based on the previous setup. The tests consisted of solving the N_S systems $(\mathbf{I}_{3N} - \mathbf{G}_D\mathbf{X})e_i^{tot} = e_i^{inc}$, and we focused on the influence of the illuminating frequency and of the contrast values in $\mathbf{X}$, since both factors impact the problem difficulty—the Born approximation, for example, is only valid for low-contrast objects with small electric size.

4.3.1. Impact of the Frequency

We first study the algorithmic performance as a function of the illuminating frequency. The object defined in Section 4.1 was used, with frequencies equal to 1, 2 and 3 GHz. The corresponding wavelengths are respectively $\lambda = 30$ cm, $\lambda = 15$ cm and $\lambda = 10$ cm, and the size of the reconstructed volume respectively corresponds to λ, 2λ and 3λ: the higher the frequency, the bigger the electric size of the object.

Table 1 gives the corresponding central processing unit (CPU) time and the number of iterations required for BiCGStab and block-BiCGStab to converge. First, in accordance with the discussion in Section 3.1, the CPU time is almost proportional to the number of iterations performed by both algorithms. Second, block-BiCGStab always yields a shorter computation time than BiCGStab, and the improvement becomes more significant as the frequency increases: at 3 GHz, block-BiCGStab is 33% faster than BiCGStab for solving the 160 systems.

Table 1. CPU time and number of iterations required for computing the 160 forward simulations with BiCGStab and block-BiCGStab for three different frequencies. Results are averaged over the 160 resolutions for BiCGStab and over 20 realizations of the random initialization for block-BiCGStab.

Frequency	CPU Time		Number of Iterations [min, Average, max]		CPU Time Ratio Block/BiCGStab
	BiCGStab	Block	BiCGStab	Block	
1 GHz	116 s	110 s	[7 8.0 10]	[7 7.9 9]	0.95
2 GHz	187 s	138 s	[13 13.4 15]	[10 10 10]	0.74
3 GHz	245 s	164 s	[17 17.9 21]	[12 12 12]	0.67

4.3.2. Impact of the Contrast

We now evaluate the computation time as a function of the contrast. A scaling factor ranging from 0.25 to 2.5 was applied to the synthetic object defined in Section 4.1. The illuminating frequency was set to 3 GHz. Table 2 gives the corresponding CPU time and the number of iterations required for BiCGStab and block-BiCGStab. Here again, the CPU time is almost proportional to the number of iterations. In all cases, block-BiCGStab is faster than BiCGStab and, more importantly, the relative saving in CPU time increases with the contrast of the object: for a contrast factor of 2.5, block-BiCGStab runs approximately twice as fast as BiCGStab.

The results reported in Sections 4.3.1 and 4.3.2 therefore suggest that block-BiCGStab is best suited for the resolution of difficult problems, i.e., involving large and/or highly contrasted objects.

Table 2. CPU time and number of iterations required for computing the 160 forward simulations with BiCGStab and block-BiCGStab for different contrast levels. Results are averaged over the 160 resolutions for BiCGStab and over 20 realizations of the random initialization for block-BiCGStab.

Contrast	CPU Time		Number of Iterations [min, Average, max]		CPU Time Ratio Block/BiCGStab
	BiCGStab	Block	BiCGStab	Block	
0.25	105 s	86 s	[7 7.01 8]	[6 6 6]	0.82
0.5	152 s	120 s	[10 10.3 12]	[8 8.5 9]	0.79
1	245 s	164 s	[17 17.9 21]	[12 12 12]	0.67
1.5	361 s	217 s	[25 26.9 29]	[15 16.1 18]	0.60
2	482 s	271 s	[34 36.1 39]	[19 20.2 21]	0.56
2.5	610 s	331 s	[44 45.9 50]	[23 24.8 26]	0.54

4.4. Inversion Procedure and Object Reconstruction

We now consider the reconstruction problem with the simulation setup described in Section 4.1. Since the object was highly contrasted, illuminations at frequencies 1, 2 and 3 GHz were used and the following frequency hopping technique was implemented (see [28,61,62] for example):

- the initial solution was set to zero, and N_{it} iterations of the inversion algorithm were performed with the 1 GHz data;
- another N_{it} iterations were then performed with the 2 GHz data, with an initial solution set to the output of the previous step;
- the reconstruction algorithm was finally applied to 3 GHz data, with an initial solution set to the output of the previous step, and iterations are run until the ℓ_∞-norm of the gradient becomes lower than 10^{-6}.

The number N_{it} of iterations for each intermediate frequency is a trade-off between reconstruction quality and computation time. In this experiment, we found that $N_{it} = 20$ iterations were sufficient to achieve satisfactory results (that is, more iterations at intermediate frequencies would not improve the final reconstruction and would unnecessarily increase the computation time). The regularization weight in (9) was set to $\gamma = 10^{-6}$ for the three frequencies.

Reconstruction was performed using BiCGStab and block-BiCGStab, following the procedure proposed in Section 3.3. Since both algorithms used the same convergence thresholds, evaluations of the objective function and of its gradient at any point by the two algorithms were nearly identical; therefore, the minimization process followed roughly the same sequence, and the reconstructed objects obtained with both approaches are comparable. Reconstruction results are shown in Figure 3, together with the true object. The shapes of the two cubes are well retrieved and the contrast of the outer cube ($\chi_1 = 0.6 + j0.8$) is very close to that of the true object. The boundaries of the reconstructed cubes slightly exceed the true ones, especially for the inner cube. The contrast value in the inner cube is also slightly misestimated (underestimation of the real part and overestimation of the imaginary part). Note that the inner cube is highly contrasted ($\chi_2 = 1.2 + j0.4$), which makes it very difficult to reconstruct.

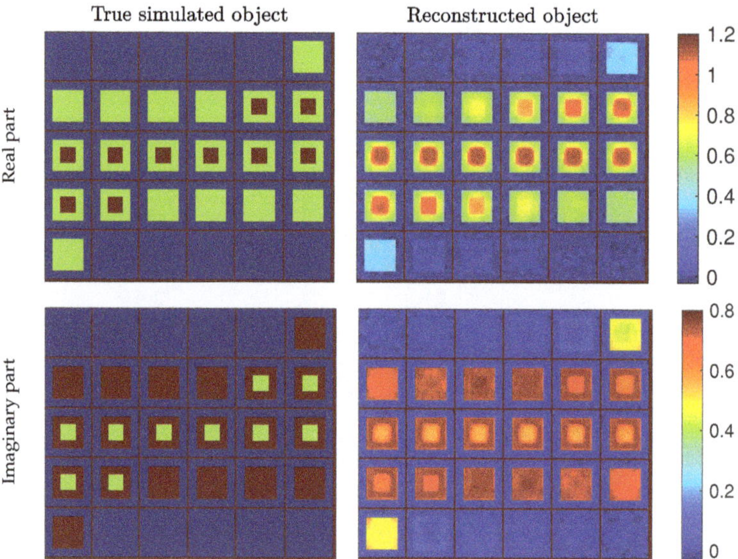

Figure 3. Reconstruction results (contrast values) for the simulated object defined in Section 4.1. Each panel represents the 30 slices (30 × 30 pixel images) of the three-dimensional object. True (**left**) and reconstructed (**right**) objects, with real (**top**) and imaginary (**bottom**) parts of the contrast function.

For this problem, the reconstruction lasted about 21.3 h when computations were performed with BiCGStab. The reconstruction time dropped to 16.8 hours when block-BiCGStab was used, which represents a saving of more than 20%.

5. Reconstruction of the Fresnel Database Objects

We finally consider the reconstruction of the five objects of the Fresnel database [54]. Such objects have already been reconstructed in many papers [27–29,61,63,64]—see also a summary of reconstruction results in [65]. Due to space limitations, we restrict the presentation to two objects, which can be considered as the two most difficult ones: the TwoSpheres and the Cylinder objects.

The 3-D experimental setup for the Fresnel data set is detailed in [54]. Using the reciprocity theorem and the two polarizations of the experimental data [29], we consider that the setup was equivalently composed of 36 sources placed on a circle at $z = 0$ with a 1.796-m radius. Data were equivalently acquired with 81 receivers placed on a sphere with a 1.796-m radius [54]. The incident fields were plane waves polarized along the z-axis and acquisitions were performed at frequencies ranging from 3 to 8 GHz. The two polarizations were used (co- and cross-polarization). Since the measurements were performed in the far-field zone of the scattering object, the electric field at the receivers is transverse (that is, its longitudinal component is zero). When projected into the Cartesian coordinate system, the three spatial components of the measured field are generally non-zero. Thus, each data set is made up of $36 \times 81 \times 3 = 8748$ complex-valued measurement points. For all objects, the volume of interest V was defined as a cube with a 10 cm edge size divided into $30 \times 30 \times 30$ voxels, so that the object is fully enclosed in V in each case. The maximum frequency is 8 GHz, that is, $\lambda = 37.4$ mm, so that the coarser discretization scheme corresponds to 11.3 voxel sides per wavelength. We then have 27,000 complex-valued unknowns. In all experiments, the regularization parameter γ in (9) was set to $\gamma = 10^{-6}$.

Here also, as discussed in Section 4.4, a frequency hopping procedure was used to carry out the reconstructions as follows:

- Perform N_{it} iterations of the reconstruction algorithm with the 3 GHz data and initial contrast function set to zero.
- Perform N_{it} iterations of the reconstruction algorithm with the 4 GHz data and initial contrast function set to the final iterate of the previous step.
- Repeat the previous step with the 5 GHz, 6 GHz and 7 GHz data.
- Perform reconstruction of the object with the 8 GHz data, initial contrast function set to the final iterate of the previous step and a stopping rule on the ℓ_∞-norm of the gradient set to 10^{-6}.

For all objects except `Cylinder`, computing $N_{it} = 10$ iterations at each frequency empirically led to a satisfactory compromise between reconstruction quality and time. For `Cylinder`, which was bigger and more contrasted, $N_{it} = 20$ intermediate iterations were necessary in order to achieve the best reconstruction.

The `TwoSpheres` object was composed of two identical spheres in contact embedded in air, 50 mm in diameter and with uniform, real-valued, relative permittivity equal to 2.6, that is, contrast $\chi = 1.6$. Figure 4 (top) shows 30 slices of the true object and of our corresponding reconstruction, and Figure 4 (bottom) represents the isosurface of the reconstruction at the contrast value $\chi = 1$. Note that we only present the real part of the estimated contrast function because the estimated imaginary part was very small. The latter is presented in the supplementary data files. It can be observed that the shape of the object was well reconstructed, even at the contact point. The contrast value was slightly overestimated at the contact point (the estimated contrast was about 1.8), and slightly underestimated around it (the estimated contrast was about 1.3).

The `Cylinder` object was composed of a cylinder of 80 mm in length and 80 mm in diameter, with uniform relative permittivity equal to 3.05 (contrast $\chi = 2.05$). It was the most difficult object of the database because of its large size and high contrast, and several methods failed to reconstruct this object [28,33,63,66], while many others produced poor quality results [27,29,61,62,64]. Figure 5 shows our reconstruction results. The shape and the contrast were adequately reconstructed: the contrast function was homogeneous inside the object and the estimated value at the center voxels was close to the true one, since the estimated permittivity is about 1.9. To the best of our knowledge, this reconstruction of `Cylinder` outperformed all other results previously reported in the literature.

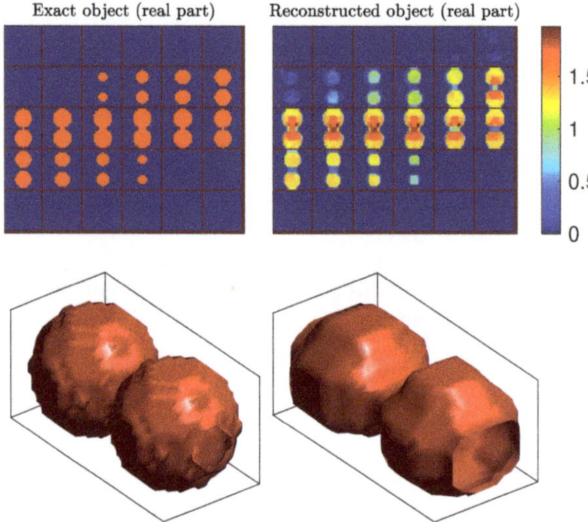

Figure 4. Reconstructed contrast function for the `TwoSpheres` object. Top: 30 horizontal slices of the true object (**left**) and the 30 slices of the reconstructed object (**right**). Bottom: isosurfaces of the true object discretized on the reconstruction grid (**left**), and of the reconstructed object (**right**), for contrast value 1.

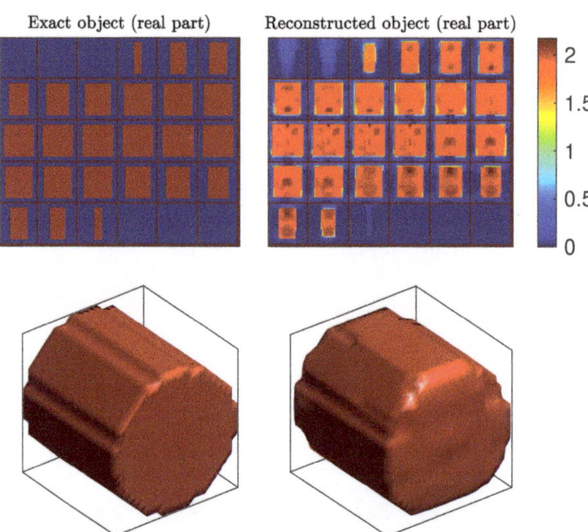

Figure 5. Reconstructed contrast function for the `Cylinder` object. Top: 30 horizontal slices of the true object (**left**) and the 30 slices of the reconstructed object (**right**). Bottom: isosurfaces of the true object discretized on the reconstruction grid (**left**), and of the reconstructed object (**right**), for contrast value 1.5.

On a broader perspective, Table 3 provides a heuristic assessment of the reconstruction quality for all objects in the Fresnel database. Each cell in the table contains a subjective comparison of our result with previously obtained ones for each object, based on the figures in the corresponding papers. We consider that the proposed method provided results at least similar to previous ones, and often better for the most difficult objects. The reader is referred to supplementary files associated with the paper, that depict graphical representations of all reconstructed objects and provide all reconstruction results as Matlab files.

Table 3. Subjective comparison of the obtained reconstruction results with previously published works on Fresnel database objects: "=", "+" and "++" respectively mean that our reconstruction is of comparable, better, and much better quality. N/A means that reconstruction of the object is not presented in the corresponding paper.

Paper Ref.	TwoCubes	IsocaSphere	CubeSpheres	TwoSpheres	Cylinder
[63]	++	=	=	++	N/A
[29]	++	=	+	++	++
[28]	+	+	=	+	N/A
[64]	++	+	+	++	++
[61]	+	=	+	=	+
[27]	+	=	=	=	++
[33]	+	=	N/A	N/A	N/A
[66]	+	N/A	N/A	=	N/A
[62]	N/A	N/A	+	=	++

Table 4 finally gives the reconstruction time for the five objects of the Fresnel database using either BiCGStab or block-BiCGStab. The time saved by block-BiCGStab becomes more important as the difficulty of the reconstruction problem increases (i.e., as objects become larger and/or more contrasted). For the most difficult `Cylinder` object, the reconstruction time drops from 60 h to 28.3 h, which means that block-BiCGStab is more than twice faster than BiCGStab.

Table 4. Reconstruction time for the five objects of Fresnel database using BiCGStab and block-BiCGStab for linear system inversions.

Object from Database	Time BiCGStab	Time Block-BiCGStab	Time Ratio Block/BiCGStab
TwoCubes	2.33 h	2.07 h	0.89
IsocaSphere	4.38 h	3.58 h	0.82
CubeSpheres	3.21 h	2.60 h	0.81
TwoSpheres	8.60 h	6.41 h	0.75
Cylinder	60.0 h	28.3 h	0.47

6. Discussion

We proposed a reconstruction strategy based on an additive regularization framework, where optimization is performed via first-order optimization. Computations of the cost function and of its gradient were performed through a dedicated implementation of the block-BiCGStab algorithm for multiple RHS systems inversion. In all our experiments, such an implementation led to a significant reduction of the computation time (up to one half) compared to BiCGStab, the gain being all the more significant that the reconstruction problem is difficult and the computation time is high. The procedure proposed in this paper is therefore particularly attractive for problems dealing with high-frequency illuminations and/or highly contrasted objects, which are of great interest: high-frequency acquisitions contain more detailed spatial information about the object under study, yielding higher resolution, and highly contrasted objects may be encountered in many application areas, such as biomedical imaging. Consequently, the proposed procedure may contribute to reducing the computational burden of microwave imaging, which still represents a crucial issue in the development of practical applications.

Improved convergence properties of block-BiCGStab over its sequential counterpart also enable block-BiCGStab to reach a low residual level for each iterative system inversion. In our experiments, the tolerance threshold on the norm of the residuals was set to 10^{-6}, whereas higher values are commonly used. More accurate solutions to the forward problem yield more efficient optimization steps; this may explain why the proposed procedure achieved satisfactory reconstruction results for the most difficult objects of the Fresnel database, for which no result of equivalent quality has been reported in the literature.

Exploiting the proposed block resolution strategy may also improve the efficiency of any other microwave imaging inversion method requiring the computation of similar quantities (i.e., the total fields from the incident fields), either based on additive [13] or multiplicative [33] regularization. This strategy can also be advantageously used whatever the choice of the optimization method, such as non-linear conjugate gradient [29,64], regularized Gauss–Newton [33], Levenberg–Marquardt [67], or distorted Born iterative method [27]. It would also be worth studying the numerical performance of the block-BiCGStab algorithm in different geometrical configurations concerning, e.g., data acquisition setups or discretization schemes. Finally, it may also be efficiently used in other applications involving the computation of the total electromagnetic field accounting for scattering, e.g., radar cross-section [48,68,69], eddy-current [26,70] or magnetotelluric [71] computations. Depending on the complexity of the problem, specific tuning of the block-BiCGStab algorithm could be required, for which the strategies proposed in this paper may help.

Further improvements may be brought by parallel implementation of the proposed procedure. In our experiments, multicore architecture was taken advantage of only by performing multithreaded operations at a low level within the block-BiCGStab loop. However, massive parallelization of the resolution of independent systems (with BiCGStab) reduces the computation time by exploiting parallelization at a higher level [33]. Indeed, the two parallelization levels could be combined into a block resolution procedure, where the multiple systems would be split into several subsystems to be solved in parallel. In [53], such an idea was successfully tested on an early version of block-BiCGStab. This remains a promising possibility to accelerate microwave imaging in the context of High Performance Computing, as well as parallelization capacities brought by GPU- or multi-GPU-based implementations.

Author Contributions: Conceptualization, J.I. and Y.G.; methodology, C.F., S.B., J.I. and Y.G.; software, C.F. and S.B.; formal analysis, C.F., S.B., J.I. and Y.G.; data curation, C.F.; writing—original draft preparation, C.F., S.B., J.I. and Y.G. All authors have read and agreed to the published version of the manuscript.

Funding: This research received no external funding.

Acknowledgments: The authors would like to thank Patrick Chaumet, from the Fresnel Institute (Marseille, France), for granting access to the Fresnel database and giving us valuable advice about data capture and post-processing.

Conflicts of Interest: The authors declare no conflict of interest.

References

1. Pastorino, M. *Microwave Imaging*; John Wiley & Sons: Hoboken, NJ, USA, 2010.
2. Gilmore, C.; Abubakar, A.; Hu, W.; Habashy, T.M.; van den Berg, P.M. Microwave Biomedical Data Inversion Using the Finite-Difference Contrast Source Inversion Method. *IEEE Trans. Antennas Propag.* **2009**, *57*, 1528–1538. [CrossRef]
3. Scapaticci, R.; Di Donato, L.; Catapano, I.; Crocco, L. A Feasibility Study on Microwave Imaging for Brain Stroke Monitoring. *Prog. Electromagn. Res. B* **2012**, *40*, 305–324. [CrossRef]

4. Abubakar, A.; Habashy, T.M.; Druskin, V.L.; Knizhnerman, L.; Alumbaugh, D.L. 2.5D Forward and Inverse Modeling for Interpreting Low-Frequency Electromagnetic Measurements. *Geophysics* **2008**, *73*, F165–F177. [CrossRef]
5. Forte, E.; Pipan, M.; Casabianca, D.; Cuia, R.D.; Riva, A. Imaging and characterization of a carbonate hydrocarbon reservoir analogue using {GPR} attributes. *J. Appl. Geophys.* **2012**, *81*, 76–87. [CrossRef]
6. Maaref, N.; Millot, P.; Pichot, C.; Picon, O. FMCW ultra-wideband radar for through-the- wall detection of human beings. In Proceedings of the 2009 International Radar Conference "Surveillance for a Safer World" (RADAR 2009), Bordeaux, France, 12–16 October 2009.
7. Asefi, M.; Jeffrey, I.; LoVetri, J.; Gilmore, C.; Card, P.; Paliwal, J. Grain bin monitoring via electromagnetic imaging. *Comput. Electron. Agric.* **2015**, *119*, 133–141. [CrossRef]
8. Isernia, T.; Pascazio, V.; Pierri, R. A nonlinear estimation method in tomographic imaging. *IEEE Trans. Geosci. Remote Sens.* **1997**, *35*, 910–923. [CrossRef]
9. Fang, Q.; Meaney, P.M.; Paulsen, K.D. Singular value analysis of the Jacobian matrix in microwave image reconstruction. *IEEE Trans. Antennas Propag.* **2006**, *54*, 2371–2380. [CrossRef]
10. Kleinman, R.E.; van den Berg, P.M. A Modified Gradient Method for Two-Dimensional Problems in Tomography. *J. Comput. Appl. Math.* **1992**, *42*, 17–35. [CrossRef]
11. Van den Berg, P.M.; Kleinman, R.E. A Total Variation Enhanced Modified Gradient Algorithm for Profile Reconstruction. *Inverse Probl.* **1995**, *11*, L5–L10. [CrossRef]
12. Abubakar, A.; van den Berg, P.M.; Mallorqui, J.J. Imaging of Biomedical Data Using a Multiplicative Regularized Contrast Source Inversion Method. *IEEE Trans. Microw. Theory Tech.* **2002**, *50*, 1761–1771. [CrossRef]
13. Barrière, P.A.; Idier, J.; Laurin, J.J.; Goussard, Y. Contrast Source Inversion Method Applied to Relatively High Contrast Objects. *Inverse Probl.* **2011**, *27*, 075012. [CrossRef]
14. Abubakar, A.; van den Berg, P.M. Iterative Forward and Inverse Algorithms Based on Domain Integral Equations for Three-Dimensional Electric and Magnetic Objects. *J. Comput. Phys.* **2004**, *195*, 236–262. [CrossRef]
15. Zhang, Z.Q.; Liu, Q.H. Three-Dimensional Nonlinear Image Reconstruction for Microwave Biomedical Imaging. *IEEE Trans. Biomed. Eng.* **2004**, *51*, 544–548. [CrossRef]
16. Catapano, I.; Crocco, L.; D'Urso, M.; Isernia, T. A Novel Effective Model for Solving 3-D Nonlinear Inverse Scattering Problems in Lossy Scenarios. *IEEE Geosci. Remote Sens. Lett.* **2006**, *3*, 302–306. [CrossRef]
17. Van den Berg, P.M.; Abubakar, A. Contrast Source Inversion Method: State of Art. *Prog. Electromagn. Res.* **2001**, *34*, 189–218. [CrossRef]
18. Abubakar, A.; van den Berg, P.M.; Habashy, T.M. Application of the Multiplicative Regularized Contrast Source Inversion Method on TM- and TE-Polarized Experimental Fresnel Data. *Inverse Probl.* **2005**, *21*, S5–S13. [CrossRef]
19. Bozza, G.; Pastorino, M. An Inexact Newton-Based Approach to Microwave Imaging Within the Contrast Source Formulation. *IEEE Trans. Antennas Propag.* **2009**, *57*, 1122–1132. [CrossRef]
20. Gilmore, C.; Mojabi, P.; LoVetri, J. Comparison of an Enhanced Distorted Born Iterative Method and the Multiplicative-Regularized Contrast Source Inversion Method. *IEEE Trans. Antennas Propag.* **2009**, *57*, 2341–2351. [CrossRef]
21. Mojabi, P.; LoVetri, J. Eigenfunction Contrast Source Inversion for Circular Metallic Enclosures. *Inverse Probl.* **2010**, *26*, 025010. [CrossRef]
22. Rubaek, T.; Meaney, P.M.; Paulsen, K.D. A Contrast Source Inversion Algorithm Formulated Using the Log-Phase Formulation. *Int. J. Antennas Propag.* **2011**, *2011*, 849894. [CrossRef]
23. Li, M.; Semerci, O.; Abubakar, A. A Contrast Source Inversion Method in the Wavelet Domain. *Inverse Probl.* **2013**, *29*, 025015. [CrossRef]
24. Ozdemir, O. Cauchy Data Contrast Source Inversion Method. *IEEE Geosci. Remote Sens. Lett.* **2014**, *11*, 858–862. [CrossRef]
25. Harada, H.; Tanaka, M.; Takenaka, T. Image Reconstruction of a Three-Dimensional Dielectric Object Using a Gradient-Based Optimization. *Microw. Opt. Technol. Lett.* **2001**, *29*, 332–336. [CrossRef]
26. Soleimani, M.; Lionheart, W.R.B.; Peyton, A.J.; Ma, X.; Higson, S.R. A Three-Dimensional Inverse Finite-Element Method Applied to Experimental Eddy-Current Imaging Data. *IEEE Trans. Magn.* **2006**, *42*, 1560–1567. [CrossRef]

27. Yu, C.; Yuan, M.; Liu, Q.H. Reconstruction of 3D Objects from Multi-Frequency Experimental Data with a Fast DBIM-BCGS Method. *Inverse Probl.* **2009**, *25*, 024007. [CrossRef]

28. De Zaeytijd, J.; Franchois, A. Three-Dimensional Quantitative Microwave Imaging from Measured Data with Multiplicative Smoothing and Value Picking Regularization. *Inverse Probl.* **2009**, *25*, 024004. [CrossRef]

29. Chaumet, P.C.; Belkebir, K. Three-Dimensional Reconstruction from Real Data Using a Conjugate Gradient-Coupled Dipole Method. *Inverse Probl.* **2009**, *25*, 024003. [CrossRef]

30. Winters, D.W.; Van Veen, B.D.; Hagness, S.C. A Sparsity Regularization Approach to the Electromagnetic Inverse Scattering Problem. *IEEE Trans. Antennas Propag.* **2010**, *58*, 145–154. [CrossRef]

31. Grzegorczyk, T.M.; Meaney, P.M.; Kaufman, P.A.; diFlorio Alexander, R.M.; Paulsen, K.D. Fast 3-D Tomographic Microwave Imaging for Breast Cancer Detection. *IEEE Trans. Med. Imaging* **2012**, *31*, 1584–1592. [CrossRef] [PubMed]

32. Abubakar, A.; Habashy, T.M.; Pan, G. Microwave Data Inversions Using the Source-Receiver Compression Scheme. *IEEE Trans. Antennas Propag.* **2012**, *60*, 2853–2864. [CrossRef]

33. Abubakar, A.; Habashy, T.M.; Pan, G.; Li, M.K. Application of the Multiplicative Regularized Gauss-Newton Algorithm for Three-Dimensional Microwave Imaging. *IEEE Trans. Antennas Propag.* **2012**, *60*, 2431–2441. [CrossRef]

34. Estatico, C.; Pastorino, M.; Randazzo, A. Microwave Imaging of Three-Dimensional Targets by Means of an Inexact-Newton-Based Inversion Algorithm. *Int. J. Antennas Propag.* **2013**, *2013*, 407607. [CrossRef]

35. Scapaticci, R.; Kosmas, P.; Crocco, L. Wavelet-Based Regularization for Robust Microwave Imaging in Medical Applications. *IEEE Trans. Biomed. Eng.* **2015**, *62*, 1195–1202. [CrossRef]

36. Zhang, Z.Q.; Liu, Q.H.; Xiao, C.; Ward, E.; Ybarra, G.; Joines, W.T. Microwave breast imaging: 3-D forward scattering simulation. *IEEE Trans. Biomed. Eng.* **2003**, *50*, 1180–1189. [CrossRef]

37. De Zaeytijd, J.; Franchois, A.; Eyraud, C.; Geffrin, J.M. Full-Wave Three-Dimensional Microwave Imaging With a Regularized Gauss-Newton Method—Theory and Experiment. *IEEE Trans. Antennas Propag.* **2007**, *55*, 3279–3292. [CrossRef]

38. Rocca, P.; Benedetti, M.; Donelli, M.; Franceschini, D.; Massa, A. Evolutionary optimization as applied to inverse scattering problems. *Inverse Probl.* **2009**, *25*, 123003. [CrossRef]

39. Cai, C.; Bilicz, S.; Rodet, T.; Lambert, M.; Lesselier, D. Metamodel-Based Nested Sampling for Model Selection in Eddy-Current Testing. *IEEE Trans. Magn.* **2017**, *53*, 6200912. [CrossRef]

40. Oliveri, G.; Zhong, Y.; Chen, X.; Massa, A. Multiresolution Subspace-Based Optimization Method for Inverse Scattering Problems. *JOSA A* **2011**, *28*, 2057–2069. [CrossRef]

41. Zaimaga, H.; Lambert, M. Soft Shrinkage Thresholding Algorithm for Nonlinear Microwave Imaging. In Proceedings of the 6th International Workshop on New Computational Methods for Inverse Problems, Cachan, France, 20 May 2016.

42. Sarkar, T.K.; Arvas, E.; Rao, S.M. Application of FFT and the Conjugate Gradient Method for the Solution of Electromagnetic Radiation from Electrically Large and Small Conducting Bodies. *IEEE Trans. Antennas Propag.* **1986**, *34*, 635–640. [CrossRef]

43. Zwamborn, P.; van den Berg, P.M. The Three-dimensional Weak Form of the Conjugate Gradient FFT Method for Solving Scattering Problems. *IEEE Trans. Microw. Theory Tech.* **1992**, *40*, 1757–1766. [CrossRef]

44. Gan, H.; Chew, W.C. A Discrete BCG-FFT Algorithm for Solving 3D Inhomogeneous Scatterer Problems. *J. Electromagnet. Wave* **1995**, *9*, 1339–1357. [CrossRef]

45. Wang, C.F.; Jin, J.M. Simple and Efficient Computation of Electromagnetic Fields in Arbitrarily Shaped Inhomogeneous Dielectric Bodies Using Transpose-Free QMR and FFT. *IEEE Trans. Microw. Theory Tech.* **1998**, *46*, 553–558. [CrossRef]

46. Ellis, R.G. Electromagnetic Inversion Using the QMR-FFT Fast Integral Equation Method. SEG Technical Program Expanded Abstracts 2002. Available online: https://library.seg.org/doi/abs/10.1190/1.1817145 (accessed on 20 October 2020).

47. Xu, X.; Liu, Q.H.; Zhang, Z.Q. The Stabilized Biconjugate Gradient Fast Fourier Transform Method for Electromagnetic Scattering. In Proceedings of the Antennas and Propagation Society International Symposium, San Antonio, TX, USA, 16–21 June 2002.
48. Wu, F.; Zhang, Y.; Oo, Z.Z.; Li, E. Parallel Multilevel Fast Multipole Method for Solving Large-Scale Problems. *IEEE Ant. Propag. Mag.* **2005**, *47*, 110–118.
49. Fang, Q.; Meaney, P.M.; Geimer, S.D.; Streltsov, A.V.; Paulsen, K.D. Microwave Image Reconstruction from 3-D Fields Coupled to 2-D Parameter Estimation. *IEEE Trans. Med. Imaging* **2004**, *23*, 475–484. [CrossRef]
50. O'Leary, D.P. The Block Conjugate Gradient Algorithm and Related Methods. *Linear Algebra Appl.* **1980**, *29*, 293–322. [CrossRef]
51. Freund, R.W.; Malhotra, M. A Block QMR Algorithm for Non-Hermitian Linear Systems with Multiple Right-Hand Sides. *Linear Algebra Appl.* **1997**, *254*, 119–157. [CrossRef]
52. El Guennouni, A.; Jbilou, K.; Sadok, H. A Block Version of BiCGSTAB for Linear Systems with Multiple Right-Hand Sides. *Electron. Trans. Numer. Anal.* **2003**, *16*, 2.
53. Friedrich, C.; Bourguignon, S.; Idier, J.; Goussard, Y. A Block Version of the BiCGStab Algorithm for 3-D Forward Problem Resolution in Microwave Imaging. In Proceedings of the 9th European Conference on Antennas and Propagation, Lisbon, Portugal, 13–17 April 2015.
54. Geffrin, J.M.; Sabouroux, P. Continuing with the Fresnel Database: Experimental Setup and Improvements in 3D Scattering Measurements. *Inverse Probl.* **2009**, *25*, 024001. [CrossRef]
55. Harrington, R.F.; Harrington, J.L. *Field Computation by Moment Methods*; Oxford University Press: Oxford, UK, 1996.
56. Friedrich, C.; Bourguignon, S.; Idier, J.; Goussard, Y. Reconstruction of 3-D Microwave Images Based on a Block-BiCGStab Algorithm. In Proceedings of the 5th International Workshop on New Computational Methods for Inverse Problems (NCMIP2015), Cachan, France, 29 May 2015.
57. Carfantan, H.; Mohammad-Djafari, A. Diffraction Tomography. *Bayesian Approach to Inverse Problems*; Idier, J., Ed.; ISTE Ltd. and John Wiley & Sons Inc.: Hoboken, NJ, USA, 2008; pp. 335–355.
58. Xu, K.; Zhong, Y.; Wang, G. A Hybrid Regularization Technique for Solving Highly Nonlinear Inverse Scattering Problems. *IEEE Trans. Microw. Theory Tech.* **2017**, *66*, 11–21. [CrossRef]
59. Nocedal, J.; Wright, S. *Numerical Optimization*; Springer: Berlin/Heidelberg, Germany, 1999.
60. Balanis, C. *Advanced Engineering Electromagnetics*, 2nd ed.; Wiley: Hoboken, NJ, USA, 2012.
61. Li, M.; Abubakar, A.; van den Berg, P.M. Application of the Multiplicative Regularized Contrast Source Inversion Method on 3D Experimental Fresnel Data. *Inverse Probl.* **2009**, *25*, 024006. [CrossRef]
62. Mudry, E.; Chaumet, P.C.; Belkebir, K.; Sentenac, A. Electromagnetic Wave Imaging of Three-Dimensional Targets Using a Hybrid Iterative Inversion Method. *Inverse Probl.* **2012**, *28*, 065007.
63. Catapano, I.; Crocco, L.; D'Urso, M.; Isernia, T. 3D Microwave Imaging via Preliminary Support Reconstruction: Testing on the Fresnel 2008 Database. *Inverse Probl.* **2009**, *25*, 024002.
64. Eyraud, C.; Litman, A.; Hérique, A.; Kofman, W. Microwave Imaging From Experimental Data Within a Bayesian Framework with Realistic Random Noise. *Inverse Probl.* **2009**, *25*, 024005.
65. Litman, A.; Crocco, L. Testing Inversion Algorithms Against Experimental Data: 3D Targets. *Inverse Probl.* **2009**, *25*, 020201.
66. Li, M.; Abubakar, A.; Habashy, T.M. A Three-Dimensional Model-Based Inversion Algorithm Using Radial Basis Functions for Microwave Data. *IEEE Trans. Antennas Propag.* **2012**, *60*, 3361–3372.
67. Franchois, A.; Pichot, C. Microwave Imaging-Complex Permittivity Reconstruction with a Levenberg-Marquardt Method. *IEEE Trans. Antennas Propag.* **1997**, *45*, 203–215.
68. Zhang, Z.Q.; Liu, Q.H.; Xu, X.M. RCS Computation of Large Inhomogeneous Objects Using a Fast Integral Equation Solver. *IEEE Trans. Antennas Propag.* **2003**, *51*, 613–618.
69. Al Sharkawy, M.; El-Ocla, H. Electromagnetic Scattering From 3-D Targets in a Random Medium Using Finite Difference Frequency Domain. *IEEE Trans. Antennas Propag.* **2013**, *61*, 5621–5626.
70. Sabbagh, H.A.; Sabbagh, L.D. An Eddy-Current Model for Three-Dimensional Inversion. *IEEE Trans. Magn.* **1986**, *22*, 282–291.

71. Liu, J.X.; Jiang, P.F.; Tong, X.Z.; Xu, L.H.; Xie, W.; Wang, H. Application of BICGSTAB algorithm with incomplete LU decomposition preconditioning to two-dimensional magnetotelluric forward modeling. *J. Cent. South Univ. (Sci. Tech.)* **2009**, *2*, 42.

Publisher's Note: MDPI stays neutral with regard to jurisdictional claims in published maps and institutional affiliations.

Article

Through-the-Wall Microwave Imaging: Forward and Inverse Scattering Modeling

Alessandro Fedeli [1], Matteo Pastorino [1], Cristina Ponti [2,3,*] Andrea Randazzo [1,*] and Giuseppe Schettini [2,3]

[1] Department of Electrical, Electronic, Telecommunications Engineering, and Naval Architecture, University of Genoa, 16145 Genoa, Italy; alessandro.fedeli@unige.it (A.F.); matteo.pastorino@unige.it (M.P.)

[2] Department of Engineering, "Roma Tre" University, via Vito Volterra 62, 00146 Rome, Italy; giuseppe.schettini@uniroma3.it

[3] National Interuniversity Consortium for Telecommunications, Roma Tre University, 00146 Rome, Italy

* Correspondence: cristina.ponti@uniroma3.it (C.P.); andrea.randazzo@unige.it (A.R.)

Received: 18 April 2020; Accepted: 16 May 2020; Published: 18 May 2020

Abstract: The imaging of dielectric targets hidden behind a wall is addressed in this paper. An analytical solver for a fast and accurate computation of the forward scattered field by the targets is proposed, which takes into account all the interactions of the electromagnetic field with the interfaces of the wall. Furthermore, an inversion procedure able to address the full underlying non-linear inverse scattering problem is introduced. This technique exploits a regularizing scheme in Lebesgue spaces in order to reconstruct an image of the hidden targets. Preliminary numerical results are provided in order to initially assess the capabilities of the developed solvers.

Keywords: electromagnetic scattering; buried objects; through-wall radar; microwave imaging; inverse scattering

---

1. Introduction

Microwave imaging of targets placed behind a wall is a topic of great interest in the remote detection of humans or objects in indoor environments, with applications in surveillance, rescue, and defense [1]. The instrument usually adopted in this class of surveys is the so-called through-wall (TW) radar, which includes a wide set of possible hardware architectures. For example, as regards the source of the interrogating field, a possible solution is to use wideband antennas radiating pulsed electromagnetic (EM) fields, or to adopt frequency-stepped sources, radiating an EM field at a limited number of frequencies.

Beyond the radar architecture, the processing of experimental data plays an important role. In most processing approaches, the goal is to localize hidden targets or produce a qualitative image of the indoor scenario [1–6], returning information about the target's shape or the presence of interfaces. Accurate forward scattering approaches may improve the reconstruction capabilities of radar surveys. First, the synthetic data obtained by the numerical modeling are helpful in providing a deeper physical insight on the fields scattered by typical TW targets, in several practical cases. Second, the theoretical solution to the forward scattering problem is a useful tool in imaging approaches, especially when aiming for a quantitative reconstruction of the target. In this respect, forward solvers find twofold applications. On the one hand, the numerical data they provide can be used as input data to validate inversion schemes. On the other hand, forward solvers can be employed as building blocks of non-linear inversion algorithms themselves. In this perspective, the forward scattering solver should develop a full-wave solution, where all the effects of the wall interfaces on the propagation of the scattered field are suitably modeled. As for the forward solvers in TW problems, the methods proposed in the literature are essentially numerical [2,3,7–10]. Due to its high flexibility, the finite-difference time-domain (FDTD) method is mainly adopted, and, being directly developed in the time-domain, its application to TW radars with pulsed sources is straightforward [7–9,11]. When modeling stepped-frequency sources, frequency domain data should be computed through an inverse fast Fourier transform [2]. However, due to the large size of investigation domains in TW settings, the FDTD modeling for frequency domain analysis may be demanding in terms of execution times and memory requirements. As for the frequency-domain techniques, other methods such as the ones based on method-of-moments (MoM) approaches may be employed, which are still numerical [3]. Asymptotic techniques are also used for modeling very large regions [10]. However, when applied to imaging approaches, forward solvers are usually implemented through linearized formulations [2,3,12–14] or by using synthetic aperture schemes [4,5,15], thus leading to qualitative images of the target. Techniques to improve the spatial resolution [16,17] and human discrimination [18,19] have also been proposed, through frequency spectral analysis. The implementation of full-wave inverse scattering approaches for quantitative TW imaging is still an open issue, as most algorithms for solving the non-linear inverse scattering problem are usually developed for free-space applications [20–30] and must be properly adapted to include the presence of the wall.

In this paper, the issues relevant to the modeling of the forward scattering problem and to the quantitative imaging are both addressed. In particular, a single-frequency non-linear inversion procedure based on a regularization scheme in Lebesgue-spaces is proposed. This kind of inversion strategy was initially developed in free-space environments [31–35] and subsequently extended to the case of targets buried in a homogeneous soil [36], showing good reconstruction capabilities. It has been found that the geometrical properties of Lebesgue-space norms lead to regularized solutions endowed with less oversmoothing than classical Hilbert-space regularization schemes, improving both target localization and their shaping. In this work, the method is extended to the case of targets hidden behind a dielectric wall, by modifying the underlying scattering model in order to include the proper Green's function for layered media [7]. The validity of this inversion scheme is assessed in its forward scattering formulation, as well as in its application to the imaging procedure. For the validation of the proposed technique, an analytical solver has been employed, called the cylindrical wave approach (CWA) [37–41]. It provides an analytical/numerical solution to a layout with buried targets, given by circular cross-section cylinders, employing cylindrical waves as basis functions of the scattered field. Due to the presence of the interface, suitable cylindrical waves are introduced, defined through spectral integrals, to deal with reflection and transmission of the scattered field by the interfaces. A preliminary implementation of the CWA for TW scattering was proposed in [38], with an approach through multiple reflection fields to accurately describe the multipath inside the wall. In [39], the CWA has been developed through a non-iterative approach, where all the multiple interactions were included in suitable reflection and transmission coefficients. The same technique has been applied in [41], where a pulsed source field has been modeled. In this work, the results provided by the analytical solver are validated considering an integral formulation based on a Green's function approach, and they are used as reference data to test the reconstruction capabilities of the inversion procedure. It is worth remarking that the novelty of the paper is both on inverse and forward modeling. From the point of view of the inversion procedure, an efficient technique working in the framework of Lebesgue spaces (which was developed for free-space and half-space scenarios) is here extended to through-the-wall configurations. Specifically, the presence of the wall is considered by inserting the proper Green's function into the scattering model. In this way, the through-the-wall propagation phenomena are fully taken into account, even under near-field conditions (differently from synthetic aperture and beamforming schemes, where far-field conditions are usually assumed). Moreover, the adopted scattering model does not rely on approximations (e.g., the Born or Kirchhoff models often adopted in TW imaging). However, the presence of multiple interfaces and the availability of few limited-view measurements (i.e., only along one side of the wall) significantly increase the difficulty of the inverse problem. Consequently, the present paper is aimed at evaluating the regularization properties of the approach even in this more involved case. Moreover, an automatic criterion for selecting the optimal Lebesgue-space norm parameter based on the entropy principle is proposed for the first time. The analytical solver used in the validation of the inversion procedure is implemented for a monochromatic line-current source and dielectric targets, applying the spectral-domain analysis developed in [39] to a more realistic source. In the proposed approach, the total field is decomposed into two different sets: scattered fields by the cylinders, and non-scattered fields, i.e., the field radiated by the line source and the fields excited by its reflection and transmission at the interfaces, in the absence of the cylinders. The non-iterative approach is applied to both sets and, through suitable reflection and transmission coefficients, all the multiple interactions of the fields inside the layer are collected in two contributions: an up-ward and a down-ward propagating wave. Therefore, a theoretical solution is developed through a very compact formulation, with the total field in each medium decomposed in a limited number of terms. This approach leads to a numerical implementation which is fast and efficient.

The paper is organized as follows: in Section 2, an overview of the proposed forward and inverse scattering approaches is reported. Forward and inverse scattering results are then presented in Section 3. Conclusions follow in Section 4.

2. Theoretical Approach to the Through-Wall Imaging Problem

The geometry of the TW imaging problem is shown in Figure 1. A two-dimensional layout is considered, with one lossless dielectric wall between two semi-infinite regions filled with air (i.e., characterized by the vacuum dielectric permittivity, ε_0). The wall has relative permittivity ε_{r1} and thickness l. The hidden investigation domain D_{inv}, highlighted by the dashed box in Figure 1, is located in the medium behind the wall, and contains one infinitely long cylinder with circular cross section having center in (x_c, y_c), radius a, and relative permittivity ε_{rc}.

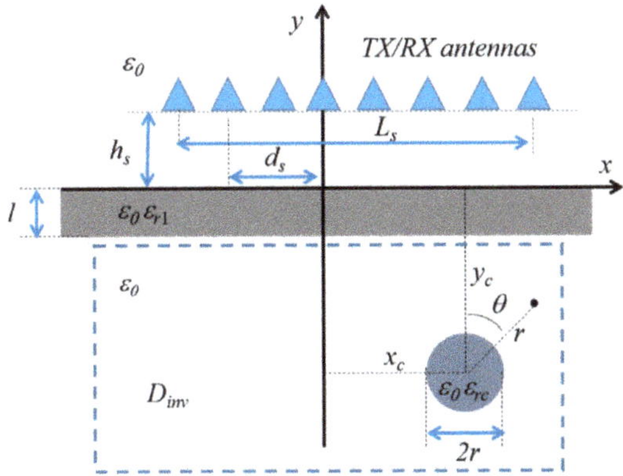

Figure 1. Configuration of the scattering problem, with one target placed behind a dielectric wall.

A set of M transmitting/receiving antennas placed along a line of length L_s parallel to the interface at a fixed distance $y = h_s$ in front of the wall is used. The transmitting antennas are modeled by monochromatic line-current sources with angular frequency ω, and it is assumed that a TM$_z$-polarized incident electric field $\mathbf{E}_{inc} = E_{inc}(x, y)\hat{\mathbf{z}}$ is excited. The expression of the field radiated by the transmitting antennas with center in (d_s, h_s) is given by [42]:

$$E_{inc}(x, y) = V_0 H_0^{(2)}\left(\sqrt{(x - d_s)^2 + (y - h_s)^2} \right) \tag{1}$$

where $H_0^{(2)}(\cdot)$ is the zero-th order second-kind Hankel function, and V_0 is the complex amplitude of the field. The $e^{j\omega t}$ term is omitted throughout the paper.

Antennas are scanned in a multi-illumination multi-view configuration, i.e., each antenna is used in turn in transmission mode to radiate the incident electric field in (1), and the total field $\mathbf{E}_{tot} = E_{tot}(x,y)\hat{\mathbf{z}}$ produced by the interaction of the EM wave with the investigation domain (including the wall and the target) is received by the remaining $M-1$ antennas. It is worth remarking that the assumed scattering model is formally exact only when dealing with cylindrical targets (i.e., ideally infinite and invariant along the z direction) under a TM_z illumination. In practical TW imaging applications, the inspected objects, as well as the wall, although not being infinite are usually elongated along the vertical direction (corresponding to the z axis in our settings). Consequently, the predicted fields are sufficiently accurate for solving the imaging problem at hand.

2.1. Forward-Scattering Problem Formulation

The theoretical method adopted to evaluate the scattered field in the layout of Figure 1 is the cylindrical wave approach [39,41]. The total field $\mathbf{E}_{tot} = E_{tot}(x,y)\hat{\mathbf{z}}$ is given by the superposition of two sets of fields. The field radiated by the transmitting antenna in (1) and the fields relevant to its reflection and transmission from the interface (in the absence of the target) belong to the first set, representing known field contributions. The second set of fields is given by the scattered field by the target in the medium behind the wall, and by the scattered-reflected and transmitted fields through the wall interfaces. In the lowest medium, the scattered electric field is found from $E_{scatt}(x,y) = E_{tot}(x,y) - E_{t2}(x,y)$, where $E_{t2}(x,y)$ is the field related to the transmission of the incident field $E_{inc}(x,y)$ in the medium behind the wall. The scattered field in the medium behind the wall is given in turn by the superposition of three contributions, $E_s(x,y)$, $E_{sr}(x,y)$, $E_{sc}(x,y)$, i.e,

$$E_{scatt}(x,y) = E_s(x,y) + E_{sr}(x,y) + E_{sc}(x,y) \tag{2}$$

where E_s represents the field scattered by the target, E_{sr} is the scattered-reflected field, describing the reflection of the field E_s by the wall, and E_{sc} is the contribution of scattered field that is transmitted inside the cylinder.

The scattered field E_s in (2) is expressed through an expansion into a series of basis functions CW_m [37]:

$$E_s(x,y) = V_0 \sum_{m=-\infty}^{+\infty} c_m CW_m(x,y) \tag{3}$$

where c_m are unknown expansion coefficients and the basis functions CW_m are cylindrical waves, proportional to m-th order Hankel functions:

$$CW_m(x,y) = H_m^{(2)}(k_0 r)e^{jm\theta} \tag{4}$$

where (r, θ) are polar coordinates centered on the cylinder.

The use of cylindrical waves as functions of expansion of the fields scattered by circular cross-section cylinders gives the analytical basis to the method. However, as the target is not in free space, but placed behind a dielectric wall, the interaction with the wall interfaces in terms of reflection and transmission must be suitably modeled. This is accomplished by expressing the cylindrical waves in (4) through an alternative definition, i.e., the plane-wave spectrum of a cylindrical wave:

$$CW_m(x,y) = \frac{1}{2\pi} \int_{-\infty}^{+\infty} F_m\big(y, k_\|\big) e^{-jk_\| x} dk_\| \tag{5}$$

where $F_m\big(y, k_\|\big)$ is the plane-wave spectrum:

$$F_m\left(y, k_\parallel\right) = -\frac{2e^{-j|y|\sqrt{1-(k_\parallel)^2}}}{\sqrt{1-\left(k_\parallel\right)^2}} \begin{cases} e^{jm\cos^{-1}n_\parallel}, & y \geq 0 \\ e^{-jm\cos^{-1}n_\parallel}, & y \leq 0 \end{cases} \tag{6}$$

The expressions (5) and (6) are used to derive the scattered-reflected field $E_{sr}(x,y)$ in (2) and the scattered fields propagating inside the wall and in the first half-space [39]. In particular, in the half-space in front of the wall, where the field is probed by the receiving antennas, the scattered field is found as $E_{scatt}(x,y) = E_{tot}(x,y) - E_{inc}(x,y) - E_{r1}(x,y)$, where $E_{r1}(x,y)$ is the contribution relevant to the reflection of incident field by the interface in $y = 0$. The scattered field $E_{scatt}(x,y)$ is defined through the following expansion [39]:

$$E_{scatt}(x,y) = V_0 \sum_{m=-\infty}^{+\infty} c_m TW_m^0(x,y;y_c) \tag{7}$$

where the basis functions $TW_m^0(x,y;y_c)$ are transmitted cylindrical waves, and they are expressed through spectral integrals:

$$TW_m^0(x,y;y_c) = \frac{1}{2\pi} \int_{-\infty}^{+\infty} T_{10}\left(k_\parallel\right)T_{21}\left(k_\parallel\right)F_m\left[n_2(y_c - l), k_\parallel\right]e^{-jy\sqrt{k_0^2-(n_2k_\parallel)^2}}e^{-jk_\parallel(x-x_c)}dk_\parallel \tag{8}$$

In (8), $T_{10}\left(n_\parallel\right)$ and $T_{21}\left(n_\parallel\right)$ are the transmission coefficients from the wall to the upper medium and from the lowest medium to the wall, respectively. In the expression (8), all the multiple reflections excited by propagation of the scattered field E_s in the wall are included through transmission and reflection coefficients related to the interaction of a plane wave with a dielectric slab [43]. A solution to the scattering problem is developed imposing the boundary conditions of continuity of the field components tangential to the cylinder's interface and deriving the unknown expansion coefficients c_m in (3) and (8) [39].

2.2. Inverse-Scattering Problem Formulation

In the inversion procedure, the space-dependent dielectric properties of the investigation domain D_{inv} are described by the contrast function $c(x,y) = \varepsilon(x,y)/\varepsilon_0 - 1$, $\varepsilon(x,y)$ being the dielectric permittivity in a generic point $(x,y) \in D_{inv}$, which represents the unknown to be retrieved. Such a quantity is related to the scattered field E_{scatt} in the measurement points by means of the following integral relationship (data equation) [21]

$$E_{scatt}(x,y) = \mathcal{G}_w^{ext}(cE_{tot})(x,y) = -k_0^2 \int_{D_{inv}} c\left(x',y'\right)E_{tot}\left(x',y'\right)g_w(x,y,x',y')dx'dy' \tag{9}$$

where $k_0 = \omega(\varepsilon_0\mu_0)^{0.5}$ is the vacuum wavenumber and $\mathcal{G}_w^{ext}$ is a linear integral operator whose kernel is the two-dimensional Green's function of the considered three-layer background, g_w, which is given by [7]

$$g_w(x,y,x',y') = \frac{j}{4\pi} \int_{-\infty}^{+\infty} \frac{e^{j\zeta(x-x')}}{\gamma_1} \begin{cases} e^{-j\gamma_0|y-y'|} + Re^{-j\gamma_0(y+y')}, & y \geq 0 \\ Te^{j\gamma_0(y+l-y')}, & y \leq -l_w \end{cases} d\zeta \tag{10}$$

where $\gamma_0 = \sqrt{k_0^2 - \zeta^2}$ and

$$R = \rho_w\frac{1-e^{-2j\gamma_1 l}}{1-\rho_w^2 e^{-2j\gamma_1 l}}, \quad T = \frac{\left(1-\rho_w^2\right)e^{-j\gamma_1 l}}{1-\rho_w^2 e^{-2j\gamma_1 l}} \tag{11}$$

with $\gamma_1 = \sqrt{k_1^2 - \zeta}$, $k_1 = k_0\varepsilon_{r1}^{0.5}$ being the wavenumber in the wall, and $\rho_w = (\gamma_0 - \gamma_1)/(\gamma_0 + \gamma_1)$. For the sake of simplicity, a single view case is considered in this Section. The total electric field $E_{tot}(x',y')$ inside the integral in (9) depends itself on the contrast function c and can be expressed by means of a second integral equation similar to (9) (the so-called state equation), i.e., $E_{tot}(x,y) = E_{inc}(x,y) + \mathcal{G}_w^{int}(cE_{tot})(x,y)$, where $\mathcal{G}_w^{int}$ is again a linear integral operator whose kernel is the Green's function for the through-wall configuration [21]. By combining the data and state equations, the inverse scattering problem can be finally formulated as [21]

$$E_{scatt}(x,y) = \mathcal{T}(c)(x,y) = \mathcal{G}_w^{ext}c\left(I - \mathcal{G}_w^{int}c\right)E_{inc}(x,y) \tag{12}$$

The non-linear problem at hand is solved in a regularized sense by using an inversion procedure developed in the framework of Lebesgue spaces L^p, i.e., function spaces endowed with the norm $\|u\|_{L^p}^p = \int |u(x,y)|^p dxdy$ (u being a generic function belonging to L^p). It is worth noting that the norm exponent p represents an additional parameter that can be tuned in order to enhance the reconstruction performance. In particular, the developed procedure is based on an iterative outer–inner Newton scheme, which can be summarized by the following steps [31,33]:

1. Set the outer iteration index to $n = 0$ and initialize the contrast function at the first outer step with $c_0 = 0$.
2. Linearize the scattering problem by computing the Fréchet derivative $\mathcal{T}'_n$ of the operator $\mathcal{T}$ around the current solution c_n. A linear problem $\mathcal{T}'_n \xi_n(x,y) = E_{scatt}(x,y) - \mathcal{T}(c_n)(x,y)$ is then obtained. It is worth remarking that, similarly to the corresponding procedures in free space [31,33], the computation of the right-hand side of the linear problem and of the Fréchet derivative $\mathcal{T}'_n$ requires the solution of a set of forward problems. To this end, a forward solver based on the MoM is adopted.
3. Solve the obtained linear problem in a regularized sense by means of the Lebesgue-space procedure detailed in [31,33]. Specifically, the solution of the linear problem obtained in step 2, i.e., ξ_n, is computed by means of the following Landweber-type iterations:

$$\xi_{n,l+1} = \mathcal{J}_q\left(\mathcal{J}_p\left(\xi_{n,l}\right) - \beta \mathcal{T}'^*_n \mathcal{J}_p\left(\mathcal{T}'_n \xi_{n,l} - E_{scatt}(x,y) + \mathcal{T}(c_n)\right)\right) \tag{13}$$

    where $\xi_{n,0} = 0$, $\beta = \|\mathcal{T}'_n\|_2^{-2}$ is a relaxation coefficient, $q = p/(p-1)$ is the Hölder conjugate of p, and the duality map $\mathcal{J}_p$ is defined as $\mathcal{J}_p(e) = \|e\|_p^{2-p}|e|^{p-1}\text{sign}(e)$, with $\text{sign}(e)= e/|e|$ (if $e \neq 0$, otherwise it is equal to zero).
4. Update the contrast function by adding the solution of the linear problem ξ_n found at step 3 to the current value, i.e., $c_{n+1} = c_n + \xi_n$
5. Iterate from step 2 until a proper stopping criterion is satisfied.

3. Numerical Results

3.1. Validation of the Forward Methods

A comparison between the analytical TW solver (Section 2.1) and the forward scattering model embedded inside the inversion procedure (Section 2.2) is reported here, for a cross-validation of the two forward approaches. A multistatic and multiview configuration has been simulated, with $M = 15$ transmitting and receiving antennas aligned in front of the wall along a line of length $L_s = 1.5$ m, with spacing $d = L_s/(M-1)$, and parallel to the wall at distance $h_s = 30$ cm. The s-th transmitting antenna ($s = 1,\ldots, M$) is placed along the horizontal axis in the following position:

$$x_s^{TX} = -\frac{L_s}{2} + (s-1)d \tag{14}$$

whereas the scattered field is probed at the remaining $M - 1$ positions along L_s. The working frequency has been fixed equal to 1 GHz. As a first case, a single dielectric cylinder with center in $(-20$ cm, 60 cm$)$, radius $a = 10$ cm, and relative permittivity $\varepsilon_{rc} = 2$, placed behind a wall of relative permittivity $\varepsilon_{r1} = 4$ and thickness $l = 20$ cm, has been considered. The actual distribution of the relative dielectric permittivity in the investigation domain is shown in Figure 2a. In the MoM solver, the target has been discretized into $N = 900$ square subdomains of side about equal to 6.7 mm. In the CWA, the order m of the cylindrical waves in Equation (7) has been truncated to $M_t = 9$, being the total number of terms in the cylindrical expansions equal to $2M_t + 1$. The truncation order has been determined applying the rule $M_t = 3\sqrt{\epsilon_{r1}}a(2\pi)/\lambda$, that allows a compromise between accuracy and computational heaviness. Figure 3 shows the amplitude and phase of the fields computed by the two approaches for some of the considered views, at the $M - 1$ measurement receiving points. Plots are evaluated for different values of the index s, which denotes the antenna used in the transmission mode, according to Equation (14). As can be seen, a very good agreement between the analytical solver used in the forward approach and the solver employed in the inversion procedure is obtained.

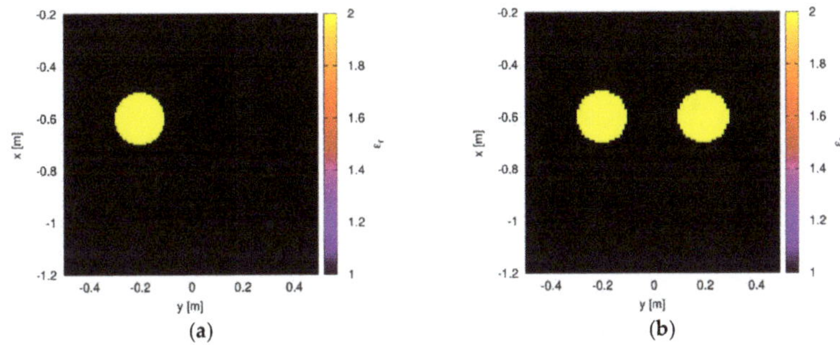

Figure 2. Actual configuration. (**a**) Single dielectric cylinder and (**b**) two separate dielectric cylinders.

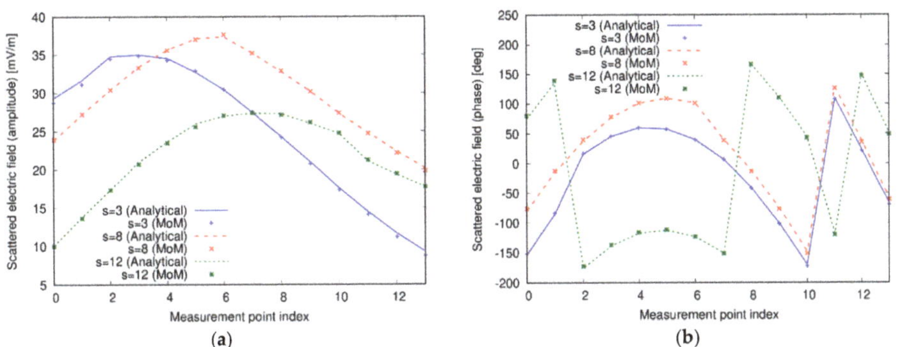

Figure 3. (**a**) Amplitude and (**b**) phase of the scattered fields in some of the considered view computed by the analytical forward solver based on the cylindrical wave approach (CWA) and by the integral equation formulation used in the inverse scattering procedure. Single dielectric cylinder.

As a second test case, two dielectric cylinders have been considered inside the investigation domain. The first one is the same considered above, whereas the second one is a dielectric cylinder with center in (20 cm, 60 cm), radius $a = 10$ cm, and relative permittivity $\varepsilon_{rc} = 2$. The corresponding distribution of the relative dielectric permittivity in the investigation domain is shown in Figure 2b. Figure 4 reports the amplitude and phase of the field computed by using the CWA and the MoM approaches. In this more complex case, too, there is a good agreement between the two solving schemes, confirming the correctness and suitability of the analytical and numerical solvers adopted in the data generation and inversion steps.

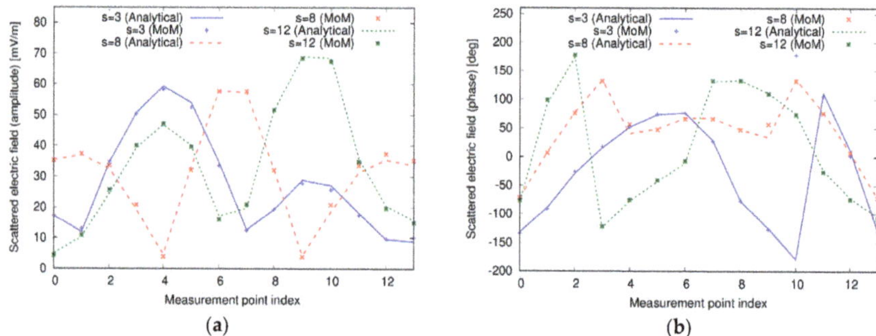

Figure 4. (**a**) Amplitude and (**b**) phase of the scattered fields in some of the considered view computed by the analytical forward solver based on the CWA and by the integral equation formulation used in the inverse scattering procedure. Two separate dielectric cylinders.

3.2. Inversion Scheme

Some preliminary examples of reconstructions provided by the previously described inversion procedure are reported in this Section. In particular, the two configurations adopted for the comparison in the previous Section are considered. In order to simulate a more realistic scenario, a Gaussian noise with zero mean value and variance corresponding to a signal-to-noise ratio of 20 dB has been added to the computed scattered electric field data. The following values of the algorithm's parameters have been used: $p \in [1.1, 2.5]$; maximum number of outer iterations, 10; maximum number of inner iterations, 50; iterations stopped when the relative variation of the residual falls below the threshold 0.005. Such values have been empirically selected, based on the previous experience on Lebesgue-space inversion in the free-space scenario. The optimal value of the norm parameter p has been found by performing a sweep in the assumed range of values and by selecting the one providing the maximum entropy. Such a choice has been made since, for this particular application, it is expected that localized targets are usually present inside the inspected scenario, and consequently the sharpness of the image, which is related to its entropy, may represent a discriminating feature [44,45].

Figure 5 shows the reconstructed distribution of the relative dielectric permittivity retrieved by the developed inverse scattering procedure. In particular, the reconstruction obtained with the optimal value of the norm parameter, i.e., $p = p_{opt} = 1.3$, is reported in Figure 5a. This result evidences a correct localization of the target. Indeed, the estimated center of the cylinder is $(-19.9\ \mathrm{cm}, -61.4\ \mathrm{cm})$, which corresponds to an average percentage error of 1.3%. Moreover, the reconstructed value of the dielectric permittivity is close to the actual one. Specifically, the maximum value of the estimated permittivity is 1.81, which compares very well with the actual value of 2. Nevertheless, the target shape, which is a circular one, is elongated along the range direction and the cross-range size is underestimated. Such a behavior can be ascribed to the use of data collected at a single frequency, as well as to their aspect-limitedness. However, it is worth noting that even with such a small number of available data and considering just a single working frequency, the approach is able to effectively provide a quite accurate indication about the target. For comparison purposes, the reconstruction obtained by using a standard inversion procedure in Hilbert spaces (corresponding to $p = 2$) is provided in Figure 5b. In this case, the target is still visible (the average percentage error on the center position is 2.9%), but the dielectric permittivity is significantly underestimated (the maximum value is 1.35). Moreover, stronger artifacts are present in the background.

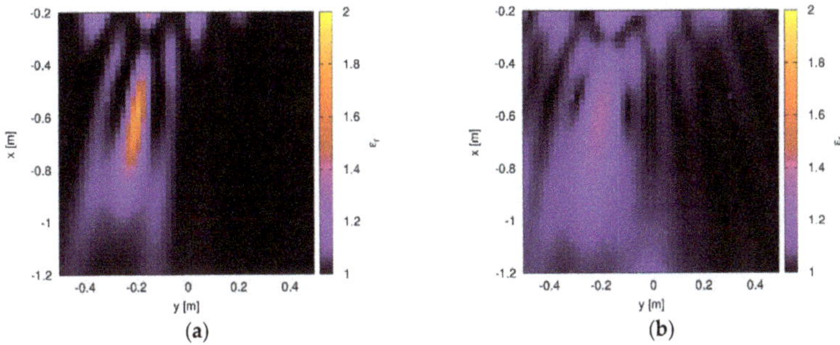

Figure 5. Reconstructed distribution of the relative dielectric permittivity inside the through-wall (TW) investigation domain. Single dielectric cylinder. (**a**) Optimal value of the norm parameter ($p_{opt} = 1.3$) and (**b**) standard Hilbert-space approach ($p = 2$).

As a second test case, the scattering data from the two dielectric cylinders considered in the previous Section have been used. In this case, too, the scattered field has been corrupted with a Gaussian noise with zero mean value and variance corresponding to $SNR = 20$ dB. The parameters of the inversion procedure are the same as in the previous case. The reconstructed distribution of the relative dielectric permittivity is shown in Figure 6. In particular, Figure 6a shows the results obtained by considering the optimal value of the norm parameter, which is equal to $p_{opt} = 1.3$ in this case. Even in this situation, the two targets are correctly localized, although the dielectric permittivity is slightly underestimated. Indeed, the estimated centers of the cylinders are $(-19.4 \text{ cm}, -57.4 \text{ cm})$ and $(-19.0 \text{ cm}, -57.0 \text{ cm})$, which correspond to the mean percentage errors of 3.7% and 5%, respectively, whereas the maximum values of the dielectric permittivity are both equal to 1.7. The corresponding reconstruction obtained by using the standard Hilbert-space reconstruction technique is shown in Figure 6b. Similarly to the preceding configuration, the two targets are visible (the mean percentage errors on the position are 11.2% and 7.3%), although their properties are strongly underestimated and with amplitude comparable to the background artifacts (the maximum value of the dielectric permittivity is equal to 1.26). The criterion for the selection of the optimal reconstruction has been assessed in this case by comparing the behavior of the scaled entropy with the one of the reconstruction errors (defined as NMSE $= \|c - c_{act}\|^2 / \|c_{act}\|^2$, c_{act} being the actual distribution of the contrast function). As can be seen from Figure 7, the scaled entropy (defined as in [45]) has a maximum corresponding to the value of the norm parameter for which the lowest reconstruction error is obtained.

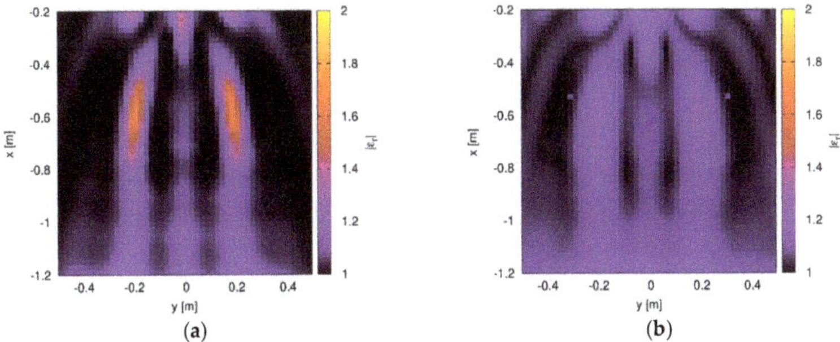

Figure 6. Reconstructed distribution of the relative dielectric permittivity inside the TW investigation domain. Two dielectric cylinders. (**a**) Optimal value of the norm parameter ($p_{opt} = 1.3$) and (**b**) standard Hilbert-space approach ($p = 2$).

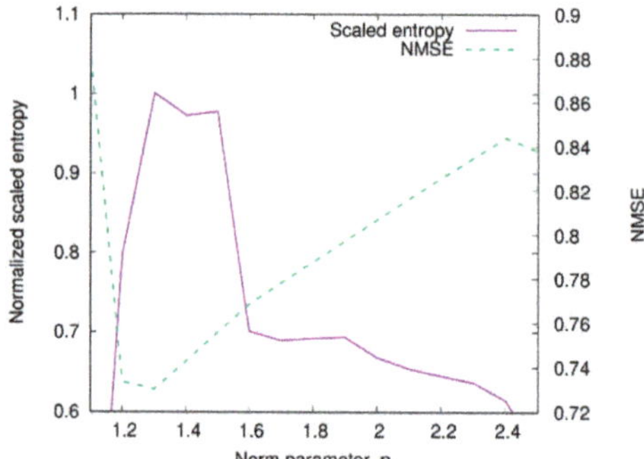

Figure 7. Behavior of the scaled entropy and of the reconstruction error versus the norm parameter.

4. Conclusions

In this work, the forward scattering problem modeling and the quantitative imaging in through-wall scenarios have been addressed. For the solution of the forward EM problem, an analytical solver based on the cylindrical wave approach has been presented. The inverse problem is solved by using a non-linear regularization technique developed in the framework of Lebesgue-spaces. The suitability of the forward and inverse solvers for the problem at hand has been evaluated with preliminary numerical simulations. Future works will be mainly devoted to the extension to multi-frequency processing, in order to increase the reconstruction accuracy, and to three-dimensional configurations. Moreover, the developed forward and inverse schemes will be validated by considering experimental measurements obtained with a real hardware setup.

Author Contributions: Conceptualization, A.F., M.P., C.P., A.R., and G.S.; formal analysis A.F., M.P., C.P., A.R., and G.S.; methodology A.F., M.P., C.P., A.R., and G.S.; validation, A.F., M.P., C.P., A.R., and G.S. All authors have read and agreed to the published version of the manuscript.

Funding: This research was partly supported by the Italian Ministry for Education, University, and Research under the project PRIN2015 U-VIEW, grant number 20152HWRSL.

Conflicts of Interest: The authors declare no conflict of interest.

References

1. Amin, M.G. *Through-The-Wall Radar Imaging*; CRC Press: Boca Raton, FL, USA, 2011. ISBN 978-1-4398-1476-5.
2. Soldovieri, F.; Solimene, R. Through-Wall imaging via a linear inverse scattering algorithm. *IEEE Geosci. Remote Sens. Lett.* **2007**, *4*, 513–517. [CrossRef]
3. Gennarelli, G.; Vivone, G.; Braca, P.; Soldovieri, F.; Amin, M.G. Multiple extended target tracking for through-wall radars. *IEEE Trans. Geosci. Remote Sens.* **2015**, *53*, 6482–6494. [CrossRef]
4. Yektakhah, B.; Sarabandi, K. All-directions through-the-wall radar imaging using a small number of moving transceivers. *IEEE Trans. Geosci. Remote Sens.* **2016**, *54*, 6415–6428. [CrossRef]
5. Yektakhah, B.; Sarabandi, K. All-Directions Through-the-Wall Imaging Using a Small Number of Moving Omnidirectional Bi-Static FMCW Transceivers. *IEEE Trans. Geosci. Remote Sens.* **2019**, *57*, 2618–2627. [CrossRef]
6. Boudamouz, B.; Millot, P.; Pichot, C. Through the Wall Radar Imaging with MIMO beamforming processing—Simulation and Experimental Results. *Am. J. Remote Sens.* **2013**, *1*, 7–12. [CrossRef]
7. Chew, W.C. *Waves and Fields in Inhomogeneous Media*; IEEE Press Series on Electromagnetic Wawes; IEEE Press: Piscataway, NY, USA, 1995. ISBN 978-0-7803-4749-6.
8. Dehmollaian, M.; Sarabandi, K. Hybrid Fdtd and Ray Optics Approximation for Simulation of Through-Wall Microwave Imaging. In Proceedings of the IEEE Antennas and Propagation Society International Symposium, Albuquerque, NM, USA, 9–14 July 2006; pp. 249–252.

9. Warren, C.; Giannopoulos, A.; Giannakis, I. gprMax: Open source software to simulate electromagnetic wave propagation for Ground Penetrating Radar. *Comput. Phys. Commun.* **2016**, *209*, 163–170. [CrossRef]

10. Chang, P.C.; Burkholder, R.J.; Volakis, J.L.; Marhefka, R.J.; Bayram, Y. High-Frequency EM Characterization of Through-Wall Building Imaging. *IEEE Trans. Geosci. Remote Sens.* **2009**, *47*, 1375–1387. [CrossRef]

11. Antonopoulos, C.S.; Kantartzis, N.V.; Rekanos, I.T. FDTD Method for Wave Propagation in Havriliak-Negami Media Based on Fractional Derivative Approximation. *IEEE Trans. Magn.* **2017**, *53*. [CrossRef]

12. Guo, Q.; Li, Y.; Liang, X.; Dong, J.; Cheng, R. Through-the-Wall Image Reconstruction via Reweighted Total Variation and Prior Information in Radio Tomographic Imaging. *IEEE Access* **2020**, *8*, 40057–40066. [CrossRef]

13. Doğu, S.; Akıncı, M.N.; Çayören, M.; Akduman, İ. Truncated Singular Value Decomposition for Through-the-Wall Microwave Imaging Application. *IET Microw. Antennas Propag.* **2020**, *14*, 260–267. [CrossRef]

14. Charnley, M.; Wood, A. A Linear Sampling Method for Through-the-Wall Radar Detection. *J. Comput. Phys.* **2017**, *347*, 147–159. [CrossRef]

15. Aamna, M.; Ammar, S.; Rameez, T.; Shabeeb, S.; Naveed, I.R.; Safwat, I. 2D Beamforming for Through-the-Wall Microwave Imaging applications. In Proceedings of the 2010 International Conference on Information and Emerging Technologies, Karachi, Pakistan, 14–16 June 2010; pp. 1–6.

16. Mizrahi, M.; Holdengreber, E.; Schacham, S.E.; Farber, E.; Zalevsky, Z. Improving Radar's Spatial Recognition: A Radar Scanning Method Based on Microwave Spatial Spectral Illumination. *IEEE Microw. Mag.* **2016**, *17*, 28–34. [CrossRef]

17. Mizrahi, M.; Holdengreber, E.; Farber, E.; Zalevsky, Z. Frequency Multiplexing Spatial Super-Resolved Sensing for RADAR Applications. *Microw. Opt. Technol. Lett.* **2016**, *58*, 831–835. [CrossRef]

18. Zhao, M.; Li, T.; Alsheikh, M.A.; Tian, Y.; Zhao, H.; Torralba, A.; Katabi, D. Through-Wall Human Pose Estimation Using Radio Signals. In Proceedings of the 2018 IEEE/CVF Conference on Computer Vision and Pattern Recognition, Salt Lake City, UT, USA, 18–22 June 2018; pp. 7356–7365.

19. Keith, S.R. *Discrimination Between Child and Adult Forms Using Radar Frequency Signature Analysis*; AFIT-ENP-13-M-20; Air Force Institute of Technology: Wright-Patterson Air Force Base, OH, USA, 2013.

20. Nikolova, N.D. *Introduction to Microwave Imaging*; Cambridge University Press: Cambridge, UK, 2017. ISBN 978-1-316-08426-7.

21. Pastorino, M.; Randazzo, A. *Microwave Imaging Methods and Applications*; Artech House: Boston, MA, USA, 2018. ISBN 978-1-63081-348-2.

22. Crocco, L.; Catapano, I.; Di Donato, L.; Isernia, T. The linear sampling method as a way to quantitative inverse scattering. *IEEE Trans. Antennas Propag.* **2012**, *60*, 1844–1853. [CrossRef]

23. Bozza, G.; Pastorino, M. An inexact Newton-based approach to microwave imaging within the contrast source formulation. *IEEE Trans. Antennas Propag.* **2009**, *57*, 1122–1132. [CrossRef]

24. De Zaeytijd, J.; Franchois, A.; Eyraud, C.; Geffrin, J.-M. Full-wave three-dimensional microwave imaging with a regularized Gauss-Newton method - Theory and experiment. *IEEE Trans. Antennas Propag.* **2007**, *55*, 3279–3292. [CrossRef]

25. Oliveri, G.; Lizzi, L.; Pastorino, M.; Massa, A. A nested multi-scaling inexact-Newton iterative approach for microwave imaging. *IEEE Trans. Antennas Propag.* **2012**, *60*, 971–983. [CrossRef]

26. Li, M.; Semerci, O.; Abubakar, A. A contrast source inversion method in the wavelet domain. *Inverse Probl.* **2013**, *29*, 025015. [CrossRef]

27. Semnani, A.; Rekanos, I.T.; Kamyab, M.; Moghaddam, M. Solving Inverse Scattering Problems Based on Truncated Cosine Fourier and Cubic B-Spline Expansions. *IEEE Trans. Antennas Propag.* **2012**, *60*, 5914–5923. [CrossRef]

28. Monleone, R.D.; Pastorino, M.; Fortuny-Guasch, J.; Salvade, A.; Bartesaghi, T.; Bozza, G.; Maffongelli, M.; Massimini, A.; Carbonetti, A.; Randazzo, A. Impact of background noise on dielectric reconstructions obtained by a prototype of microwave axial tomograph. *IEEE Trans. Instrum. Meas.* **2012**, *61*, 140–148. [CrossRef]

29. Zeitler, A.; Migliaccio, C.; Moynot, A.; Aliferis, I.; Brochier, L.; Dauvignac, J.-Y.; Pichot, C. Amplitude and phase measurements of scattered fields for quantitative imaging in the W-band. *IEEE Trans. Antennas Propag.* **2013**, *61*, 3927–3931. [CrossRef]

30. Ye, X.; Chen, X. Subspace-Based Distorted-Born Iterative Method for Solving Inverse Scattering Problems. *IEEE Trans. Antennas Propag.* **2017**, *65*, 7224–7232. [CrossRef]

31. Estatico, C.; Pastorino, M.; Randazzo, A. A novel microwave imaging approach based on regularization in Lp Banach spaces. *IEEE Trans. Antennas Propag.* **2012**, *60*, 3373–3381. [CrossRef]
32. Estatico, C.; Fedeli, A.; Pastorino, M.; Randazzo, A. Microwave imaging of elliptically shaped dielectric cylinders by means of an Lp Banach-space inversion algorithm. *Meas. Sci. Technol.* **2013**, *24*, 074017. [CrossRef]
33. Estatico, C.; Pastorino, M.; Randazzo, A.; Tavanti, E. Three-dimensional microwave imaging in Lp Banach spaces: Numerical and experimental results. *IEEE Trans. Comput. Imaging* **2018**, *4*, 609–623. [CrossRef]
34. Estatico, C.; Fedeli, A.; Pastorino, M.; Randazzo, A. Quantitative microwave imaging method in Lebesgue spaces with nonconstant exponents. *IEEE Trans. Antennas Propag.* **2018**, *66*, 7282–7294. [CrossRef]
35. Estatico, C.; Fedeli, A.; Pastorino, M.; Randazzo, A. Microwave imaging by means of Lebesgue-space inversion: An overview. *Electronics* **2019**, *8*, 945. [CrossRef]
36. Estatico, C.; Fedeli, A.; Pastorino, M.; Randazzo, A. A multifrequency inexact-Newton method in Lp Banach spaces for buried objects detection. *IEEE Trans. Antennas Propag.* **2015**, *63*, 4198–4204. [CrossRef]
37. Ponti, C.; Santarsiero, M.; Schettini, G. Electromagnetic scattering of a pulsed signal by conducting cylindrical targets embedded in a half-space medium. *IEEE Trans. Antennas Propag.* **2017**, *65*, 3073–3083. [CrossRef]
38. Frezza, F.; Pajewski, L.; Ponti, C.; Schettini, G. Through-wall electromagnetic scattering by N conducting cylinders. *J. Opt. Soc. Am. A Opt. Image Sci. Vis.* **2013**, *30*, 1632–1639. [CrossRef]
39. Ponti, C.; Vellucci, S. Scattering by conducting cylinders below a dielectric layer with a fast noniterative approach. *IEEE Trans. Microw. Theory Tech.* **2015**, *63*, 30–39. [CrossRef]
40. Ponti, C.; Schettini, G. Direct scattering methods in presence of interfaces with different media. In Proceedings of the 11th European Conference on Antennas and Propagation, Paris, France, 19–24 March 2017; pp. 1707–1710.
41. Ponti, C.; Schettini, G. The cylindrical wave approach for the electromagnetic scattering by targets behind a wall. *Electronics* **2019**, *8*, 1262. [CrossRef]
42. Balanis, C.A. *Advanced Engineering Electromagnetics*, 2nd ed.; John Wiley & Sons: Hoboken, NJ, USA, 2012. ISBN 978-0-470-58948-9.
43. Orfanidis, S.J. Electromagnetic Waves and Antennas. 2016. Available online: http://www.ece.rutgers.edu/~{}orfanidi/ewa/ (accessed on 18 May 2020).
44. Wei, X.; Zhang, Y. Autofocusing Techniques for GPR Data from RC Bridge Decks. *IEEE J. Sel. Top. Appl. Earth Obs. Remote Sens.* **2014**, *7*, 4860–4868. [CrossRef]
45. Giannakis, I.; Tosti, F.; Lantini, L.; Alani, A.M. Diagnosing Emerging Infectious Diseases of Trees Using Ground Penetrating Radar. *IEEE Trans. Geosci. Remote Sens.* **2020**, *58*, 1146–1155. [CrossRef]

Article

Experimental Validation of a Microwave Imaging Method for Shallow Buried Target Detection by Under-Sampled Data and a Non-Cooperative Source

Adriana Brancaccio [1,2,*], Giovanni Leone [1,2], Rocco Pierri [1,2] and Raffaele Solimene [1,2,3]**

[1] Dipartimento di Ingegneria, Università degli Studi della Campania Luigi Vanvitelli, 81031 Aversa, Italy; giovanni.leone@unicampania.it (G.L.); rocco.pierri@unicampania.it (R.P.); raffaele.solimene@unicampania.it (R.S.)

[2] Consorzio Nazionale Interuniversitario per le Telecomunicazioni-CNIT, 43124 Parma, Italy

[3] Department of Electrical Engineering, Indian Institute of Technology, Chennai 600036, India

[*] Correspondence: adriana.brancaccio@unicampania.it; Tel.: +39-081-5010-2070

Abstract: In microwave imaging, it is often of interest to inspect electrically large spatial regions. In these cases, data must be collected over a great deal of measurement points which entails long measurement time and/or costly, and often unfeasible, measurement configurations. In order to counteract such drawbacks, we have recently introduced a microwave imaging algorithm that looks for the scattering targets in terms of equivalent surface currents supported over a given reference plane. While this method is suited to detect shallowly buried targets, it allows one to independently process all frequency data, and hence the source and the receivers do not need to be synchronized. Moreover, spatial data can be reduced to a large extent, without any aliasing artifacts, by properly combining single-frequency reconstructions. In this paper, we validate such an approach by experimental measurements. In particular, the experimental test site consists of a sand box in open air where metallic plate targets are shallowly buried a (few *cm*) under the air/soil interface. The investigated region is illuminated by a fixed transmitting horn antenna, whereas the scattered field is collected over a planar measurement aperture at a fixed height from the air-sand interface. The transmitter and the receiver share only the working frequency information. Experimental results confirm the feasibility of the method.

Keywords: radar imaging; target detection; experimental measurements; microwave imaging

Citation: Brancaccio, A.; Leone, G.; Pierri, R.; Solimene, R. Experimental Validation of a Microwave Imaging Method for Shallow Buried Target Detection by Under-Sampled Data and a Non-Cooperative Source. *Sensors* **2021**, *21*, 5148. https://doi.org/10.3390/s21155148

Academic Editor: Giacomo Oliveri

Received: 21 June 2021
Accepted: 27 July 2021
Published: 29 July 2021

Publisher's Note: MDPI stays neutral with regard to jurisdictional claims in published maps and institutional affiliations.

1. Introduction

Microwave imaging, and in general radar imaging, is a mature research field that finds application in a number of different contexts where pursuing non-destructive investigation is convenient or mandatory [1–11].

Ground Penetrating Radar (GPR) is a radar system that is properly conceived to address non-destructive imaging. Generally, GPRs work in contact with the interface between the air and the medium under investigation. However, there is great interest in achieving target detection through non-contact measurement layouts, for example, with GPRs mounted on a flying platform [12,13]. Indeed, stand-off distance configurations allow for the investigation of regions that are not easily (or safely) accessible, as it happens, for instance, when one has to deal with mine or unexploded device detection [14]. Moreover, a flying GPR can allow for inspecting large areas quickly [15,16].

In this framework, however, the system cost and the achievable performance must be traded-off [17]. This requires finding a compromise between the time needed to collect data, the number of sensors to be simultaneously deployed, and the way transmitter and receivers "communicate". In this regard, a single-view/multistatic configuration seems convenient, since only one sensor transmits and the others act as mere passive receivers,

53

hence with reduced weight and cost. However, synchronization between TX and RXs is a critical issue when they are mounted on different platforms, since, differently from multi-monostatic arrangement, TX and RXs are no more co-located and hence do not share the same electronic system. In addition, the number of measurement points is directly linked to the number of flying platforms and hence must be reduced as much as possible.

Actually, a number of different processing schemes have been proposed in the literature to address subsurface imaging [18]. Recently, we have introduced a reconstruction scheme that allows one to mitigate the previously mentioned drawbacks [19,20]. This method relies on a certain scattering formulation where, according to the equivalence principle [21], the scattered field is modeled as being radiated by equivalent surface currents that are supported over the air/soil interface, or at some other reference plane whose depth is chosen. A related method that uses equivalence principles for target shape reconstruction is reported in [22], where, however, the multiple experiments arise from using different illuminations (i.e., a multistatic/multiview configuration is employed).

In our approach, basically, the main idea is that if the reference plane is close to the scattering target, then the spatial support of the surface current gives information concerning the transverse location of the target. If it is a priori known that the targets are in close proximity to the air/soil interface, e.g., for detection of a mine [23] or unexploded improvised device [24] (where the targets are often very shallowly buried just to hide from sight), then the reference plane can be set just at the air/soil interface. In this case, the surface currents radiate in free-space, and the related simple Green function can be considered as a propagator. The reconstruction is cast as a 2D inverse problems, since only the targets detection and their transverse locations are looked for, and hence single-frequency data can be employed. Accordingly, RXs do not need information about the TX, except the working frequency. Hence, the source can be considered as being non-cooperative (it does not share information with the receivers). However, it is not opportunistic as in most passive radar, since it is deliberately deployed in the scene. Multi-frequency data can be processed separately (i.e., incoherently) and then combined to counteract aliasing artifacts that can arise if the spatial measurement points are reduced (not properly sampled) [20]. Finally, depth can be explored by performing the reconstructions at different reference planes.

In previous contributions (see [19,20]), we have shown the feasibility of the method by employing synthetic numerical data and some measured data collected under lab conditions for a free-space scattering scenario. However, the method still need to be validated in a realistic scattering scenario.

In this contribution, we aim at pursuing such a task. To this end, the method and the related achievable performances are checked for a more realistic scattering scenario where the background medium is actually not homogeneous. Indeed, the test site mimics a realistic on-field situation, as it consists of an open-air sand box. As to the RXs and the Txs, they are not mounted on flying platforms. However, they are at a stand-off distance from the air/soil interface, and the TX signal is unknown to the RXs but the working frequency is known.

In sum, the advancements that we are conveying in this contribution concern:

- The generalization of the scattering model for a near-field configuration (previous results refer to a far-field case);
- The experimental validation of the approach for a realistic scattering scenario.

The rest of the paper is organized as follows. In Section 2 the scattering model is generalized to deal with the new scattering configuration, whereas in Section 3, the related reconstruction algorithm is briefly described. In Section 4, the experimental set-up as well as a few experimental results are presented and discussed. Conclusions end the paper.

2. Scattering Model

In this section, we introduce the adopted scattering model and the necessary notation that is required in the following. A detailed derivation of the model can be found in [19]. Here, we adapt that derivation to the configuration used in the experimental set up (described later).

The background medium is two-layered with the upper half-space being air, while the lower one models the soil. The two half-spaces are separated by a planar interface (i.e., the air/soil interface) located at $z = 0$. The scattering targets are located in the lower half-space (i.e., for $z < 0$) and buried in close proximity to the separation interface. Moreover, the "transverse" investigation domain (i.e., the spatial region where the targets can belong to) is denoted as $D = [-x_M, x_M] \times [-y_M, y_M]$. The 2D investigation domain is considered at a fixed depth z_T. Generally, we will consider $z_T = 0$ (which is just at the separation interface). Reconstructions at different $z_T < 0$ can be considered as well in order to explore the depth.

The scattering scene is probed by a single source located in the upper half-space at some stand-off distance from the air/soil interface, h_t, whereas the field scattered by the buried targets is collected over a set of sensors still located on the air side and all at the same height h_O. Accordingly, the spatial measurement positions lay over a plane; say $\mathbf{r}_n = (x_n, y_n, h_O)$, $n = 1, 2, \ldots, N_O$, with their positions, N_O being the number of spatial measurements. Figure 1 shows a pictorial view of the scattering configuration along with the adopted reference frame.

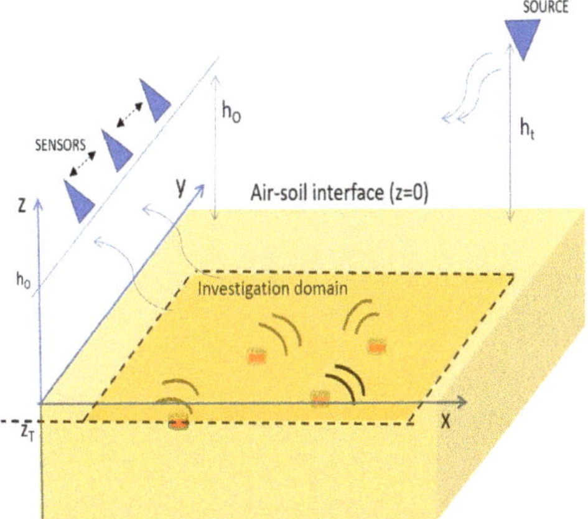

Figure 1. Reference system: the air/soil interface is at $z = 0$, the investigation plane at z_T.

After the scene is illuminated by the incident field, the scattered field arises. Since all the targets are located in the half-space $z < z_T$, by invoking the equivalence theorem [21], the scattered field can be considered as being radiated by equivalent surface currents supported over the plane at $z = z_T$. In particular, by filling the half-space $z < z_T$ with a perfect electric conductor, only the magnetic surface current survives. Such a current can be expressed as

$$\mathbf{J}_m(\mathbf{r}; k) = \mathbf{J}_{eq}(x, y; k)\delta(z - z_T). \tag{1}$$

Note that the magnetic equivalent current depends on the scattering scene and on the incident field. As such, it depends on the working frequency, which is indeed highlighted in (1) in terms of the wavenumber k. In particular, if $z_T = 0$ (i.e., just at the air/soil interface), the current in (1) radiates in free space, and hence the scattered field (in the upper half-space $z > 0$) can be written as

$$\mathbf{E}_S(\mathbf{r}; k) = \int \nabla_\mathbf{r} g(\mathbf{r}, \mathbf{r}'; k) \times \mathbf{J}_m(\mathbf{r}'; k) d\mathbf{r}' = \int_D \underline{G}(\mathbf{r}, \mathbf{r}'; k) \cdot \mathbf{J}_{eq}(\mathbf{r}'; k) d\mathbf{r}'_t, \tag{2}$$

where $\mathbf{r}' = \mathbf{r}'_t + z_T\hat{z}$ is the source point, with $\mathbf{r}'_t = (x', y')$, $\mathbf{r}$ is the field point and

$$g(\mathbf{r}, \mathbf{r}'; k) = -\frac{e^{-jk|\mathbf{r}-\mathbf{r}'|}}{4\pi|\mathbf{r}-\mathbf{r}'|}, \tag{3}$$

is the free-space 3D scalar Green function. $\underline{G}(\mathbf{r}, \mathbf{r}'; k)$ is the magnetic to electric dyadic Green function, whose expression is given as

$$\underline{G}(\mathbf{r}, \mathbf{r}'; k) = \begin{bmatrix} 0 & -\frac{\partial g}{\partial z}(\mathbf{r}, \mathbf{r}'; k) & \frac{\partial g}{\partial y}(\mathbf{r}, \mathbf{r}'; k) \\ \frac{\partial g}{\partial z}(\mathbf{r}, \mathbf{r}'; k) & 0 & -\frac{\partial g}{\partial x}(\mathbf{r}, \mathbf{r}'; k) \\ -\frac{\partial g}{\partial y}(\mathbf{r}, \mathbf{r}'; k) & \frac{\partial g}{\partial x}(\mathbf{r}, \mathbf{r}'; k) & 0 \end{bmatrix}. \tag{4}$$

It must be remarked that, to be rigorous, surface integration in (2) should run over the entire plane $z = 0$. However, since the targets are very close to the air/soil interface (or in general to the reference plane z_T), it can reasonably be assumed that the current support is very similar to the target's cross section. Hence, D is chosen according to the size of the spatial region to be investigated. Again, in (2), we considered the free-space Green function. This is correct for $z_T = 0$. When this is not the case, because the reconstruction at a different depths is required, we will still use the same Green function. Indeed, using the free-space Green function avoids dealing with the computation of the Green function pertaining to a layered background medium. What is more, the layered medium Green function requires the knowledge of the electromagnetic features (dielectric permittivity and conductivity) of the soil, which are in general not available. Herein, such background medium electromagnetic parameters are assumed not to be known. By contrast, using the free-space Green function leads to the targets appearing more deeply located, because soil is electromagnetically denser than air. However, this is not a serious drawback if the targets of interest are shallowly buried.

The magnetic surface current $\mathbf{J}_{eq}$ has no component along $\hat{z}$. Accordingly, (2) particularizes as

$$\mathbf{E}_S(\mathbf{r}; k) = \int_D \begin{bmatrix} 0 & -\frac{\partial g}{\partial z}(\mathbf{r}, \mathbf{r}'; k) \\ \frac{\partial g}{\partial z}(\mathbf{r}, \mathbf{r}'; k) & 0 \\ -\frac{\partial g}{\partial y}(\mathbf{r}, \mathbf{r}'; k) & \frac{\partial g}{\partial x}(\mathbf{r}, \mathbf{r}'; k) \end{bmatrix} \cdot \mathbf{J}_{eq}(\mathbf{r}'_t; k) d\mathbf{r}'_t. \tag{5}$$

It is seen that E_{Sx} is solely linked to J_{eqy} and E_{Sy} to J_{eqx}. Therefore, if one collects separately such field components, then the inverse problem in (5) splits in two identical scalar inverse problems from which one can reconstruct the two source components independently. Then, these reconstructions can be combined as in [20]. However, in general, this is not the case, even in view of the receiving antenna response. More precisely, what one can actually measure is the antenna output voltage. Therefore, in place of (5), the following equation should be considered

$$V(\mathbf{r}; k) = \mathcal{T}(\mathbf{E}_S)(\mathbf{r}; k), \tag{6}$$

where V is the voltage data, and $\mathcal{T}$ is a linear operator schematizing the antenna response. Eventually, this is the scattering model from which to start in order to perform the current reconstructions. A few details concerning the reconstruction algorithm along with some further simplifications to achieve such a task are provided in the next section.

3. Reconstruction Algorithm

According to (6), the magnetic current components cannot in general be separately reconstructed. Moreover, the antenna response should be known and put into the model. It would be useful to avoid both these issues. Indeed, looking for simultaneously both the source components means to deal with a doubled number of unknowns. Furthermore, antenna response must be measured/known in advance.

As to the first question, if the receiving antenna is linearly polarized, for example, in the $x - z$ plane, then its plane-wave spectrum vector belongs to the same plane. Accordingly, the main contribution to the voltage is due to J_{eqy}, which is the one that contributes to E_{Sx}. Similar considerations apply if the other source component is considered. Therefore, we make the problem scalar by approximating (6) as

$$V(\mathbf{r}_n; k) = \int_D H(\mathbf{r}_n, \mathbf{r}'; k) J_{eqy}(\mathbf{r}'_t; k) d\mathbf{r}'_t, \tag{7}$$

where $\mathbf{r}_n$ are the measurement positions (antenna phase center) and $H(\mathbf{r}_n, \mathbf{r}'; k)$ is the kernel of the integral operator in (7), which is actually the approximation of the composition between the antenna response operator and the propagator, once the contribution due to J_{eqx} has been neglected. However, the antenna response is still there.

It is noted that both the data and the unknown depend on the frequency k. Hence, employing all the available multiple-frequency data to perform the reconstruction is not formally allowed. On the one hand, single-frequency reconstructions do not allow one to estimate targets' depths. This is a minor drawback for shallowly buried targets. Moreover, depth at which reconstruction is achieved can be changed. On the other hand, at a single frequency, the antenna response basically introduces a complex weight, which is the same for each measurement position. This means that it does not affect source localization, which is instead related to the phase term that depends on $\mathbf{r}_n - \mathbf{r}'$. Hence, since we are mainly interested in the detection and the localization of the targets (i.e., we do not aim at quantitative reconstructions, even in view of the other approximations), antenna response is neglected in (7) while achieving single-frequency reconstructions. Note that this would not be the case for multi-frequency reconstructions, since the complex weight in general changes with frequency.

In order to perform the reconstruction, the presented model (at a given frequency k_l) is discretized by representing the unknown current component J_{eqy} as a truncated Fourier series, that is,

$$J_{eqy}(x, y; k_l) = \sum_{n=-N}^{N} \sum_{m=-M}^{M} I_{mn} \exp\left[-j\pi\left(\frac{mx}{x_M} + \frac{ny}{y_M}\right)\right]. \tag{8}$$

where I_{nm} are the Fourier coefficients and the exponentials represent the two-dimensional Fourier spatial harmonics over extent of the domain of investigation . Accordingly, the unknowns of the problem now become the expansion coefficients I_{nm}. The choice of N and M is linked to the so-called number of degrees of freedom of the problem, reflects the ill-posedness of the inverse problem at hand, and depends on the operating frequency as well as the investigation and the observation domain extensions [19,20]. In general, however, this discretization scheme allows one reduce to a large extent the number of unknowns, as compared to a pixel based representation. The corresponding discrete model is then obtained as

$$\mathbf{V}(k_l) = \mathbf{H} \cdot \mathbf{I}, \tag{9}$$

where $\mathbf{V}(k_l) \in C^{N_O}$ is the data vector at the l-th frequency, $\mathbf{H} \in C^{N_O \times (2N+1)(2M+1)}$ is the matrix version of the scattering operator, and $\mathbf{I} \in C^{(2N+1)(2M+1)}$ is the (vectorized) expansion coefficient matrix.

Equation (9) is inverted for $\mathbf{I}$ by a standard Truncated Singular Value Decomposition (TSVD) [25] of the relevant matrix operator. This allows to counteract the ill-posedness of the problem and to obtain a stable reconstruction.

Once the coefficients I_{nm} have been recovered, the corresponding equivalent current $J_{eqy}(x, y; k_l)$ is computed by means of (8). Then, the support of such an equivalent surface current is provided by the image in the x, y investigation domain,

$$I(x, y; k_l) = |J_{eqy}(x, y; k_l)|. \tag{10}$$

In order to limit the system complexity, the number of spatial data must be reduced. This in general can lead to a reconstruction that is corrupted by aliasing artifacts that are difficult to distinguish from the actual current. To cope with this drawback, a simple strategy based on the combination of single-frequency reconstructions has been introduced in [20]. In more detail, suppose that N_k is the number of adopted frequencies; then the final reconstruction is obtained as

$$I(x,y) = \Pi_{l=1}^{N_k} I(x,y;k_l). \tag{11}$$

The very basic idea behind (11) is that aliasing artifacts change positions with the working frequency, whereas the actual source reconstruction does not. Therefore, (11) tends to mitigate all those peaks in the reconstruction that do not overlap (or overlap only partially) while the frequency changes. A criterion for the choice of the frequencies is provided in [20].

In sum, the algorithm presents the following steps:

1. Fix one frequency value;
2. Compute the scattering matrix model **H**;
3. Compute the SVD of **H**;
4. Fix a regularizing threshold for the normalized singular values of **H** (in the following experimental results, 20 dB is used) and compute the unknown vector **I** via a TSVD inversion by retaining the data projection over the singular vectors corresponding to the singular values above the threshold;
5. Calculate $I(x,y;k_l)$ by (8) and (10);
6. Repeat from point 1 by changing the frequency value;
7. Compute $I(x,y)$ from (11).

4. Experimental Results

In this section, we check the proposed algorithm against some experimental measurements collected in a semi-controlled scattering scenario. In particular, we first describe the test site and then show some reconstructions aiming to highlight the role of the number of spatial measurements and of the employed frequencies.

4.1. Test Site

The test site consisted of a tank full of sand of about 3.5 m (length) 2.5 m (width) and 1.5 m (depth) in size. The tank was placed in the open air so that the sand appeared wet, apart from the very surface layer, which was dried by sun. The electromagnetic features of the sand were unknown and wer not estimated for detection purposes.

The transmitting antenna was a horn positioned at a $h_t = 1.5$ m height from the sand floor and located at one of the end sides of the tank. It was tilted to point to the spatial region under investigation. The receiving antenna was still a horn and was located on the other side of the tank. In particular, it was mounted on a wooden slide that allowed it to synthesize a planar measurement aperture at a fixed height from the air/sand interface (in the following examples $h_O = 80$ cm or $h_O = 130$ cm). Furthermore, the receiving antenna was tilted toward the investigated spatial region and was linearly polarized. Figure 2 shows a schematic view of the measurement configuration along with some pictures of the test site. As a target, a metallic rectangular plate 17.5 cm $\times$ 48 cm in size is considered.

A vector network analyzer was connected to the antennas by means of coaxial cables. Standard calibration at the end of each channel was performed at the beginning of each measurement stage in order to avoid mismatch between VNA and cables. Data have been acquired in the frequency band [2–9] GHz (201 equispaced frequencies) for each different position of the receiving antenna.

We performed measurements under two different conditions: *flat* and *rough* air/sand interface. Flatness was obtained by manually using a shovel for smoothing the sand floor. Of course, the obtained sand surface, though smooth, was far from "ideally" flat.

Roughness interface was instead obtained by turning over the sand. For such scenarios, we took measurements with and without targets (background measurements) for comparison purposes.

It is worth remarking that the measurement scenario actually contained many features that are not accounted for in the scattering model used to develop the detection algorithm. For example, though antennas were tilted towards the scattering scene, they still presented a direct link, which implies direct coupling between the receiving and the transmitting antennas. Furthermore, because of the finite dimensions of the tank, the two-half-space medium assumption clearly does not correspond to the actual background medium. This entails that the received signal actually consisted of different contributions besides the one expected from the targets. In addition, note that, even though the medium was a perfect two-layered medium with a flat separation interface, the air/sand interface reflection always superimposes the target signals. Nonetheless, in the following reconstructions, we did not mitigate such unwanted contributions by data pre-processing.

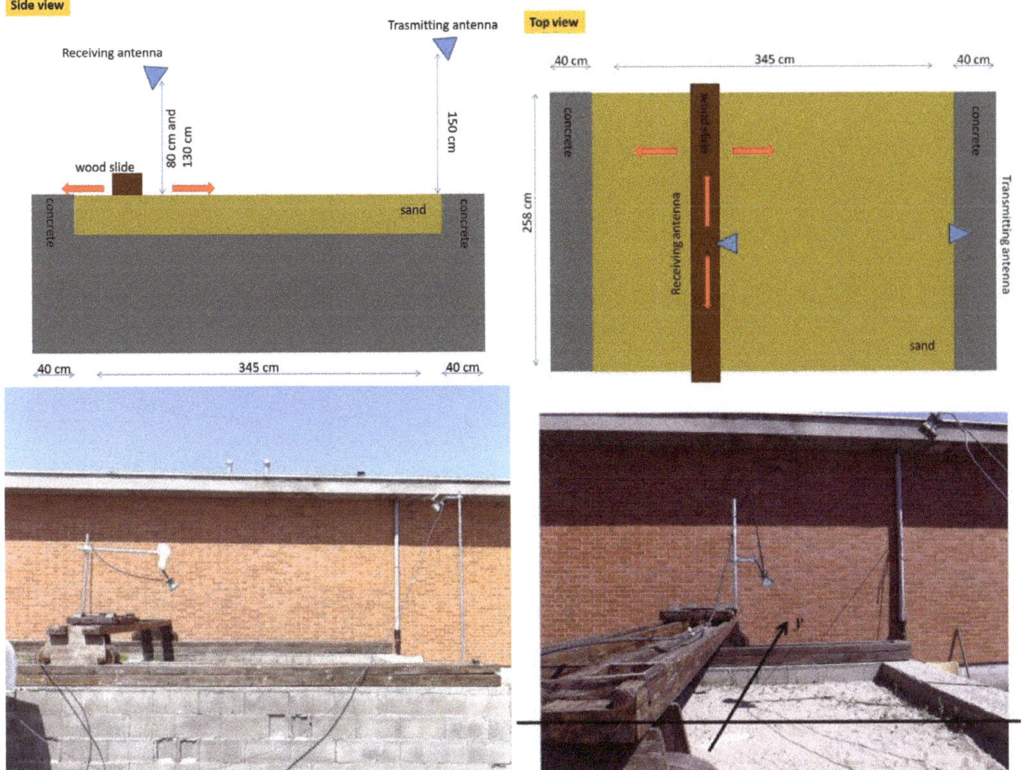

Figure 2. Schematic view of the measurement configuration (**top**) and photos of the test site (**bottom**).

4.2. Detection Results for Flat Air/Soil Interface

We start by considering the case of flat air/sand interface in the sense clarified above.

For this case, we collected data over a grid of 5×7 positions. To this end, the receiving antenna scanned the measurement aperture with a spatial step (along both the x and y directions) of 20 cm at the height and $h_O = 80$ cm.

The investigation domain is a rectangle of 160 cm along the x and 100 cm along the y direction, and it is located at $z_T = 0$. We also introduce the two parameters *offsetx* and *offsety*, which indicate the displacement, along the x and y directions, respectively, of the center of the investigation domain with respect to the central point of the first measurement line (see Figure 3a for the reference system). Basically, after acquiring the data, changing the investigation domain center location (by varying the offset parameters) entails looking for the targets in different spatial regions. Accordingly, the same target will appear at different relative positions. This can be considered as a way to check reconstructions' consistency and stability. A detection can be considered successful if the target localization point moves coherently with the change of the investigation domain center position.

It must be remarked that the number of employed spatial data is already below the one required if the field has to be properly spatially sampled (see [20]). This, of course, limits the highest frequency that can be used in the reconstructions. Indeed, we tested the inversion algorithm against different bands inside the available one of [2–9] GHz. As expected, when dealing with higher frequencies, even by our approach, detection is impaired because the reconstruction results were crowded by a number of artifacts. Hence, we ended up processing data collected only within the band [2–4] GHz. In particular, in such a band, $N_k = 11$ frequencies were considered in order to be as close as possible to frequency selection criterion provided in [20].

Finally, the target was located as shown in Figure 3b and roughly buried 2 cm below the air/sand interface.

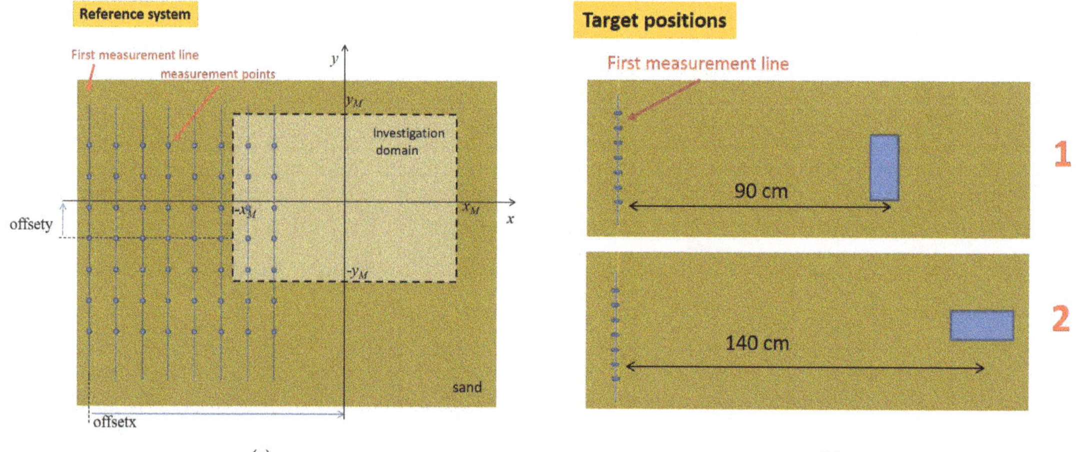

Figure 3. (**a**) Investigation domain and measurement points; the reference system is centred on the investigation domain. (**b**) Different positions of the target with respect to the first (numbered from the left) measurement line.

In particular, Figure 4 reports the reconstructions for a target located as in position 1 sketched in Figure 3b, for different investigation domain center offsets. As can be seen, a good detection was achieved with no significant artifacts corrupting the reconstructions. What is more, the reconstructed spot moved coherently with the investigation domain displacement.

Some comments are in order here.

Firstly, we remark that the inversion algorithm does not aim at providing the shape of the targets, but rather, it is intended for target detection. This is mainly due to the strategy we adopted to combine the different single frequency reconstructions. Indeed, the proposed multiplicative combination highlighted in (11) allows us to mitigate aliasing artifacts but at the same time tends to enhance the strongest part of the reconstructions so that targets actually appear as hot spots.

Secondly, as we mentioned above, we did not process data to counteract the direct link (from the transmitting antenna to the receiving one) or the air/sand reflection. However, reconstructions do not suffer from such spurious signals. This can be justified by observing that we have performed the reconstruction over a given spatial region: the investigation domain. Hence, the direct link should appear located outside such a region. As to the air/sand interface reflection, it actually enters in the investigation domain. However, it is not localized as the target contribution and tends to be spread over the whole investigation domain. The multiplicative combination strategy hence also enhances the target reconstruction against such a contribution. In Figures 5 and 6, reconstructions corresponding to the target (still approximately buried at 2 cm) located as in position 2 (see Figure 3b) are reported. In particular, while Figure 5 has been obtained using the same source polarization as in the previous case, in Figure 6, the transmitting antenna has been rotated 90° so that the scene is illuminated by an orthogonal polarization. As can be seen, in this case as well, the target is clearly detected, regardless of the transmitting antenna features (in this case polarization). Actually, according to the formulation and related approximations presented in Section 2 and under Equation (7), what matters is that the equivalent source contributes to the polarization and that the receiving antenna is sensitive. Hence, even by changing the transmitting antenna polarization, it is expected that the method works as long as the previous statement holds true. This is basically what happened in Figure 6. In other words, unless the equivalent current has rigorously no component to which the receiving antenna is sensitive, the method is expected to work.

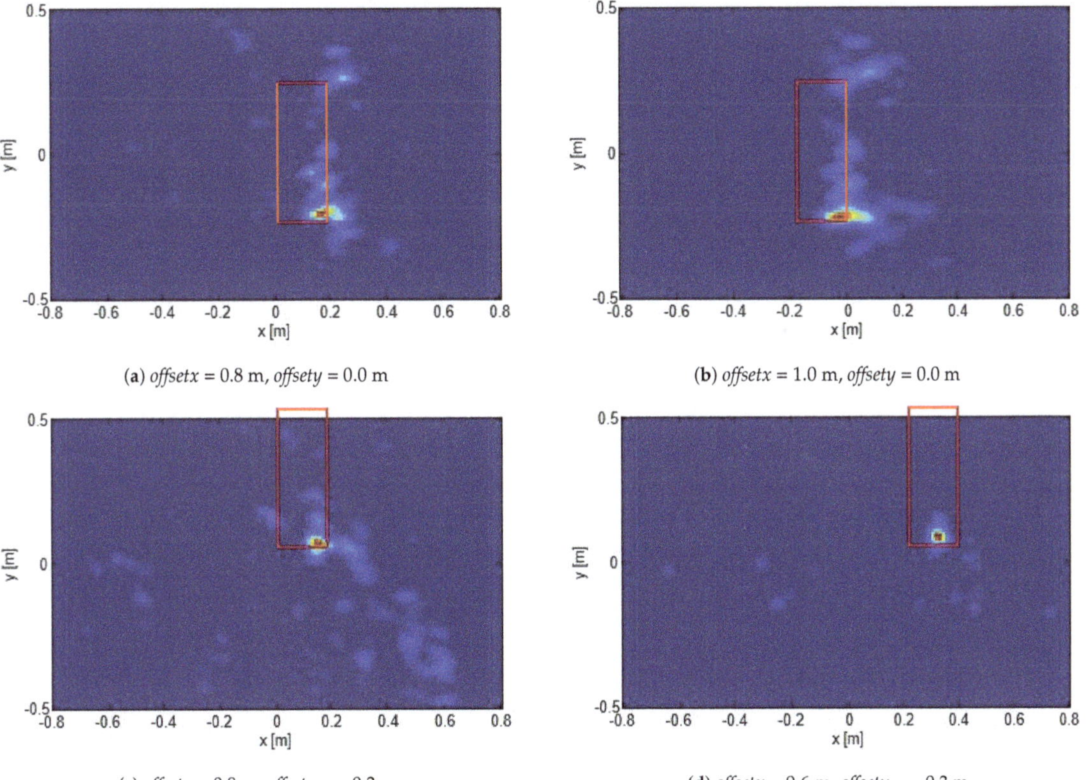

(**a**) *offsetx* = 0.8 m, *offsety* = 0.0 m

(**b**) *offsetx* = 1.0 m, *offsety* = 0.0 m

(**c**) *offsetx* = 0.8 m, *offsety* = −0.3 m

(**d**) *offsetx* = 0.6 m, *offsety* = −0.3 m

Figure 4. Normalized reconstructions for target located at position 1 shown in Figure 3b for different investigation domain offsets (reported in the figures). The red rectangles show the actual target positions.

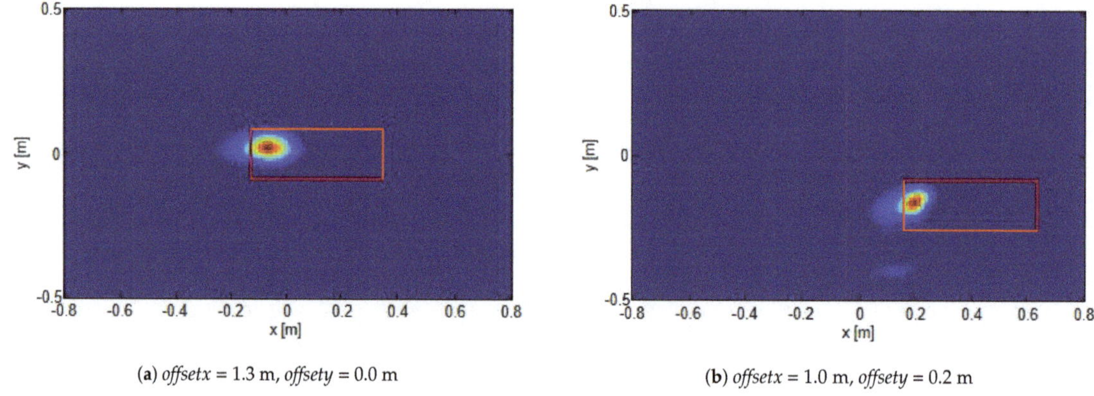

(**a**) *offsetx* = 1.3 m, *offsety* = 0.0 m   (**b**) *offsetx* = 1.0 m, *offsety* = 0.2 m

Figure 5. Normalized reconstructions for target located at position 2 shown in Figure 3b for different investigation domain offsets (reported in the figures). The red rectangles show the actual target positions.

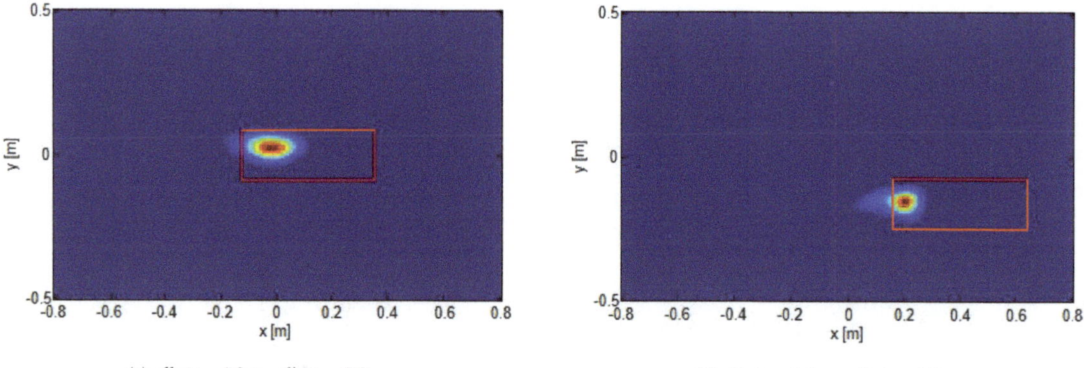

(**a**) *offsetx* = 1.3 m, *offsety* = 0.0 m   (**b**) *offsetx* = 1.0 m, *offsety* = 0.2 m

Figure 6. The same as in Figure 5 but incident field polarized orthogonal with respect to the previous case.

4.3. Detection Results for Rough Air/Sand Surface

We now turn to consider the case in which the air/sand interface was not smoothed. In this case, we consider data collected over a grid of 8×7 positions with the same spatial step as above but at a height $h_O = 130$ cm. Note that the spatial data are slightly greater than the previous case but still under-sampled [20].

First, we show the reconstruction of a shallowly buried target (the target depth and type are the same as above) by employing all the available frequencies. These results are reported in the left column of Figure 7. As can be appreciated, the target is clearly detected and the related hot spot indicator changes position accordingly to the investigation domain center offset. On the same figure (right column), instead, we report the reconstructions obtained by processing background data, i.e., in absence of the target. Differently from the target case, now the reconstructions do not exhibit a clear hot spot. Moreover, the reconstruction changes as the investigation domain center offset varies. This confirms the previous discussion that processing air/soil interface reflection does not return a focused hot spot. Furthermore, the reconstruction corresponding to this contribution is in general different when the investigated spatial region changes. This is in particularly true here because roughness entails that the air/soil interface has different spatial details. Eventually, these results suggest a possible strategy to recognize actual targets against surface clutter. Indeed, comparing images obtained using different investigation domain offsets, the actual targets are those ones for which the reconstructions "move" coherently with the change in the investigation domain center.

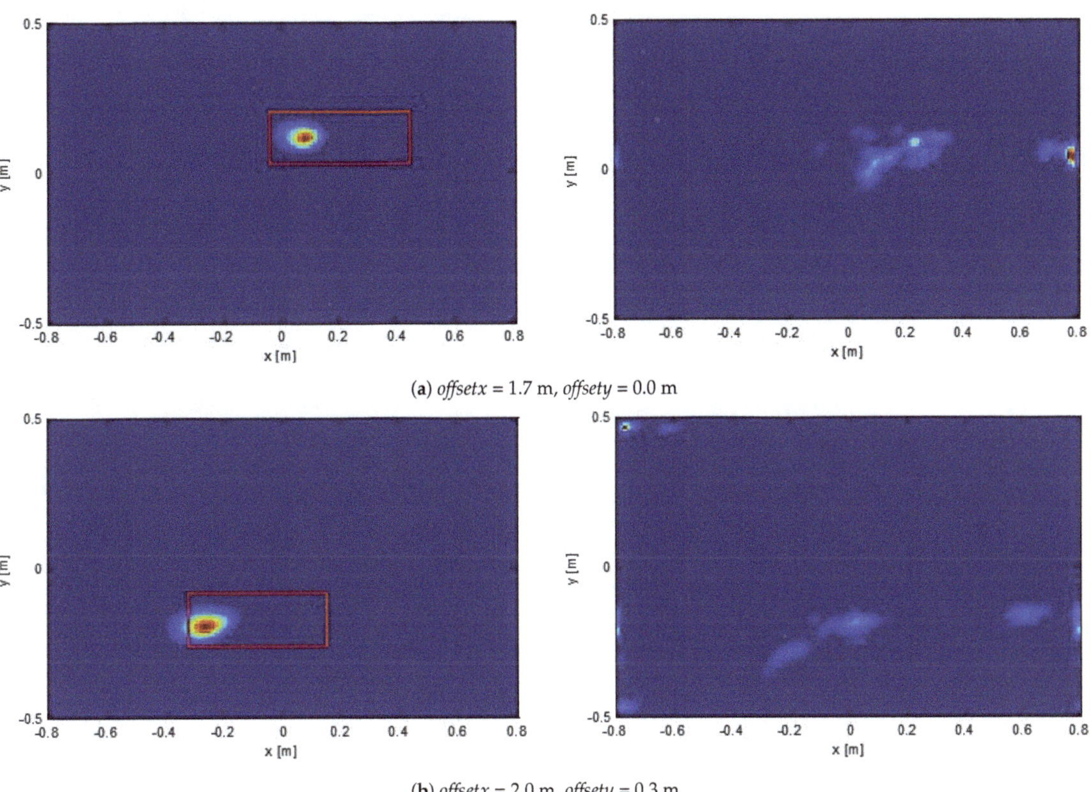

(**a**) *offsetx* = 1.7 m, *offsety* = 0.0 m

(**b**) *offsetx* = 2.0 m, *offsety* = 0.3 m

Figure 7. Normalized reconstructions of a shallowly buried targets for two different investigation domain offsets along with the actual target location denoted as red rectangles (figures un the left column). Normalized reconstructions background medium data, i.e., in absence of target, (figures on the right column).

In Figure 8, as well as in Figure 7, by keeping fixed the investigation domain with *offsetx* = 2.0 m, *offsety* = 0.3 m is considered. However, different numbers of frequencies are employed. As can be seen, when the number of frequencies is increased, the aliasing artifacts actually tend to disappear and the hot spots narrows. This is, of course, expected and perfectly consistent with the theoretical arguments discussed above.

The simple proposed strategy for reducing spatial data hence works very well. To further check this procedure, we consider an even more challenging case by reducing the spatial data employed to achieve the reconstructions. In more detail, we collected data over a 4×4 measurement grid, with twice the spatial step, that is, 40 cm. The height of the measurement aperture was still $h_O = 130$ cm, and the frequencies were taken within the same band as above. Note that in this case the number of spatial data is even lower than the ones used in the first example reported at the beginning of this section.

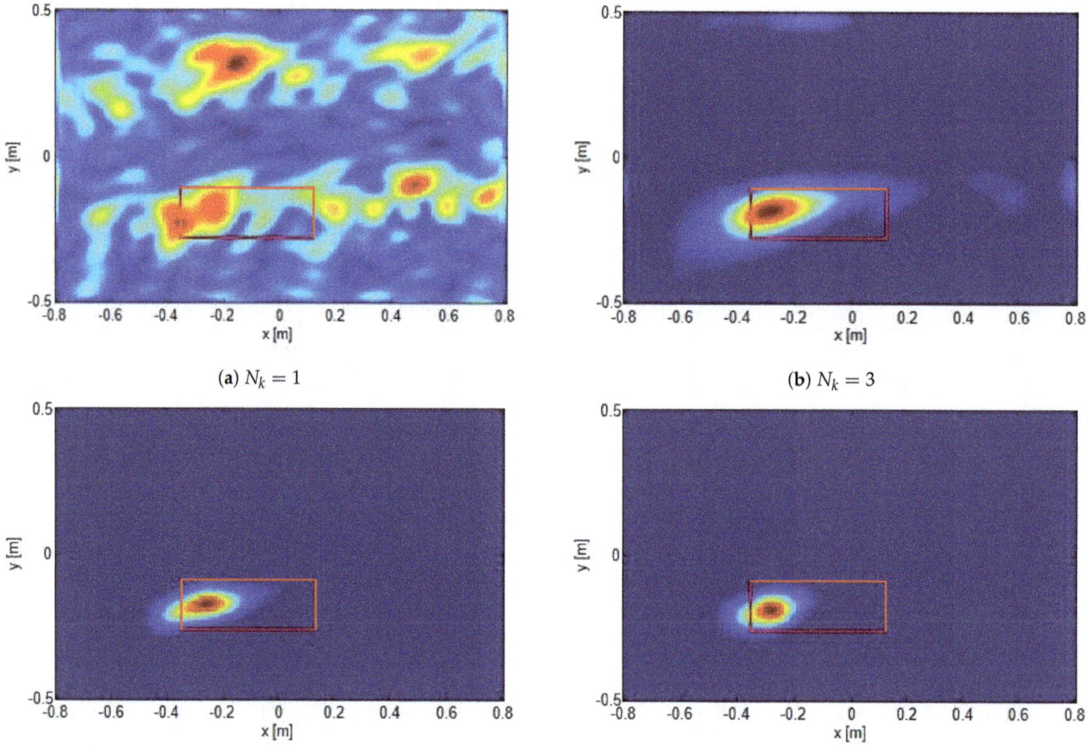

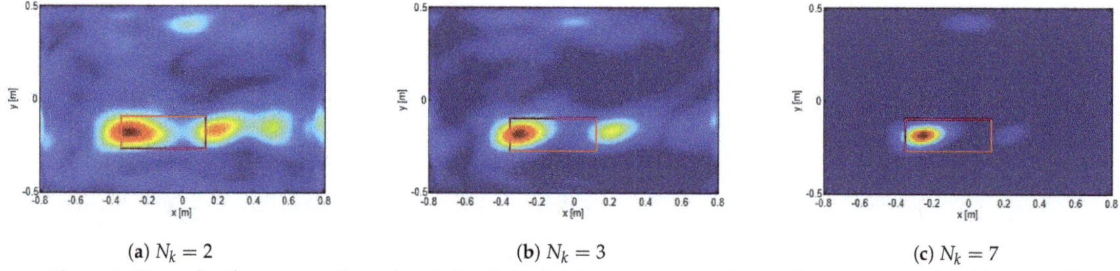

(a) $N_k = 1$

(b) $N_k = 3$

(c) $N_k = 5$

(d) $N_k = 7$

Figure 8. The same case as in Figure 7 with *offsetx* = 2.0 m; *offsety* = 0.3 m. Comparison of reconstructions obtained by employing different number of frequencies N_k.

It can be seen in Figure 9 that even under this more challenging situation where the spatial data have been reduced further, the proposed method works very well in detecting and localizing the target and in counteracting aliasing artifacts, though the number of frequencies is actually low as well.

(a) $N_k = 2$

(b) $N_k = 3$

(c) $N_k = 7$

Figure 9. Normalized reconstructions of a shallowly buried target corresponding to the case reported in Figure 7, *offsetx* = 2.0 m, *offsety* = 0.3 m but with only 4 × 4 measurement grid collected with a spatial step of 40 cm. The actual target location is denoted as red rectangles. Comparison of reconstructions obtained for different values of N_k.

As a final example, we consider the case the target is buried more deeply. In particular, in this case, the plate target is buried approximately 10 cm below the air/sand interface. Figure 10 shows the reconstructions obtained by considering the image plane at different depths. In all the examples, $N_k = 7$ frequencies (which were shown to be enough for mitigating aliasing artifacts) have been employed, and the same spatial grid measurement as in Figure 9 is considered. These results show that the proposed method is still effective in detecting the target. Moreover, as discussed above, changing the depth at which reconstruction is achieved allows one to get information about the target depth as the one for which the reconstruction is more focused. Of course, since the soil permittivity is not known (and hence not accounted for in the reconstruction algorithm), the estimated target depth does not coincide with the actual one. In particular, since the relative dielectric permittivity of a wet sand likely stands between 4 and 9, the "apparent" target depth can range from 20 cm to 30 cm, which is perfectly consistent with the result shown in Figure 10.

A brief comment is in order about the actual location of the reconstructed spots that, in most of the figures above, appear focused at the edges of the true shape. As a matter of fact, one can expect that the detected hot spot should roughly appear in correspondence with the target center. Indeed, this circumstance has been observed in [19,20], where synthetic data were employed. However, in those cases, we have considered a reflection-mode configuration (i.e., the TX (plane wave incidence) and the RXs were on the same side), the investigation domain was centered with respect to the measurement aperture, and the target was precisely parallel to the reconstruction plane. For the present case, because of the experimental set up, of course, the target position and its orientation are not precisely known; uncertainty is a few cm. In addition, the configuration is quite different since a transmission-mode set up is under consideration and the measurement aperture is actually side-looking the investigation plane. All these aspects could have contributed to the obtained results. However, we believe that such reconstructions are most likely due to the illumination, which is not uniform across the target and depends on the relative position between the TX and the target itself. In order to obtain a clear understanding of this effect, one should go through the study of the equivalent current behavior. This can be numerically achieved and is beyond the aim of the paper.

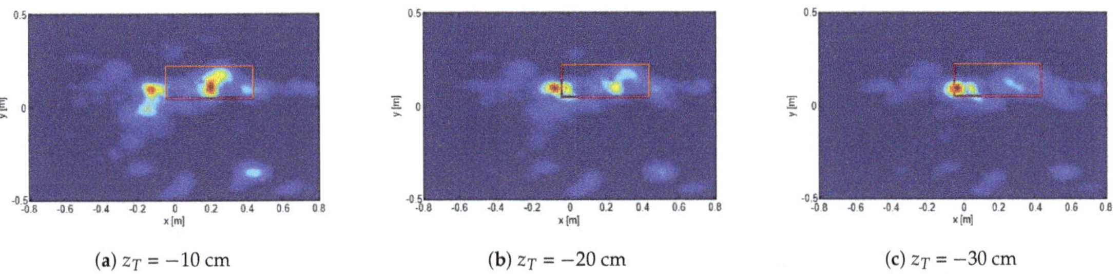

(a) $z_T = -10$ cm   (b) $z_T = -20$ cm   (c) $z_T = -30$ cm

Figure 10. Normalized reconstructions of a target approximately buried 10 cm below the air/sand interface. Data have been collected over 4×4 measurement grid with a spatial step of 40 cm and *offsetx* = 1.7 m, *offsety* = 0.0 m. The actual target location is denoted as red rectangles. Comparison of reconstructions obtained using $N_k = 7$ at different depths.

5. Conclusions

In this paper, we experimentally validated a method for detecting and localizing shallowly buried targets from under-sampled multi-frequency data. The method relies on two main ingredients: a suitable scattering model and a simple procedure to process multi-frequency under-sampled data. The scattering model is based on the equivalence theorem, which allows one to cast the detection as the reconstruction of equivalent sources supported over a reference plane. This way, detection becomes a 2D problem. Furthermore, the incident field is embodied into the unknown equivalent sources. Therefore, coherence between the TXs and Rxs is not necessary when single-frequency data are employed. Reconstructions obtained at different frequencies are then suitably combined. This allows one to mitigate aliasing artifacts and hence to achieve the reconstructions with a very reduced number of spatial measurements.

Experimental results confirm the feasibility of the method even under a rather complex scattering scenario, such as the one addressed herein. Indeed, it is shown that the targets can be detected by using very few spatial measurement points and a number of properly selected frequencies. In addition, the results confirm the robustness of the method against the reflection occurring at the air/soil interface, even when this exhibits some degree of roughness.

The obtained results can be meant as a proof of concept. However, they encourage the application of the proposed measurement configuration and of the related inversion algorithm to more challenging scenarios, where many targets may be present inside a non-homogeneous terrain and the sensors may be deployed at a larger stand-off distance.

Author Contributions: Conceptualization, A.B., R.S., and G.L.; methodology, R.S.; software, A.B. and R.S.; validation, A.B.; formal analysis, A.B. and R.S.; investigation, A.B. and G.L.; resources, A.B.; data curation, A.B.; writing—original draft preparation, A.B. and R.S.; writing—review and editing, A.B.; visualization, A.B.; supervision, G.L. and R.P. All authors have read and agreed to the published version of the manuscript.

Funding: This research was partially funded by VALERE: VAnviteLli pEr la RicErca research program by Universita degli Studi della Campania Luigi Vanvitelli.

Institutional Review Board Statement: Not applicable.

Informed Consent Statement: Not applicable.

Data Availability Statement: No new data were created or analyzed in this study. Data sharing is not applicable to this article.

Conflicts of Interest: The authors declare no conflict of interest.

References

1. Daniels, D.J. *Ground Penetrating Radar*, 2nd ed.; IEE Radar, Sonar, Navigation and Avionics Series 15; Stewart, N., Griffiths, H., Eds.; IEE PRESS: London, UK, 2004; ISBN 0 86341 3609.
2. Pastorino, M.; Randazzo, A. *Microwave Imaging Methods and Applications*; Artech House: Boston, MA, USA, 2018; ISBN 978-1-63081-348-2.
3. Benedetto, A.; Pajewsky, L. *Civil Engineering Applications of Ground Penetrating Radar*; Springer: Cham, Switzerland, 2015; ISBN 978-3-319-04813-0.
4. Ivashov, S.I.; Razevig, V.V.; Vasiliev, I.A.; Andrey, A.V.; Bechtel, T.D.; Capineri, L. Holographic subsurface Radar of RASCAn type: Development and applications. *IEEE J. Sel. Top. Appl. Earth Observ. Remote Sens.* **2011**, *4*, 763–778. [CrossRef]
5. Soldovieri, F.; Brancaccio, A.; Prisco, G.; Leone, G.; Pierri, R. A Kirchhoff-based shape reconstruction algorithm for the multi-monostatic configuration: The realistic case of buried pipes. *IEEE Trans. Geosci. Remote Sens.* **2008**, *46*, 3031–3038. [CrossRef]
6. Soliman, M.; Wu, Z. Buried object location based on frequency domain UWB measurements. *J. Geophys. Eng.* **2008**, *5*, 221–231. [CrossRef]
7. Urbini, S.; Cafarella, L.; Marchetti, M.; Chiarucci, P. Fast geophysical prospecting applied to archaeology: Results at «villa ai Cavallacci» (Albano Laziale, Rome) site. *Ann. Geoph.* **2007**, *50*, 291–299. [CrossRef]
8. Catapano, I.; Crocco, L.; Di Napoli, R.; Soldovieri, F.; Brancaccio, A.; Pesando, F.; Aiello, A. Microwave tomography enhanced GPR surveys in Centaur's Domus—Regio vI of Pompeii. *J. Geophys. Eng.* **2012**, *9*, S92–S99. [CrossRef]
9. Sato, M. GPR evaluation test for humanitarian demining in Cambodia. In Proceedings of the IEEE International Geoscience and Remote Sensing Symposium (IGARSS), Honolulu, HI, USA, 25–30 July 2010. [CrossRef]
10. Randazzo, A.; Salvadè, A.; Sansalone, A. Microwave tomography for the inspection of wood materials: Imaging system and experimental results. *IEEE Trans. Microw. Theory Tech.* **2018**, *66*, 3497–3510. [CrossRef]
11. Brancaccio, A.; Migliore, M.D. A Simple and Effective Inverse Source Reconstruction with Minimum a Priori Information on the Source. *IEEE Geosci. Remote Sens. Lett.* **2017**, *14*, 454–458. [CrossRef]
12. Fu, L.; Liu, S.; Liu, L.; Lei, L. Development of an airborne ground penetrating radar system: Antenna design, laboratory experiment, and numerical simulation. *IEEE J. Sel. Top. Appl. Earth Obs. Remote Sens.* **2014**, *7*, 761–766. [CrossRef]
13. Catapano, I.; Crocco, L.; Krellmann, Y.; Triltzsch, G.; Soldovieri, F. A Tomographic Approach for Helicopter-Borne Ground Penetrating Radar Imaging. *IEEE Geosci. Remote Sens. Lett.* **2012**, *9*, 378–382. [CrossRef]
14. Fernández, M.G.; López, Y.Á; Arboleya, A.A.; Valdés, B.G.; Vaqueiro, Y.R.; Andrés, F.L.H.; García, A.P. Synthetic aperture radar imaging system for landmine detection using a ground penetrating radar onboard an unmanned aerial vehicle. *IEEE Access* **2018**, *6*, 45100–45112. [CrossRef]
15. Noviello, C.; Esposito, G.; Fasano, G.; Renga, A.; Soldovieri, F.; Catapano, I. Small-UAV radar imaging system performance with GPS and CDGPS based motion compensation. *Remote Sens.* **2020**, *12*, 3463. [CrossRef]
16. Gennarelli, G.; Rosen, P.A.; Beatty, R.; Freeman, T.; Gim, Y. A low frequency airborne GPR system for wide area geophysical surveys: The case study of Morocco Desert. *Remote Sens. Environ.* **2019**, *233*. [CrossRef]

17. Catapano, I.; Gennarelli, G.; Ludeno, G.; Noviello, C.; Esposito, G.; Soldovieri, F. Contactless Ground Penetrating Radar Imaging: State of the Art, Challenges, and Microwave Tomography-Based Data Processing. *IEEE Geosci. Remote Sens. Mag.* **2021**. [CrossRef]
18. Ludeno, G.; Gennarelli, G.; Lambot, S.; Soldovieri, F.; Catapano, I. A comparison of linear inverse scattering models for contactless GPR imaging. *IEEE Trans. Geosci. Remote Sens.* **2020**, *58*, 7305–7316. [CrossRef]
19. Brancaccio, A.; Leone, G.; Solimene, R. Single-frequency Subsurface Remote Sensing via a Non-Cooperative Source. *J. Electrom. Waves Appl.* **2016**, *30*, 1147–1161. [CrossRef]
20. Brancaccio, A.; Dell'Aversano, A.; Leone, G.; Solimene, R. Subsurface Detection of Shallow Targets by Undersampled Multifrequency Data and a Non-Cooperative Source. *Appl. Sci.* **2019**, *9*, 5383. [CrossRef]
21. Balanis, C. *Advanced Engineering Electromagnetics*; Wiley and Sons: New York, NY, USA, 1989.
22. Bevacqua, M.T; Isernia, T. Shape Reconstruction via Equivalence Principles, Constrained Inverse Source Problems and Sparsity Promotion. *Prog. Electromagn. Res.* **2017**, *158*, 37–48. [CrossRef]
23. Potin, D.; Vanheeghe, P.; Duflos, E.; Davy, M. An abrupt change detection algorithm for buried landmines localization. *IEEE Trans. Geosci. Remote Sens.* **2006**, *44*, 260–272. [CrossRef]
24. Pasternak, M.; Karczewski, J.; Silko, D. Stepped frequency continuous wave GPR unit for unexploded ordnance and improvised explosive device detection. In Proceedings of the International Radar Symposium (IRS), Leipzig, Germany, 7–9 September 2011; pp. 105–109.
25. Bertero M.; Boccacci, P. *Introduction to Inverse Problems in Imaging*; Institute of Physics: Bristol, UK, 1998; ISBN 9780750304351.

Article

Low Cost, High Performance, 16-Channel Microwave Measurement System for Tomographic Applications

Paul Meaney [1,*], **Alexander Hartov** [1], **Timothy Raynolds** [1], **Cynthia Davis** [2], **Sebastian Richter** [3], **Florian Schoenberger** [3], **Shireen Geimer** [1] and **Keith Paulsen** [1]

[1] Thayer School of Engineering, Dartmouth College, Hanover, NH 03755, USA;
    alex.hartov@dartmouth.edu (A.H.); timothy.raynolds@dartmouth.edu (T.R.);
    shireen.geimer@dartmouth.edu (S.G.); keith.paulsen@dartmouth.edu (K.P.)
[2] GE Global Research, Niskayuna, NY 12309, USA; davisc@ge.com
[3] German Federal Ministry of Defense, 2E1202 Hamburg, Germany; richter.sebastian@hotmail.de (S.R.);
    flo3003@web.de (F.S.)
*   Correspondence: paul.meaney@dartmouth.edu; Tel.: +1-603-646-3939

Received: 17 August 2020; Accepted: 17 September 2020; Published: 22 September 2020

Abstract: We have developed a multichannel software defined radio-based transceiver measurement system for use in general microwave tomographic applications. The unit is compact enough to fit conveniently underneath the current illumination tank of the Dartmouth microwave breast imaging system. The system includes 16 channels that can both transmit and receive and it operates from 500 MHz to 2.5 GHz while measuring signals down to −140 dBm. As is the case with multichannel systems, cross-channel leakage is an important specification and must be lower than the noise floors for each receiver. This design exploits the isolation inherent when the individual receivers for each channel are physically separate; however, these challenging specifications require more involved signal isolation techniques at both the system design level and the individual, shielded component level. We describe the isolation design techniques for the critical system elements and demonstrate specification compliance at both the component and system level.

Keywords: microwave imaging; breast; multipath; dynamic range; software defined radio; leakage

1. Introduction

Interest in microwave imaging for medical applications has grown significantly since the early 1980s, but more prominently over the past decade. Much of the recent expansion has resulted from advances in technology that facilitate implementation of novel ultrawideband approaches [1–3] and the advent of powerful computing that can accommodate time-consuming inverse problems [4–6]. Microwave imaging offers important advantages over other emerging technologies. For instance, microwave methods achieve far greater penetration depths relative to optical imaging techniques, which allow broader applications to be pursued [7–9]. Compared to lower frequency electrical impedance approaches, microwave imaging systems provide superior spatial resolution [10–13]. The list of medical microwave imaging applications is substantial and growing, and currently includes breast cancer imaging [14,15], stroke diagnosis [16,17], bone health assessment [18], and thermal therapy monitoring [19–21], among others. In all of these cases, microwave methods either exploit significant tissue dielectric property contrast [16,17,22–25], or inherent property dependencies [26,27]. While some debate over breast tumor/normal tissue property contrast remains, most studies have demonstrated that substantial contrast exists between tumor properties and those of both the associated adipose and fibroglandular tissue [22–29]. Further, several clinical publications indicate effective tumor detection, diagnosis, and treatment monitoring based on endogenous dielectric property contrast between tumor and normal tissue [14,15,30]. Although the potential of medical microwave imaging is significant, tangible progress in terms of physical implementations has been slow.

Hardware challenges arise from the need to devise a near-field configuration that illuminates the body from multiple directions and receives scattered waves in either monostatic or multistatic modes. In multistatic cases, the difficulties are compounded by antenna mutual coupling, multipath signal distortion within the illumination array, and possible cross-channel leakage within the measurement electronics, themselves [31]. Various strategies have been developed to contend with these signal corruption problems, for example, by multiplexing high-end vector network analyzers (e.g., [32,33]), utilizing lossy coupling media (e.g., [6]), and/or translating small numbers or single source–detector pairs (e.g., [1]).

An alternative and increasingly popular hardware option [34,35] is centered around a measurement system design based on inexpensive, commercially available software defined radios (SDRs). We have described a 4-channel prototype using the B210 USRP boards developed by Ettus Research (Santa Clara, California) [36]. These units incorporate the major functional elements of a sophisticated transmitter and receiver into a single chip (Analog Devices AD9361, Norwood, MA), and operate over an impressively large bandwidth of 70 MHz to 6 GHz. They are finding use in a range of radar applications [34]. Measurement requirements for our log transformed tomographic image reconstruction algorithm along with the associated calibration process have been stable and robust over several generations of hardware systems. Our microwave tomography implementation that does not require a priori information for convergence to the desired solution, and does so with less measurement data relative to competing implementations [37]. These innovations are integral to our success in translating this technology into the clinic [14,15]. Our software advances are described elsewhere, and a comprehensive summary of the most recent innovations appears in Meaney et al. [37]. The system design includes a synchronization strategy, a technique to enhance dynamic range, and a calibration process that compensates for uneven amplitude and phase steps associated with internal variable gain amplifiers [36]. In this paper, we focus on the significant challenges associated with channel-to-channel leakage, packaging SDRs into a compact and light form factor that fits underneath the antenna mounting fixture for our existing microwave breast imaging system, and instrumentation control through existing software platforms. In the case of channel-to-channel leakage, we investigate the problem in terms of (a) leakage through existing transmission lines, and (b) ambient leakage from signals escaping their immediate structures and propagating along the overall structure as surface waves.

In this context, the novel aspects of the system include: (a) enhanced dynamic range, (b) multichannel operation with high degree of channel-to-channel isolation, and (c) data acquisition acceleration via reordering of the data recording sequence. Integration of these attributes enables the realization of an SDR-based acquisition system with data frames rates short enough for use in clinical breast examinations. We demonstrate a compact and efficient packaging of a complete multichannel system, and present results that confirm excellent dynamic range and cross-channel isolation.

2. Methods

2.1. Overall System Configuration

Figure 1 shows a schematic diagram for a 16 channel system. It includes: (a) a single B210 USRP board operating as a dedicated transmitter, (b) a single-pole, 16 throw switch (SP16T) for multiplexing the transmission channel, (c) a 6-bit, digital attenuator (0–63 dB) for controlling the reference signal amplitude, (d) a series of 16 single-pole, double-throw switches (SPDT) for toggling individual channels between transmit and receive modes, (e) a series of 16 single-pole, single-throw switches (SPST) for extra isolation between the SP16T and SPDT's (not shown), (f) a series of low noise amplifiers (LNAs) for increasing received signal strength, (g) a series of 16 monopole antennas, (h) eight B210 USRP boards for dedicated reception, (i) a SPST in combination with an 8-way power divider for synchronizing reference signals to each receiver board, and (j) an OctoClock-G (Ettus

Research, Santa Clara, CA) for providing coherent clock signals and an accurate reference for B210 signal synthesis. Control, bias lines, USB communication lines, power supply, and controlling computer are not shown. Since our imaging system approach only utilizes transmission data, incorporation of reflection measurements is unnecessary, and would increase system complexity substantially. Substantially greater channel-to-channel isolation was achieved by separating the transmit and receive functions physically and utilizing a dedicated B210 (Ettus Research, Santa Clara, CA) for transmission. The signal from Channel 1 of the transmit board drove individual antennas as each radiated into the illumination tank (illumination tank shown later in Figure 5). Part of the signal was separated via a power divider and passed through a digital attenuator after which it acted as a variable reference signal for synchronization—a process described in more detail in Meaney et al. [36].

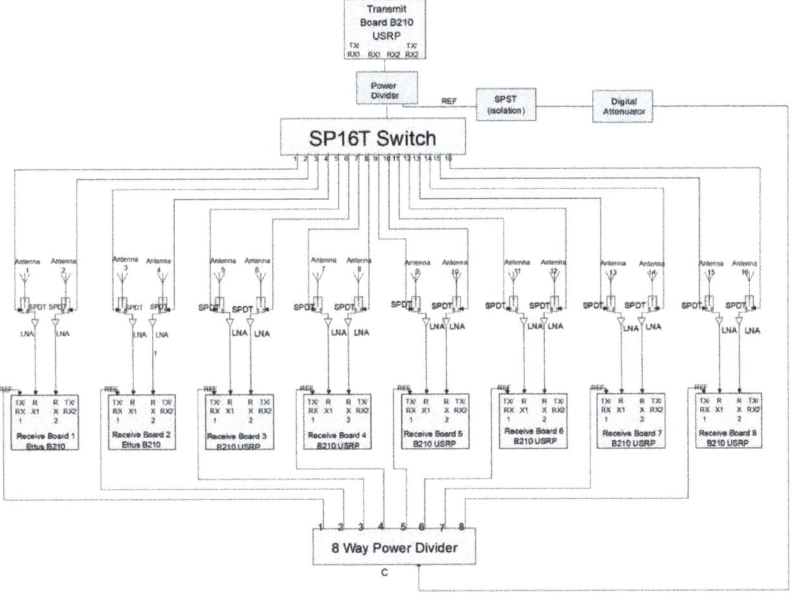

Figure 1. Schematic diagram of the complete system illustrating: (**a**) transmitting B210 board, (**b**) transmitting SP16T switch, (**c**) 16 switch/amplifier modules, (**d**) eight dedicated receive B210s, (**e**) switch module for the reference signal, and (**f**) 1-by-8 power dividers for the reference signal, respectively.

Sets of SP16T, SPDT, and SPST switches direct the transmission signal to a single antenna (one at a time) while allowing the reception of response signals by the remaining 15 channels. Transmitting signals must not leak into receiver ports where they will be amplified and detected erroneously by the receive B210 modules. Here, maintaining adequate isolation is challenging because transmitted signals are on the order of 0 dBm while received signals on the order of −140 dBm need to be detected. This level of isolation is uncommon in most measurement systems and demands significant attention. In particular, signal attenuation from transmitters to receivers directly opposing the signal source often exceed −120 dBm and can reach −140 dBm or more at higher frequencies. Lossiness in the coupling liquid confers substantial benefits, specifically, in suppression of surface waves that can corrupt desired signals [31]. Achieving a large dynamic range with concomitant channel-to-channel isolation is challenging but essential for high fidelity data acquisition. As a result, measurement of signals at these low power levels is critical to system success, especially in the breast imaging context. The SP16T switch was purchased in a connectorized configuration from Universal Microwave Components Corporation (UMCC SR-J010-16S, Alexandria, VA) while the SPDT and SPST switches

(Peregrine Semiconductor PE4240 and PE4246, San Diego, CA) were integrated into a custom housing (one per channel) with the LNA.

LNAs (Mini-Circuits TSS-53LNB+, Brooklyn, NY) were added to the design to improve the receiver noise figure and dynamic range. The dynamic range of the B210's by themselves is limited by the internal A/D discretization error at the low end instead of the actual noise level. By adding extra gain in front of these boards, the low end of the dynamic range is determined by the noise floor, which can be controlled by altering the sampling bandwidth—i.e., increasing the sampling time decreases the noise floor. This design approach was discussed in depth in Meaney et al. [36].

The B210 SDRs are driven primarily by the AD9361 (Analog Devices, Norwood, MA) as an agile transceiver, which incorporates two channels of transmission and four ports of reception, respectively (only two of the latter can be used simultaneously). As discussed in detail in Meaney et al. [36], multiport synchronization is possible and leads to a coherent, multichannel system. Variable gain was included in order to measure signals at very low received power depending on sampling bandwidth. In this configuration, Rx ports were utilized for signal reception and one of the Tx/Rx ports was used for reference signal handling. Since the on-board LO oscillators were synchronized, sampling the reference signal on only one of the two remaining ports was required. Furthermore, reference signal sampling could not be performed simultaneously with the signal measurements from the antenna on the same channel, since the Tx/Rx port shared the same pin on the AD9361 chip with the corresponding Rx port. In fact, as discussed in Section 2.5.3., the reference signal was sampled simultaneously with the antenna detection signal from the complementary channel on the same board (only their phase differences were coherent), and then a second scan samples both antenna signals (multiple subtractions produced coherent versions of both antenna—see Section 2.5.3.). In addition, more B210 shielding is critical relative to cross channel leakage, and is discussed in detail below.

The coherent reference signal was fed into a module with three SPSTs placed in series after a digital attenuator that was inserted for extra isolation, which was necessary because the reference signal is not turned OFF during antenna transmission and becomes a potential multipath propagation conduit. The signal was then fed into an 8-way power divider (Pulsar Microwave PS8-12-454/5S, Clifton, NJ) from which the Channel 1 Tx/Rx ports of the receive B210s were driven. Feeding eight receive B210s allowed reception to be performed simultaneously on all 15 channels. The system synchronization utilizing this external reference signal configuration is described in detail in Meaney et al. [36].

The OctoClock-G CDA-2990 (Ettus Research, Santa Clara, CA, USA) provided eight pairs of outputs to the B210 SDR. The first signal is a stable, 10 MHz reference that was used to synchronize the B210s, and the second is a 1 Hz PPS clock signal that was utilized for B210 coordination. In our system configuration, eight signal pairs service nine B210s: one pair is fed into two low frequency power splitters that subsequently drive two B210s.

The list below summarizes the overall system costs. Costs for instruments with comparable performance (such as the Keysight M9800A and Rohde and Schwarz ZNBT) are in excess of $120,000. Utilizing the Ettus B210 SDRs as the foundational building block offers a substantial price advantage.

Ettus OctoClock	$2030
Ettus B210 (9)	$11,538
UMCC 1 × 16 Switch	$4620
Pulsar 1 × 8 Power Divider	$750
Miscellaneous cables, connectors, and wires	$1200
Shielded housings	$3500
Switch/amplifier circuit boards & components (16)	$1200
NI USB digital I/O boards (3)	$370
USB hubs (2)	$150
Total	$25,358

2.2. Isolation and Shielding

Two components of the hardware require significant shielding to attenuate signals propagating into the environment and back into the system—the B210 USRP circuit boards and the switch/amplifier modules.

2.2.1. B210 USRP Housing Design

B210 USRP circuit boards can be purchased with no shield or one provided by the manufacturer as shown in Figure 2a,b. Our system was comprised of nine B210 boards—one serving as a dedicated transmitter and the remaining eight acting as dedicated receivers (two channels per board to produce a 16 channel system).

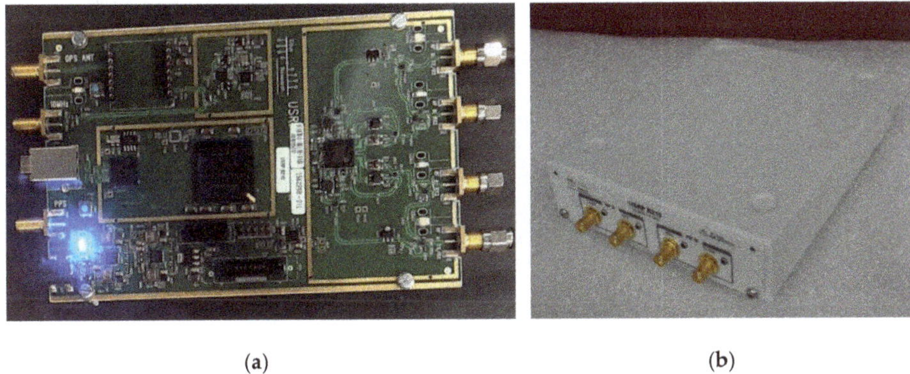

(**a**) (**b**)

Figure 2. Photographs of the B210 USRP circuit board (**a**) without and (**b**) with a commercial cover.

In order to develop a suitable shield, we explored several options before converging on the final design shown in Figure 3. An initial design, based on housing walls configured to accommodate existing B210 connectors, was unsuccessful (signals easily escaped through gaps between connector flanges and the housing wall). Instead, coaxial SMA connectors were removed and panel mount flange connectors were utilized to achieve a significantly better seal. The power connector (lower left in Figure 2) was also replaced with a panel mount coaxial cable connector to increase isolation further. The USB3.0 connector was retained on the board and was accommodated through a cutout in the housing wall around its casement.

Figure 3. Photograph of the B210 USRP circuit board mounted inside its custom shielded housing and associated cover, which exhibits a central ridge that isolates RF fields from the digital portion of the circuitry.

The shielded housing cover was designed with a raised surface that extended 0.5 mm into the main housing chamber, and fit snugly within the bottom of the housing—adding isolation from ambient fields escaping through seals between housing parts and aligning components to minimize damage during assembly. The board shown in Figure 3 was logically divided into two functional areas: (a) the RF componentry (right) and (b) the digital interface, power supplies, and FPGA controlling unit (left). A gold band was printed on the top surface with plated-through vias to the ground plane below. We incorporated a ridge into the housing cover that closely contacted the gold band to prevent leakage of RF signals from the right portion of the board.

2.2.2. Switch/Amplifier Housing Design

The switch/amplifier module (see Figure 4) incorporated a single-pole/single-throw (SPST) switch (lower left), a single-pole/double-throw (SPDT) switch (lower right), and a low noise amplifier (LNA; top right). The SPDT selects whether a particular antenna operates in transmit or receive mode. In the transmit mode, the signal from the transmit B210 passes through both switches directly to the antenna port for propagation into the illumination tank. In the receive mode, the signal passes from the antenna port, through the SPDT and LNA to the receive port. This SPDT selects the receiver mode, and the LNA added overall gain to the composite receive channel (including the receiver portion of the B210 boards) and reduced the effective channel noise figure [36]. The SPST also added extra isolation so that signals did not leak from the transmitter to the receiver through the circuitry, but instead propagated along the desired path into the tank and through intervening object. The design goal for minimizing leakage from the transmit to receive ports was 80 dB attenuation. The design compartmentalized components such that openings between each acted as quasi-cutoff waveguides, which attenuated unwanted signal propagation between partitions—leaving the coplanar waveguide transmission line as the only viable signal path. Bias and control lines were connected via resin sealed, bolt-in filters (API Technologies Corporation—Spectrum Control—part # 51-729-312, Schwabach, Germany) threaded into the housing floor. Previous analysis of connectorized, single component devices demonstrated that bias and control lines created significant leakage pathways. The extra filtering impacted ON/OFF switching speed; however, elimination of signal corruption was deemed more important even though the design degraded data acquisition speed slightly.

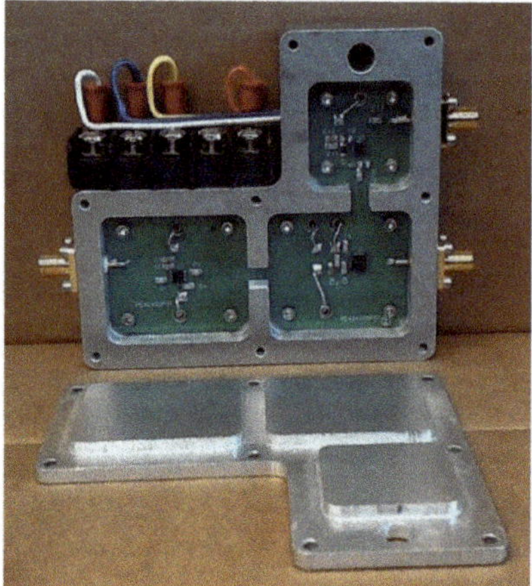

Figure 4. Photograph of the switch/amplifier module illustrating compartmentalization of the single-pole/single-throw (SPST; left), single-pole/double-throw (SPDT; lower right), and low noise amplifier (LNA; upper right), and associated cover with raised surfaces.

2.2.3. Antenna Mutual Coupling

While not part of the microwave transmit/receive electronics, antenna mutual coupling warrants some discussion. The system feeds an array of 16 monopole antennas submerged in a glycerin: water bath. Figure 5 shows (a) a photograph of a test illumination chamber and (b) a schematic diagram of the imaging field-of-view with a test object present. In our measurement system, a complement of 13 antennas directly opposite the transmitter receives signals from each transmitting antenna. We have studied the effects of the presence of adjacent antennas on the signals received. In Paulsen and Meaney [38] and Meaney et al. [39], we identified the signal perturbations caused by adjacent antennas and developed a strategy for compensating for the effects by modeling each adjacent antenna as an electromagnetic sink. The compensation is effective and produces substantially improved images. Interestingly, at frequencies above 1000 MHz, these perturbations were not observed, primarily because the lossiness of the glycerin-based coupling liquid increased substantially with frequency, and effectively suppressed the interantenna interactions.

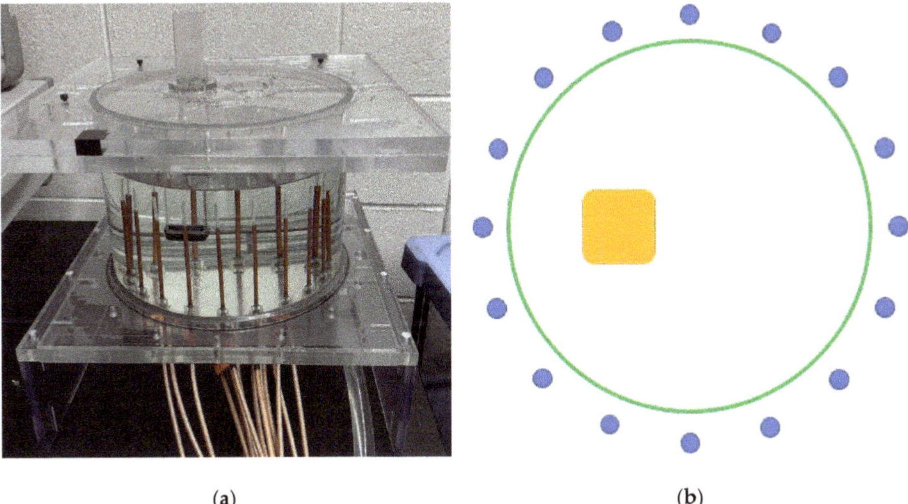

(a)   (b)

Figure 5. (**a**) Photograph of a test illumination chamber and (**b**) schematic diagram of the imaging field-of-view with 16 monopole antennas and the presence of a yellow test object.

2.3. Packaging

Key features that directly addressed expansion of our 4-channel prototype [36] were: (1) physical separation of receive modules, (2) separation of the transmit module from receive modules, (3) SPST addition to the switch/amplifier modules to minimize signal leakage from the transmitter when operating as a receiver, and (4) component compartmentalization in the switch/amplifier modules to reduce internal multipath propagation of unwanted surface waves.

Figure 6a,b shows two views of the microwave subsystem as (a) fully assembled and in a (b) separated configuration. The unit was mounted below the antenna array in the operational mode as shown in Figure 6c. Components were organized to minimize wiring and cabling lengths and complexity. For example, groups of eight switch/amp modules had their associated digital I/O cards and power supply screw terminals located directly above to keep bias and control lines as short as possible. Connections to external computers and power supplies were limited to two USB cables, a single 110 V 60 Hz power cable, and two DC power supply wires.

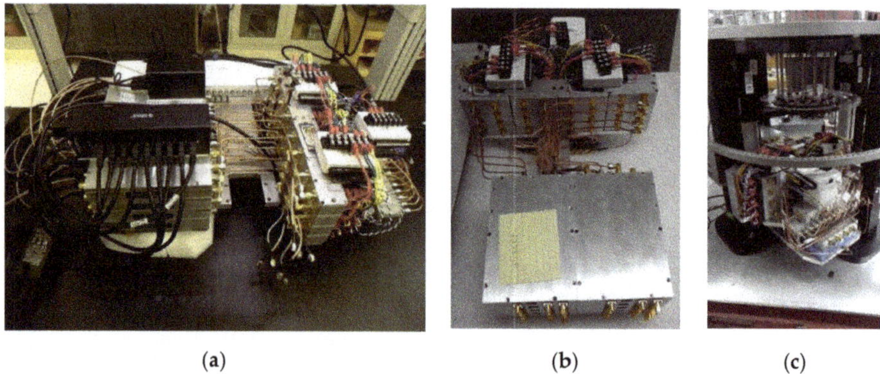

(a)   (b)   (c)

Figure 6. Photographs of the microwave electronic subsystem showing: (**a**) a complete system fully assembled external to the imaging system, (**b**) electronics with the grouping of eight shielded B210s (bottom) and switch/amplifier modules (top; USB hubs removed to expose componentry), respectively, and (**c**) a complete system integrated below the imaging tank and supported by the antenna array mounting plate.

2.4. Data Acquisition Sequencing and Time Considerations

Several techniques were combined to accomplish efficient data acquisition: (1) transmission runs continuously, (2) time stabilization occurs at start-up, and (3) transmission antenna switching occurs through the SP16T whose rise time is well under 1 μsec.

The data acquisition sequence was arranged so that a single receive module was selected first and then transmission was incremented over the remaining antennas. Delays because receiver gain levels need to be adjusted did occur (depending on which transmitting antenna was broadcasting), but only involved one or two acquisitions before settling. Here, a single data set was comprised of 10,000 measurements obtained with a sampling rate of 10 MHz for an acquisition time of 8.0 msec. Since stabilization time for each module was identical, the process was initiated for a subsequent module several seconds before the data acquisition was completed from the current module. Data acquisition time for a single receive module from the 14 complementary transmitting antennas and for 5 frequencies required 15 s—total time to acquire data from all 8 receive modules was 2.2 min for a given antenna array position. For each transmitting antenna, a closely coupled channel was always available within the system design, which allowed associated switch/amplifier modules to be connected to the same receive B210 circuit board. Since complete isolation of the transmitting signal from neighboring receive channels was difficult, we omitted the adjacent data points, leaving 14 receive signals for each transmitter (7 boards with two receivers per board). Our previous experience has shown that image quality is not reduced significantly, if at all, when 13 receiver antennas are used instead of the full complement of 15 channels (i.e., the two receivers closest to the transmitter are omitted).

2.5. Software and Performance Considerations

Software challenges associated with maintaining data acquisition performance included: (1) retaining coherence between transmit channels on a single board—especially when changing operating frequency, (2) utilizing the receive function on Tx/Rx ports, (3) acquiring signals from different ports on the same board coherently, and (4) delay times associated with each receive B210 board when switching between boards.

2.5.1. Transmit Channel Coherence

To maintain transmit channel coherence, only signals from one of the transmitting ports were used by incorporating a power divider to create two coherent waveforms, and one of which was attenuated digitally and served as the reference signal, which guaranteed coherence (Figure 1). In general, maintaining coherence overcomes limitations that occur when the two transmit signals are not phase locked and different boards, themselves, are not phase locked. Without coherence, the measurement phase becomes arbitrary. We exploited the B210 design feature that the receive channels within a single board are coherent (i.e., differences in their phase are constant), and performed combinations of signal acquisitions involving a transmitter reference signal to maintain coherence between the transmitter board and all receiver boards. The details of this design strategy are discussed in Meaney et al. [36].

2.5.2. Tx/Rx Port Receive Function

Each Ettus B210 board was comprised of two channels—an Rx port for receiving signals and a Tx/Rx port for both transmitting and receiving signals. Signal leakage occurred within the B210 board, itself, from an unconnected Tx/Rx port to its associated Rx port such that a reference signal could be connected to the Tx/Rx port, and sampled on the Rx port. Respective signal power levels were optimized to mitigate cross-channel contamination and associated interference through a calibration process.

2.5.3. Coherent Signal Acquisition Across Same-Board Ports

Receive signals from the two channels (Rx and Tx/Rx) on each board were coherent only when sampled simultaneously. Accordingly, reference signal amplitude was tailored to minimize signal corruption from channel crosstalk.

2.5.4. Set-up Time Minimization

Relative to earlier prototypes [40], data acquisition was reconfigured to loop over receive modules, and then loop over associated transmitting antennas for each module. Transmitter channel selection was performed by an external SP16T switch (not by the B210 boards), which offered rise times on the order of 10 nsec, which were negligible compared to receive board start up times. Only a few receive boards can be operational at any given time without overloading the control computer. Board set up times were minimized by starting each subsequent board before its previous counterpart had completed its data acquisition cycle. Under these conditions, the majority of next-board set up times was performed while data acquisition continued, which allowed sets of data consisting of 5 frequencies and 7 vertical antenna positions to be completed in approximately 15 min per breast.

2.5.5. System Calibration

Our imaging algorithm utilizes measurement data formed as differences between field values acquired when an object is present in the illumination chamber relative to when the tank is empty [40]. The subtraction was performed in terms of the logarithm of field values, and therefore, processed both log magnitudes and phases. It also cancelled phase and amplitude variations associated with the individual antennas, cables, and measurement channels, and preserved measurement coherence. Coherence for this configuration was described and discussed in detail in Meaney et al. [36]. Processing log magnitudes and phases required phase unwrapping, and unwrapping strategies are described in detail in Meaney et al. [37].

3. Results

3.1. Isolation of Individual B210s

To test isolation characteristics of representative configurations, all B210 coaxial connectors were terminated with shielded matched loads and Channel 1 of each board was programmed to transmit 0 dBm signals at frequencies from 0.9 to 1.7 GHz. Configurations under the test included (a) B210 with no shielding and (b) B210 mounted in the custom shielded housing (see Figure 3). Measurements were recorded at representative locations (Figure 7) to assess shielding benefits, and are reported in Tables 1 and 2 for transmission from TX/Rx Channel 1 and 2 ports, respectively. An Electro-Metrics

(Johnstown, NY, USA) hand-held probe (EM-6992) connected to Keysight Technologies (formerly HP) 8563E spectrum analyzer (26.5 GHz) sensed emissions outside of the housing at selected locations.

Figure 7. Photograph of the enclosed B210 with labels indicating probe measurement sites.

Table 1. Measured leakage signal values at 900, 1300, and 1700 MHz for circuit board locations with: (**a**) no shield and (**b**) shielded housing, respectively, for Channel 1.

Location of Measurement Points	No Shielding (dBm)			Custom Shield (dBm)			Difference (dBm)		
	900 MHz	1300 MHz	1700 MHz	900 MHz	1300 MHz	1700 MHz	900 MHz	1300 MHz	1700 MHz
Location 1	−74	−80	−83	−108	−106	−109	−34	−24	−26
Location 2	−66	−70	−77	−99	−98	−107	−33	−28	−30
Location 3	−86	−75	−74	−112	−111	−107	−26	−36	−33
Location 4	−84	−80	−73	−106	−112	−109	−22	−32	−36
Location 5	−94	−99	−92	−105	−112	−115	−11	−13	−23
Location 6	−94	−109	−94	−94	−102	−112	0	+7	−18
Location 7	−98	−97	−85	−95	−102	−114	+3	−5	−29
Location 8	−100	−92	−78	−101	−106	−114	−1	−14	−36

Table 2. Measured leakage signal values at 900, 1300, and 1700 MHz for circuit board locations with: (**a**) no shield and (**b**) shielded housing, respectively, for Channel 2.

Location of Measurement Points	No Shielding (dBm)			Custom Shield (dBm)			Difference (dBm)		
	900 MHz	1300 MHz	1700 MHz	900 MHz	1300 MHz	1700 MHz	900 MHz	1300 MHz	1700 MHz
Location 1	−81	−89	−90	−108	−112	−109	−27	−23	−19
Location 2	−85	−81	−84	−114	−113	−102	−29	−32	−18
Location 3	−82	−78	−80	−96	−99	−102	−14	−21	−22
Location 4	−78	−87	−81	−92	−94	−102	−14	−7	−21
Location 5	−100	−89	−94	−106	−111	−112	−6	−22	−18
Location 6	−97	−89	−102	−99	−108	−114	−2	−19	−12
Location 7	−92	−90	−102	−92	−102	−114	0	−12	−12
Location 8	−90	−88	−92	−99	−107	−116	−9	−19	−24

Predictably, leakage values were reduced for shielded B210s relative to unshielded boards, and advantages were more pronounced at the higher frequencies. For channel transmissions with no shield, leakage signals were high relatively at measurement points 1–4 compared to 5–8 (see Figure 7), likely due to the RF circuitry being immediately adjacent to these connectors compared to the more distant measurement sites (5–8). For points 1–4, shielded values were uniformly superior. For positions 5–8, leakage was also superior uniformly with shielding for the highest frequency, but more uneven for lower frequencies. Interestingly, instances occurred where unshielded values were nearly identical or slightly better than their shielded counterparts. In cases where shielded and unshielded values were similar, individual measurements were promising (above −94 dBm once).

Predictably, leakage values were reduced for shielded B210s relative to unshielded boards, and advantages were more pronounced at the higher frequencies. For channel transmissions with no shield, leakage signals were high relatively at measurement points 1–4 compared to 5–8 (see Figure 7), likely due to the RF circuitry being immediately adjacent to these connectors compared to the more distant measurement sites (5–8). For points 1–4, shielded values were uniformly superior. For positions 5–8, leakage was also superior uniformly with shielding for the highest frequency, but more uneven for lower frequencies. Interestingly, instances occurred where unshielded values were nearly identical or slightly better than their shielded counterparts. In cases where shielded and unshielded values were similar, individual measurements were promising (going above −94 dBm only once).

3.2. Performance of the Switch/Amplifier Module

Figure 8 shows plots of insertion loss (or gain) for cases where (a) signals were transmitted to an antenna, (b) signals were received from an antenna and amplified prior to being measured by the B210 receiver, and (c) leakage signals from a transmitter port to receiver port when the switch/amplifier module operated in the receive mode. The latter leakage signal is also shown when the housing cover was removed (i.e., no shielding). For (a), insertion loss was mild and less than 2.4 dB up to 2 GHz, which is nominally the highest frequency used in the imaging system. The result implies that the SPST and SPDT switches were operating efficiently and power loss was minimal in the system transmission mode. Similarly, gain for signals received by antennas (and directed to receiver boards) was relatively flat and ranged from 20.0 to 21.3 dB with modest monotonically decreasing gain up to 2 GHz. Isolation was greater than 80 dB over the frequency range of interest when the housing enclosure was available, and remained above 70 dB up to 3 GHz. In the unshielded case, values were substantially worse and uneven across the frequency range likely due to excitation of surface wave modes that propagated over the circuit board and housing surfaces and recombined in unpredictable patterns with the desired signals at the output. For the enclosed case, small channels between compartments acted as cutoff waveguides and provided isolation for all modes other than the desired Quasi-TEM transmission line mode that led to well-behaved transfer characteristics.

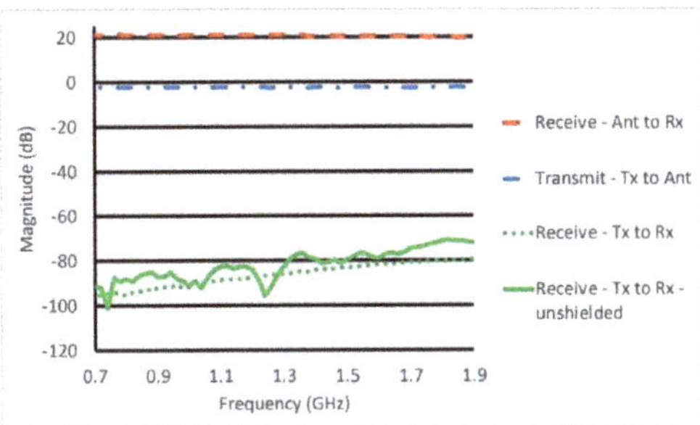

Figure 8. Plots of the switch/amplifier insertion loss (or gain) for the transmission mode (between the Tx and Ant ports), and receive mode (between the Ant and Rx ports). Leakage between transmit and receive ports while operational in the receive mode is shown for the completely shielded housing and the compartmentalized housing without cover, respectively.

We also tested housing isolation for ambient signals. Figure 9 shows a SolidWorks rendering of the switch/amplifier module with four associated measurement sites: (a) connected to the antenna (Ant), (b) connected to the transmission network (Tx), (c) connected to the receive B210 boards (Rx), and (d) not directly associated with a connector. In one test, a 0 dBm signal was supplied to the Tx port while the other ports were terminated with shielded 50 ohm loads. Switch control lines were set so that the signal was transmitted to the antenna port. Measurement data are presented in Table 3. Interestingly, no significant difference occurred in shielded and unshielded values for the three frequencies reported in the table. Overall, measurements were lower than −90 dBm. Signals measured near Rx ports were noticeably lower. While signals normally emanating from this port were amplified by the low noise amplifier by about 20 dB, the SPDT switch restricted signal propagation to the port, which more than compensated for the additional gain.

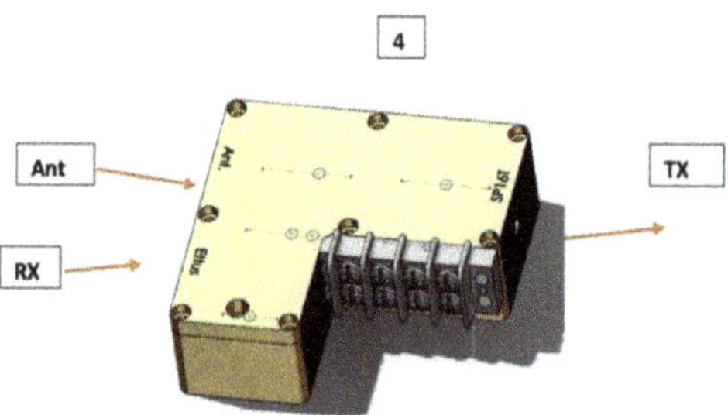

Figure 9. SolidWorks 3D rendering of the switch amplifier housing with four measurement sites.

Table 3. Average transmission network (Tx) mode leakage signal measurements at designated locations for 900, 1300, and 1700 MHz, respectively.

Location of Measurement Points in TX Mode	No Shielding (dBm)			Custom Shield (dBm)			Difference (dBm)		
	900 MHz	1300 MHz	1700 MHz	900 MHz	1300 MHz	1700 MHz	900 MHz	1300 MHz	1700 MHz
Ant	−104	−107	−107	−97	−106	−106	+7	+1	+1
Tx	−101	−99	−99	−91	−99	−87	+10	0	+12
Rx	−109	−105	−115	−109	−100	−107	0	+5	+8
Location 4	−95	−90	−92	−97	−95	−103	−2	−5	−11

In a second test, a −10 dBm signal was fed into the antenna port while the remaining ports were terminated with shielded matched loads. The SPDT control line was set to direct a signal coming from the antenna port to pass through the low noise amplifier and Rx port. The measurement data are presented in Table 4. Substantial improvement in leakage occurred with the shielded housing relative to the unshielded case for all measurements. Interestingly, signals measured at the Rx site were about 10 dB higher than those from the other sites, which most likely reflected the fact that these signals have been amplified previously by the LNA (which does not occur for the other ports).

Table 4. Average receive B210 board (Rx) mode leakage signal measurements at designated locations for 900, 1300, and 1700 MHz, respectively.

Location of Measurement Points in TX Mode	No Shielding (dBm)			Custom Shield (dBm)			Difference (dBm)		
	900 MHz	1300 MHz	1700 MHz	900 MHz	1300 MHz	1700 MHz	900 MHz	1300 MHz	1700 MHz
Ant	−91	−87	−96	−115	−116	−112	−24	−29	−16
Tx	−105	−88	−87	−114	−117	−118	−9	−29	−31
Rx	−83	−82	−84	−102	−102	−107	−19	−20	−23
Location 4	−104	−86	−80	−115	−115	−118	−11	−29	−38

3.3. System Isolation Specifications

Antenna ports in the switch/amplifier modules were terminated with 50 ohm loads and a 0 dBm signal was the output from the transmit B210 and response signals were measured at each of 16 channels with receiver gain levels set to their maximum (76 dB). Measurements were repeated for frequencies from 700 to 1900 MHz in 200 MHz increments. Figure 10 reports the isolation level measured at each port when the transmit channel was selected to be Channel 1, and remaining ports were activated in the receive mode. Under these conditions, isolation levels were largely below −140 dB and closer to −150 dB. A few measurements for the 700 MHz case rose above the −130 dB level. A few of the 1100 MHz values were also above the −140 dB level.

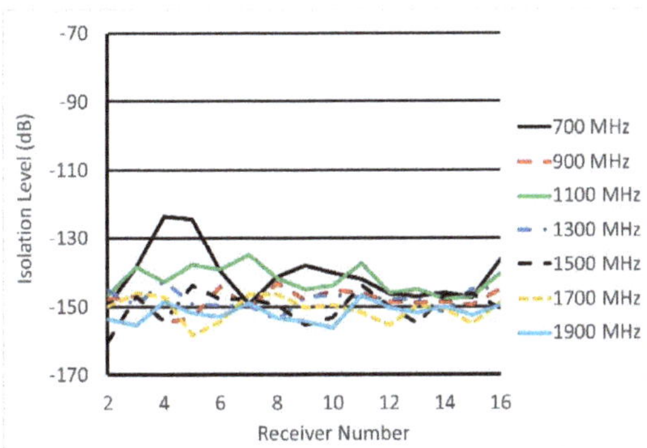

Figure 10. Isolation levels measured at receivers for signal transmission from Channel 1 at 7 frequencies when remaining channels were activated in the receive mode and antenna ports were terminated with a 50 Ω matched load. Except for the 700 MHz case, all values are less than or equal to -135 dB.

3.4. Images Reconstructed from Measurement Data

For the imaging experiment, we utilized a 2.7 cm $\times$ 2.7 cm square cylindrical phantom filled with 60% glycerin and offset 3.2 cm from the center. Its properties were ε_r = 51.53 and σ = 1.18 S/m at 1100 MHz. The background medium was an 80% glycerin bath having ε_r = 29.87 and σ = 1.25 S/m at 1100 MHz. The height of the liquid was adjusted to be the same as when the object is present and when it was absent. Figure 11a shows measured magnitudes for a single transmitting antenna when the object signals were calibrated, i.e., they had the measurement values for the homogeneous bath subtracted (phases were also shown for completeness). Here, leakage was assumed to arise from any one of the situations described above. Measured data were converted from the log magnitude/phase format to the associated complex numbers. We generated random numbers between -1 and 1 for the real and imaginary leakage signal components and multiplied them by amplitude factors, which were equivalent to those associated with the prescribed leakage simulation levels utilized in the results (in these cases, phases and amplitudes of multipath contributions are unknown. Accordingly, we utilized this random number approach to control the amplitude within a worst-case scenario for the phase. Ultimately, the experiment assesses trends in the presence of progressively distorting signals). These numbers were then added to the complex-valued versions of the original data. Once these terms were summed, they were transformed back to their log magnitude and phase forms, and used directly in the reconstruction algorithm. Here, deviations between the non-leakage case and the background were more pronounced (because these values were normalized to those of the background, the background values were effectively zero). Plots confirm the observations in (a), i.e., that the field values varied most for the most remote antennas. They also show signal deviations increased as leakage levels increased for both magnitude and phase.

Figure 12a reports reconstructed relative permittivity and conductivity images at 1100 MHz with no leakage signal. Images were reconstructed with a Gauss–Newton iterative algorithm under log transformation [37]. Regularization consisted of a standard Levenberg–Marquardt scheme, and the initial image estimate corresponded to the properties of the homogeneous bath. The algorithm completed 50 iterations with an iteration step size of 0.2, which required 5.5 min of computation time to produce each image pair. The permittivity inclusion was recovered at the correct location and with the right size. Artifact levels were minimal in the permittivity images but higher in the conductivity case. In addition, the squared shape of the inclusion cross section is also evident. Conductivity images have the inclusion recovered with the correct size and location, approximately, and the recovered inclusion property values were lower than the background as expected.

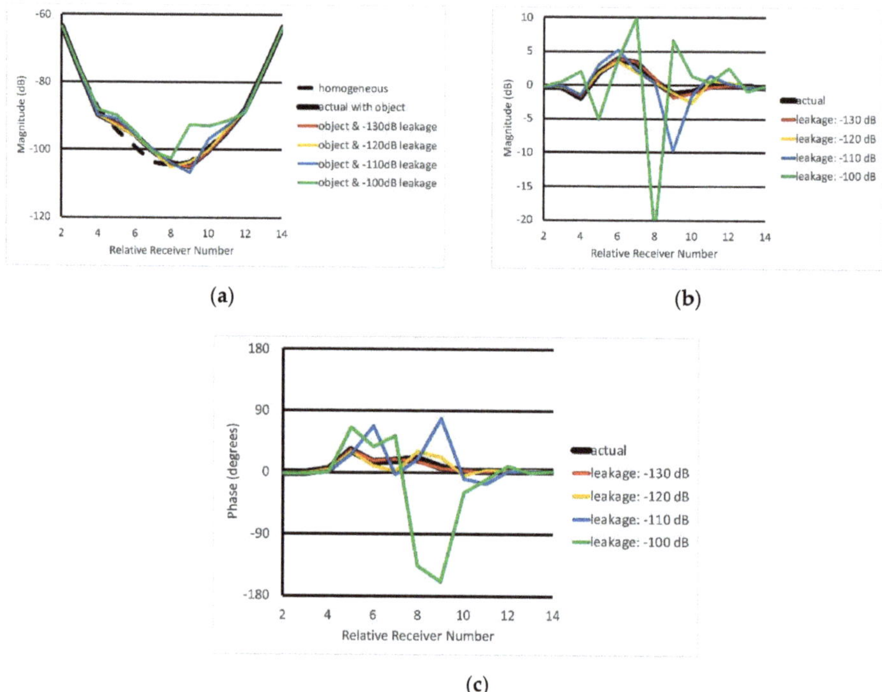

Figure 11. Plots of measurement data for antenna 1 transmission including cases with varying levels of added leakage signal: (**a**) raw magnitude, (**b**) calibrated magnitude, and (**c**) calibrated phases. The dashed black line represents the measured signal for the homogeneous bath whereas the solid black line indicates results when the object is present, respectively. Colored lines symbolize signals when the object is present but with progressively increasing leakage added.

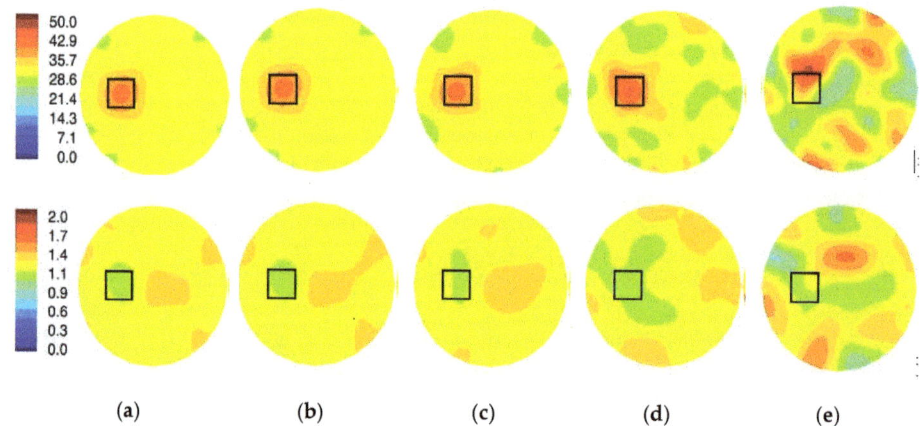

Figure 12. Reconstructed relative permittivity (**top**) and conductivity (**bottom**) at 1100 MHz for (**a**) no signal leakage, and (**b–e**) signal leakage of −130 dB, −120 dB, −110 dB, and −100 dB, respectively, of the square object depicted in Figure 5 for the 14.2 cm diameter field of view.

Figure 12b–e shows images from the same experiment but with data with increased levels of leakage signals. For the −130 through to −110 dB leakage, the permittivity object was recovered properly but with increasing degrees of artifacts. For the −100 dB case, the image reconstruction diverged. The conductivity object was visible in the −130 dB case, and not discernable at all once signal leakage reached −110 dB. Artifacts increased with increasing leakage signal.

Figure 13a,b shows the plots of permittivity and conductivity values along transects through the center of the imaging zone (the actual property distributions are shown for reference). These results confirm previous observations that the algorithm recovers correct trends in the actual permittivity distribution, although peak values in the object are attenuated because of smoothing inherent in the regularization process. The conductivity profile is flatter as expected because contrast between the object and background is low, and the algorithm recovers values slightly lower than the background in the proper location, even though the object properties are only 5.6% lower than the background. Artifact levels are higher in the conductivity images, which is a consistent feature of our approach as well as other systems [6]. Corresponding plots are also shown for the different leakage levels. These results confirm that when leakage is suppressed sufficiently, quality images can be recovered from the data. As leakage levels increase, image quality progressively decreases.

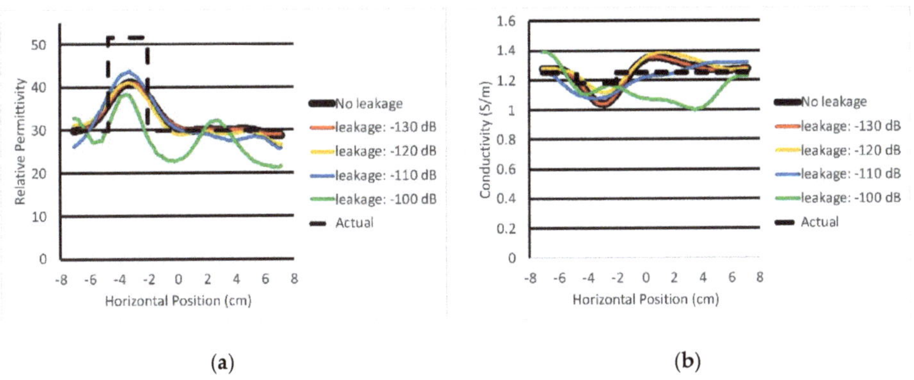

(a)   (b)

Figure 13. Horizontal transects through the 1100 MHz (**a**) permittivity and (**b**) conductivity images shown in Figure 12.

4. Discussion

We examined leakage sources from critical elements in the system—the B210 transceivers and the switch/amplifier modules—because they were the most likely to cause signal contamination. In the case of B210 transceivers, we examined their behavior in the transmission mode while we tested the switch/amplifier modules in both transmit and receive modes. Leakage tests compared instances in which boards were operated with and without the shielding. We also measured leakage effects when boards were isolated with a commercial enclosure supplied by Ettus Research, and these results were not noticeably different from the non-shielded case. Leakage results obtained with custom shields were superior and the leakage signals that did appear were most likely derived from seams between the top and bottom housing parts. While overall system design specifications called for cross-channel isolation greater than 140 dB, the Electro-Metrics probe we used measured levels near −100 dB, which were considered acceptable because these data only reflect a one-way loss through the shielded housing.

Results for the switch/amplifier module exhibited similar trends. Here, we examined both unwanted signals propagating through existing transmission lines, and those caused by surface waves that leaked between shielded housing structures. For the former, leakage signals from the Tx to Rx port were generally below −80 dB. Given this loss would be combined with about 65 dB of attenuation from the SP16T switch associated with the dedicated transmit B210 channel, the maximum unwanted signal reaching the receive B210 was expected to be on the order of −145 dBm, which is acceptable because the lowest level signals getting to the B210 were expected to be on the order of −120 dBm.

Corresponding surface wave-related leakage results were similar. Interestingly, leakage did not appear to be significantly better in the shielded case in the transmit mode. Nonetheless, overall leakage levels were low and suggest these signals were not problematic. When the probe was placed above uncovered circuits at random positions, we observed transient values often 30–40 dB higher than those measured near connectors. In the receive mode, the shielded housing offered a clear advantage and isolation data reported here were informative relative to overall system performance.

Field measurement power levels varied when no breast was present relative to when breasts of different sizes and densities were placed in the illumination tank. For example, the magnitude plot in Figure 11 for the case when no object or breast was present represented a lower bound on signal amplitude for the particular liquid and operating frequency given the lossy coupling medium. When the breast was present, signal strengths were higher, and could be as much as 30 dB higher for a fatty breast and 5–10 dB higher for denser breasts. The example emphasizes the fact that central antenna signals were impacted most by leakage levels, similarly to the phantom results, and that the data acquisition needs to handle a broad range of signal levels. Our imaging algorithm was able to recover images of quality similar to those reported in the past for standard phantoms with relatively high property contrast relative to the background. These results were consistent with image reconstruction performance obtained with data acquired with previous hardware implementations of our imaging system, and confirmed the multichannel software defined radio approach described here was operating properly.

While classical imaging defines resolution in terms of the Rayleigh criteria, we were typically able to recover objects considerably smaller than the $\lambda/2$ standard. Our experience indicates shows we could detect objects as small as $\lambda/12$, although we could not characterize their size or electrical properties, accurately. Experiments with varying levels of noise indicate that resolution was inversely related to the noise level.

5. Conclusions

We developed a high performance, multichannel measurement system that could be used for microwave imaging of biological targets. Novel aspects of the system include: (a) an unusually high dynamic range that is critical in medical applications involving lossy coupling media; (b) multichannel SDR-based design with multipath signals reduced to levels having minimal impact on reconstructed image quality; (c) reorganization of data acquisition sequencing to reduce data recording time significantly; and (d) complete integration of these features that makes clinical microwave exams possible. The system operated from 500 MHz to 2.5 GHz, allowed transmission and reception from 16 channels, had a dynamic range of roughly 140 dB, and achieved excellent cross-channel signal isolation. The system design is compact and can be located below the associated antenna array. It also included design features required to ensure coherence between transmit and receive signals. By exploiting emerging technology designed into commercially available software defined radios as a fundamental system building blocks, overall system costs and complexity were reduced dramatically while still leading to a high-performance measurement system.

Author Contributions: The following summarizes the contributions for each author: Conceptualization, P.M., C.D. and K.P.; methodology, P.M., T.R. and S.G.; software, A.H., T.R., S.R., and F.S.; validation, T.R., C.D., S.R., F.S. and S.G.; formal analysis, P.M. and S.G.; writing—original draft preparation, P.M. and K.P.; writing—review and editing, P.M. and K.P.; supervision, P.M. All authors have read and agreed to the published version of the manuscript.

Funding: This work was supported by NIH grant # R01-CA240760.

Acknowledgments: We thank Dr. Selaka Bulumulla for his technical support in designing the microwave electronics while he worked at GE Global Research in Niskayuna, NY, USA.

Conflicts of Interest: The authors declare no conflict of interest.

References

1. Kurrant, D.J.; Fear, E.C. An improved technique to predict the time-of-arrival of a tumor response in radar-based breast imaging. *IEEE T Bio-Med. Eng.* **2009**, *56*, 1200–1208. [CrossRef] [PubMed]
2. Klemm, M.; Craddock, I.J.; Leendertz, J.A.; Preece, A.; Benjamin, R. Radar-based breast cancer detection using a spherical antenna array—experimental results. *IEEE Trans. Antenn. Propag.* **2009**, *57*, 1692–1704. [CrossRef]
3. Fear, E.C.; Li, X.; Hagness, S.C.; Stuchly, M.A. Confocal microwave imaging for breast tumor detection: Localization of tumors in three dimensions. *IEEE T Bio-Med. Eng.* **2002**, *49*, 812–822. [CrossRef] [PubMed]
4. Ong, C.; Weldon, M.; Cyca, D.; Okoniewski, M. Acceleration of large-scale FDTD simulations on high performance GPU clusters. In Proceedings of the 2009 IEEE Antennas and Propagation Society International Symposium, Charleston, SC, USA, 1–5 June 2009.
5. Baran, A.; Kurrant, D.J.; Zakaria, A.; Fear, A.C.; LoVetri, J. Breast imaging using microwave tomography with radar-based tissue-regions estimation. *Prog. Electromagn. Res.* **2014**, *149*, 161–171. [CrossRef]
6. Semenov, S.Y.; Bulyshev, A.E.; Abubakar, A.; Posukh, V.G.; Sizov, Y.E.; Souvorov, A.E.; van den Berg, P.M.; Williams, T.C. Microwave-tomographic imaging of the high dielectric-contrast objects using different image-reconstruction approaches. *IEEE Trans. Microw. Theory* **2005**, *53*, 2284–2294. [CrossRef]
7. Chen, F.C.; Chew, W.C. Integrated optical coherence tomography (OCT) and fluorescence laminar optical tomography (FLOT). *IEEE J. Sel. Top. Quant.* **2010**, *16*, 755–766. [CrossRef]
8. Jiang, S.; Pogue, B.W.; Carpenter, C.M.; Poplack, S.P.; Wells, W.A.; Kogel, C.A.; Forero, J.A.; Muffly, L.S.; Schwartz, G.N.; Paulsen, K.D.; et al. Evaluation of breast tumor response to neoadjuvant chemotherapy with tomographic diffuse optical spectroscopy: Case studies of tumor region-of-interest changes. *Radiology* **2009**, *252*, 551–560. [CrossRef]
9. Yalavarthy, P.K.; Pogue, B.W.; Dehghani, H.; Paulsen, K.D. Weight-matrix structured regularization provides optimal generalized least-squares estimate in diffuse optical tomography. *Med. Phys.* **2007**, *34*, 2085–2098. [CrossRef]
10. Halter, R.J.; Hartov, A.; Paulsen, K.D. A broadband high-frequency electrical impedance tomography system for breast imaging. *IEEE Trans. Bio-Med. Eng.* **2008**, *55*, 650–659. [CrossRef]
11. Jam, H.; Isaacson, D.; Edic, P.M.; Newell, J.C. Electrical impedance tomography of complex conductivity distributions with noncircular boundary. *IEEE Trans. Bio-Med. Eng.* **1997**, *44*, 1051–1060.
12. Holder, D. *Clinical and Physiological Applications of Electrical Impedance Tomography*; UCL Press: London, UK, 1993.
13. Woo, E.J.; Hua, P.; Webster, J.G.; Tompkins, W.J. A robust image reconstruction algorithm and its parallel implementation in electrical impedance tomography. *IEEE Trans. Med. Imaging* **1993**, *12*, 137–146. [CrossRef] [PubMed]
14. Poplack, S.P.; Paulsen, K.D.; Hartov, A.; Meaney, P.M.; Pogue, B.W.; Tosteson, T.; Grove, M.; Soho, S.; Wells, W.A. Electromagnetic breast imaging: Pilot results in women with abnormal mammography. *Radiology* **2007**, *243*, 350–359. [CrossRef] [PubMed]
15. Meaney, P.M.; Kaufman, P.A.; Muffly, L.S.; Click, M.; Wells, W.A.; Schwartz, G.N.; di Florio-Alexander, R.M.; Tosteson, T.D.; Li, Z.; Poplack, S.P.; et al. Microwave imaging for neoadjuvant chemotherapy monitoring: Initial clinical experience. *Breast Cancer Res.* **2013**, *15*, R35. [CrossRef]
16. Persson, M.; Fhager, A.; Trefna, H.D.; Yu, Y.; McKelvey, T.; Pegenius, G.; Karlsson, J.-E.; Elam, M. Microwave-based stroke diagnosis making global pre-hospital thrombolytic treatment possible. *IEEE Trans. Bio-Med. Eng.* **2014**, *61*, 2806–2817. [CrossRef] [PubMed]
17. Semenov, S.Y.; Corfield, D.R. Microwave tomography for brain imaging: Feasibility assessment for stroke detection. *Int. J. Antenn. Propag.* **2008**, *2008*, 254830. [CrossRef]
18. Meaney, P.M.; Goodwin, D.; Zhou, T.; Golnabi, A.; Pallone, M.; Geimer, S.D.; Burke, G.; Paulsen, K.D. Clinical microwave tomographic imaging of the calcaneus: Pilot study. *IEEE T Bio-Med. Eng.* **2012**, *59*, 3304–3313. [CrossRef]

19. Meaney, P.M.; Fanning, M.W.; Li, D.; Pendergrass, S.A.; Fang, Q.; Moodie, K.L.; Paulsen, K.D. Microwave thermal imaging: Initial in vivo experience with a single heating zone. *Int. J. Hyperther.* **2003**, *19*, 617–641. [CrossRef]

20. Stauffer, P.R.; Snow, B.W.; Rodrigues, D.B.; Salahi, S.; Oliveira, T.R.; Reudink, D.; Maccarini, P.F. Non-invasive measurement of brain temperature with microwave radiometry: Demonstration in a head phantom and clinical case. *J. Neuroradiol.* **2014**, *27*, 3–12. [CrossRef]

21. Haynes, M.; Stang, J.; Moghaddam, M. Real-time microwave imaging of differential temperature for thermal therapy monitoring. *IEEE Trans. Bio-Med. Eng.* **2014**, *61*, 1787–1797. [CrossRef]

22. Joines, W.T.; Zhang, Y.; Li, C.; Jirtle, R.L. The measured electrical properties of normal and malignant human tissue from 50 to 900 MHz. *Med. Phys.* **1994**, *41*, 547–550. [CrossRef]

23. Sugitani, T.; Kubota, S.-I.; Kuroki, S.-I.; Sogo, K.; Arihiro, K.; Okada, M.; Kadoya, T.; Hide, M.; Oda, M.; Kikkawa, T. Complex permittivities of breast tumor tissues obtained from cancer surgeries. *Appl. Phys. Lett.* **2014**, *104*, 253702. [CrossRef]

24. Surowiec, A.; Stuchly, S.; Barr, J.; Swarup, A. Dielectric properties of breast carcinoma and the surrounding tissues. *IEEE Trans. Bio-Med. Eng.* **1988**, *35*, 257–263. [CrossRef] [PubMed]

25. Lazebnik, M.; Popovic, D.; McCartney, L.; Watkins, C.B.; Lindstrom, M.J.; Harter, J.; Sewall, S.; Ogilvie, T.; Magliocco, A.; Breslin, T.M.; et al. A large-scale study of the ultrawideband microwave dielectric properties of normal, benign and malignant breast tissues obtained from cancer surgeries. *Phys. Med. Biol.* **2007**, *52*, 6093–6115. [CrossRef] [PubMed]

26. Duck, F.A. *Physical Properties of Tissue: A Comprehensive Reference Book*; Academic Press: London, UK, 1990; pp. 167–223.

27. Schwan, H.P.; Foster, K.R. RF-field interactions with biological systems: Electrical properties and biophysical mechanisms. *Proc. IEEE* **1980**, *68*, 104–113. [CrossRef]

28. Martellosio, A.; Pasian, M.; Bozzi, M.; Perregrini, L.; Mazzanti, A.; Svelto, F.; Summers, P.E.; Renne, G.; Preda, L.; Bellomi, M. Dielectric properties characterization from 0.5 to 50 GHz of breast cancer tissues. *IEEE Trans. Microw. Theory* **2017**, *65*, 998–1011. [CrossRef]

29. Cheng, Y.; Fu, M. Dielectric properties for non-invasive detection of normal, benign, and malignant breast tissues using microwave theories. *Thorac. Cancer* **2018**, *9*, 459–465. [CrossRef]

30. Preece, A.W.; Craddock, I.; Shere, M.; Jones, L.; Winton, H.L. MARIA M4: Clinical evaluation of a prototype ultrawideband radar scanner for breast cancer detection. *J. Med. Imaging* **2016**, *3*, 033502. [CrossRef]

31. Meaney, P.M.; Schubitidze, F.; Fanning, M.W.; Kmiec, M.; Epstein, N.R.; Paulsen, K.D. Surface wave multi-path signals in near-field microwave imaging. *Int. J. Biomed. Imaging* **2012**, *2012*, 697253. [CrossRef]

32. Ostadrahimi, M.; Zakaria, A.; LoVetri, J.; Shafai, L. A near-field dual polarized (TE-TM) microwave imaging system. *IEEE Trans. Microw. Theory* **2013**, *61*, 1376–1384. [CrossRef]

33. Ciocan, R.; Jiang, H. Model-based microwave image reconstruction: Simulations and experiments. *Med. Phys.* **2004**, *31*, 3231–3241. [CrossRef]

34. Hershberger, J.; Pratt, T.; Kossler, R. A software-defined, dual-polarized radar system. In Proceedings of the 2016 IEEE Conference on Antenna Measurements & Applications (CAMA), Syracuse, NY, USA, 23–27 October 2016.

35. Marimuthu, J.; Bialkowski, K.S.; Abbosh, A.M. Software-defined radar for medical imaging. *IEEE Trans. Microw. Theory* **2016**, *64*, 643–652. [CrossRef]

36. Meaney, P.M.; Hartov, A.; Bulumulla, S.; Raynolds, T.; Davis, C.; Schoenberger, F.; Richter, S.; Paulsen, K.D. 4-channel, vector network analyzer microwave imaging prototype based on software defined radio technology. *Rev. Sci. Instrum.* **2019**, *90*, 044708. [CrossRef]

37. Meaney, P.M.; Geimer, S.D.; Paulsen, K.D. Two-step inversion in microwave imaging with a logarithmic transformation. *Med. Phys.* **2017**, *44*, 4239–4251. [CrossRef]

38. Paulsen, K.D.; Meaney, P.M. Compensation for nonactive array element effects in a microwave imaging system: Part I—Forward solution vs. measured data comparison. *IEEE Trans. Med. Imaging* **1999**, *18*, 496–507. [CrossRef]

39. Meaney, P.M.; Paulsen, K.D.; Chang, J.T.; Fanning, M. Compensation for nonactive array element effects in a microwave imaging system: Part II—Imaging results. *IEEE Trans. Med. Imaging* **1999**, *18*, 508–518. [CrossRef]
40. Meaney, P.M.; Paulsen, K.D.; Chang, J.T. Near-field microwave imaging of biologically based materials using a monopole transceiver system. *IEEE Trans. Microw. Theory* **1998**, *46*, 31–45. [CrossRef]

Article

A Prototype Microwave System for 3D Brain Stroke Imaging

Jorge A. Tobon Vasquez [1], Rosa Scapaticci [2], Giovanna Turvani [1], Gennaro Bellizzi [3], David O. Rodriguez-Duarte [1], Nadine Joachimowicz [4], Bernard Duchêne [5], Enrico Tedeschi [6], Mario R. Casu [1], Lorenzo Crocco [2] and Francesca Vipiana [1,*]

[1] Department of Electronics and Telecommunications, Politecnico di Torino, 10129 Torino, Italy; jorge.tobon@polito.it (J.A.T.V.); giovanna.turvani@polito.it (G.T.); david.rodriguez@polito.it (D.O.R.-D.); mario.casu@polito.it (M.R.C.)

[2] Institute for the Electromagnetic Sensing of the Environment, National Research Council of Italy, 80124 Naples, Italy; scapaticci.r@irea.cnr.it (R.S.); crocco.l@irea.cnr.it (L.C.)

[3] Department of Electric Engineering and Information Technologies, University of Naples Federico II, 80125 Naples, Italy; gbellizz@unina.it

[4] Group of Electrical Engineering-Paris (GeePs), CNRS, CentraleSupélec, Université Paris-Sud, Univ. Paris-Saclay, Sorbonne Univ., 91190 Gif-sur-Yvette, France; nadine.joachimowicz@paris7.jussieu.fr

[5] Laboratoire des Signaux et Systèmes (L2S), Université Paris-Saclay, CNRS, CentraleSupélec, 91190 Gif-sur-Yvette, France; bernard.duchene@l2s.centralesupelec.fr

[6] Department of Advanced Biomedical Sciences, University of Naples Federico II, 80131 Napoli, Italy; enrico.tedeschi@unina.it

* Correspondence: francesca.vipiana@polito.it

Received: 6 April 2020; Accepted: 30 April 2020; Published: 3 May 2020

Abstract: This work focuses on brain stroke imaging via microwave technology. In particular, the open issue of monitoring patients after stroke onset is addressed here in order to provide clinicians with a tool to control the effectiveness of administered therapies during the follow-up period. In this paper, a novel prototype is presented and characterized. The device is based on a low-complexity architecture which makes use of a minimum number of properly positioned and designed antennas placed on a helmet. It exploits a differential imaging approach and provides 3D images of the stroke. Preliminary experiments involving a 3D phantom filled with brain tissue-mimicking liquid confirm the potential of the technology in imaging a spherical target mimicking a stroke of a radius equal to 1.25 cm.

Keywords: microwave imaging; brain stroke; monitoring; antenna array

---

1. Introduction

Brain stroke is one of the main causes of permanent injury and death worldwide, with an incidence of over 5 million annual deaths [1]. Since prompt intervention (such as the administration of specific drugs that can affect the acute stage of the stroke) can significantly improve the prognosis, a rapid diagnosis of the disease and continuous monitoring after its onset represent important clinical goals.

Currently, the most effective tool for brain stroke diagnosis is the imaging-based diagnostics performed in an emergency department after the recognition of stroke-like symptoms. In this respect, magnetic resonance imaging (MRI) or X-ray based computerized tomography (CT) are the clinically adopted techniques. However, although they are continuously evolving, these technologies still have several limitations. In particular, despite its high resolution and accuracy, MRI is not widely available in emergency settings and is therefore actually used only as a secondary diagnostic tool [2,3]. On the other hand, non-contrast CT may be limited by the fact that the early signs of ischemia may not be easily recognizable by non-experienced personnel. Moreover, due to the use of ionizing radiation, CT is not suitable for repeated examinations, which are especially useful for post-acute monitoring purposes. Furthermore, both MRI and CT equipment is bulky and so not currently suited for ambulance use or as bedside devices.

The above circumstances have led to increased interest in the development of different diagnostic imaging techniques [3]. Among others, microwave imaging (MWI) [4] has emerged as a complementary technique which is able to address the different needs arising in stroke diagnosis and management, namely the early—possibly prehospital—diagnosis of the kind of stroke (ischemia or hemorrhage), bedside brain imaging and continuous brain monitoring for stroke in the post-acute stage. MWI takes advantage of the different electric properties (electric permittivity and conductivity) that human tissues exhibit at microwave frequencies depending on their kind (e.g., blood versus gray or white matter) and pathological status. These differences permit a functional map of the inspected anatomical region to be obtained. The benefits of MWI mainly stem from the non-ionizing nature of microwave radiation and the reduced intensity required to obtain reliable imaging (at an intensity comparable to that currently used for mobile phones), which make it completely safe and suitable for repeated applications. Moreover, MWI technology is cost-effective and benefits from a reduced size, as it makes use of miniaturized, low-cost, off-the-shelf components that are available in the microwave frequency range for signal generation and acquisition [5] and low-cost accelerators to speed up processing [6].

Recently, several MWI devices and prototypes have been proposed [7–17]. Among them, the two most prominent examples (which are already being tested on humans) are the Strokefinder, developed by Medfield Diagnostics [7,8], and the EMTensor BrainScanner [10]. The Strokefinder is a device which aims to discriminate between ischemic and hemorrhagic strokes in the early stage of patient rescue, based on an automated classification which is carried out by comparing the measured data to a database (obtained by data collected from already examined patients). This device is characterized by its very simple and compact hardware, consisting of a small number of printed antennas mounted on a support that can be adapted to the patient's head. Some initial clinical trials have been reported for the Strokefinder [7], but it should be remarked that it does not provide images; thus, its intended role is to complement standard imaging tools. The EMTensor BrainScanner aims to perform brain stroke tomography. The system is characterized by a high complexity, as it consists of a large number of radiating elements (177 truncated waveguides, loaded with ceramic material of appropriate permittivity [10]), which considerably increase its cost and size, thus reducing to some extent the advantages of its use. In addition, the image reconstruction task involves the processing of a considerable amount of measured data and has to face the pitfalls of dealing with a non-linear and ill-posed inverse problem. This entails long elaboration times and possibly results in false solutions; i.e., producing images which fulfill the underlying optimization but are different from the ground truth.

In this paper, we describe the realization, characterization and initial experimental validation of a prototype device representing a different approach to dealing with a still open issue in stroke management; that is, continuous monitoring during the hospitalization of the patient in order to evaluate the effectiveness of the administered therapies [18]. This specific application aims to image only a "small" variation occurring in the brain and not its overall structures and features. As a consequence, it is possible to keep the device complexity low, and therefore also its size and cost, as well as to rely on the Born approximation to model the scattering phenomenon, thus enabling reliable real-time imaging. Accordingly, the proposed device is based on the low-complexity architecture designed with the rigorous procedure as described in [18]. Moreover, it adopts a differential imaging approach, where data gathered at two different acquisition times are processed [19] with simple and fast imaging algorithms based on the distorted Born approximation [18,19].

The proposed system provides 3D images of the head by relying on data measured through an array of 24 printed monopole antennas organized as an anatomically conformal shape mimicking a wearable and adaptable helmet. Each antenna is enclosed in a box of graphite-silicon material, acting as the coupling medium, and connected to a two-port vector network analyzer (VNA) through a 24×24 switching matrix, which allows the whole differential scattering matrix required for imaging to be acquired. The use of a semi-solid matching medium is a distinct feature of the system which allows for an increased simplicity of operation and repeatability, as compared to other arrangements that make use of a coupling liquid [10]. Finally, as detailed below, the device presented here is equipped with a "digital twin" based on a proprietary electromagnetic (EM) solver that allows us to properly characterize and foresee its behavior, as well as to provide the building blocks needed for the imaging.

In the following sections, the different components of the device are described and discussed, and a first experimental assessment on an anthropomorphic head phantom is presented. This phantom consists of a plastic shell with the shape and size of a human head, which is filled up with a homogeneous material whose dielectric properties are equal to an average value of the properties of the different tissues present in the brain. The reported experimental results provide an initial demonstration of the capabilities of the developed device.

It is worth noting that the presented system is not the only example of a low-complexity device for brain imaging, as other devices, using a low number of antennas (8–16) arranged in a circular array, have been proposed [12–16]. However, these devices only provide 2D maps of the transverse cross-section of the head in the array plane , whereas the device herein presented provides a full 3D image of the head.

2. Material and Methods

2.1. Three-Dimensional Microwave Imaging System Design

In this section, we present the choices made in the design of the proposed imaging system (i.e., the operating frequency, coupling medium, antenna number and arrangement).

The operating frequency as well as the properties of the selected coupling medium were set according to previous findings obtained from theoretical formulations and experimentally validated [20] and [4] (Chapter 2). In particular, a working frequency of around 1 GHz and a coupling medium with a relative permittivity of around 20 were determined to be optimal and therefore chosen for the realization of the hardware. As all the simulations and measurements were made at 1 GHz, this operating frequency will be considered as implicit in the following.

To design the layout of the array of antennas (the number and position of the radiating/measuring elements), a recently proposed rigorous procedure was adopted [18,21]. This procedure is based on the analysis of the spectral properties of the discretized scattering operator [22], assuming the antennas to be located on a surface conformal to the head (a helmet) and taking into account the dynamic range and signal-to-noise ratio (SNR) of the adopted measurement device as well as the actual size of the antennas. The result of this study allowed us to identify a 24-element array as the suitable candidate to perform the desired imaging task while keeping the system complexity as low as possible. The expected performances were confirmed by a preliminary numerical analysis [18].

As far as the choice of the imaging algorithm was concerned, the key aspect was that the targeted application was to monitor the time evolution of the stroke. Hence, a differential approach was a suitable choice [18,23]. In particular, the input data of the imaging problem, denoted as ΔS in the following, were represented by the difference between the scattering matrices measured at two different times, while the output was a 3D image showing the possible variation of the electric contrast of the brain tissues—i.e., $\Delta\chi$—occurring between these two different times. The differential electric contrast function $\Delta\chi$ is defined as $\Delta\epsilon/\epsilon_b$, where ϵ_b is the complex permittivity of the non-homogeneous background at the reference instant (e.g., a map of the brain at the first diagnosis) and $\Delta\epsilon$ is the complex permittivity variation between the two time instants.

Since the contrast variation $\Delta\chi$ was localized in a small portion of the imaging domain, it was possible to take advantage of the distorted Born approximation [22], so that a linear relationship held between ΔS and $\Delta\chi$:

$$\Delta S\left(\boldsymbol{r}_p, \boldsymbol{r}_q\right) = \mathcal{S}\left(\Delta\chi\right), \tag{1}$$

where $\mathcal{S}$ is a linear and compact integral operator, whose kernel is $-j\omega\epsilon_b/4\,E_b\left(\boldsymbol{r}_m, \boldsymbol{r}_p\right) \cdot E_b\left(\boldsymbol{r}_m, \boldsymbol{r}_q\right)$, with $\boldsymbol{r}_m \in D$, $\boldsymbol{r}_p$ and $\boldsymbol{r}_q$ denote the positions of the transmitting and receiving antennas, and $\boldsymbol{r}_m$ shows the positions of the points in which the imaging domain D is discretized. E_b is the background field in the unperturbed scenario; that is, the field radiated inside the imaging domain by each element of the array. The symbol "·" denotes the dot product between vectors, $\omega = 2\pi f$ is the angular frequency, and j the imaginary unit.

As a reliable and well-established method to invert (1), we exploit the truncated singular value decomposition (TSVD) scheme [22], which allows us to obtain the unknown differential contrast function through the explicit inversion formula:

$$\Delta\chi = \sum_{n=1}^{L_t} \frac{1}{\sigma_n} \langle\Delta S, [u_n]\rangle\, [v_n], \tag{2}$$

where σ_n, $[u_n]$ and $[v_n]$ are the singular values and the right and left singular vectors of the discretized scattering operator $\mathcal{S}$, respectively. L_t is the truncation index of the SVD, which acts as a regularization parameter and was chosen to obtain a good compromise between the stability and accuracy of the reconstruction [22].

2.2. Three-Dimensional Microwave Imaging System Realization

The realized 3D microwave imaging system prototype is shown in Figure 1. It consists of several parts, described in the following sub-sections, all of which are controlled by a laptop.

Figure 1. Overview of the realized 3D microwave imaging system prototype. From left to right: vector network analyzer (VNA), switching matrix and head phantom wearing a helmet interconnected to the switching matrix by means of coaxial cables.

2.2.1. Vector Network Analyzer and Switching Matrix

First, all the signals are generated and received by a standard VNA (Keysight N5227A, 10 MHz–67 GHz) where the input power is set to 0 dBm and the intermediate frequency (IF) filter to 10 Hz. The two ports of the VNA are connected, via flexible coaxial cables, to the two input ports of the 2×24 switching matrix. The switching matrix has been realized with two single-pole-four-throw (SP4T), eight single-pole-six-throw (SP6T), and 24 single-pole-double-throw (SPDT) electro-mechanical coaxial switches. The internal connections between the switches are made with semi-rigid coaxial cables to maximize the isolation and minimize the insertion losses. Then, the 24 output ports of the switching matrix are connected, via flexible coaxial cables, to the 24 antennas placed on the helmet as supports hosting the 3D anthropomorphic head phantom. As detailed in [23], the switching matrix has been realized so that there are 24 paths from VNA port 1 to the corresponding 24 antennas, as well as 24 paths back to VNA port 2; all the paths were designed to have the same electrical length. In this way, each antenna can work as a transmitter (TX) or as a receiver (RX), and while one antenna is transmitting, the signals collected by the other 23 antennas are received in sequence by the VNA.

2.2.2. Brick Antenna Array

The antenna array is the core of the imaging system. The antenna numbers, positions and orientations were determined according to the design procedure described in Section 2.1. In the prototype shown in Figure 2, the 24 antennas are placed on a 3D printed plastic (acrylonitrile butadiene styrene, ABS) support with the shape of a helmet conformal to the head phantom. This prototype configuration allows us to easily change or remove the antennas, if needed.

Figure 2. Brick antenna array conformal to the head phantom.

Each antenna, denoted as "brick" antenna in the following, was a monopole antenna printed on a standard FR4 substrate, with a thickness equal to 1.55 mm, and embedded in a dielectric brick. The dielectric brick was made of a mixture of urethane rubber and graphite powder and was designed in order to reach a relative dielectric permittivity of $\epsilon_r \cong 20$ and to minimize the losses. The actual EM properties obtained with this mixture were $\epsilon_r = 18.3$ and $\sigma = 0.19\,\text{S/m}$. The graphite powder was needed to increase the relative dielectric constant of the urethane rubber; however, it also increased the conductivity, although still within acceptable levels compared to other matching media used in medical microwave imaging (e.g., [23,24]). Moreover, the adopted matching medium is usually liquid [10], which is inherently inconvenient for a helmet-like device. Here, instead, the implemented matching medium was solid rubber, which can be easily placed conformally to the head, as shown in Figure 2. The overall dimensions of each brick were $5 \times 5 \times 7\,\text{cm}^3$ to accommodate the need to place 24 brick antennas around the head.

Figure 3 reports the 24×24 scattering matrix measured by the microwave imaging system in the presence of the head phantom; the self-terms were set to zero in order to highlight the range variation of the measured coupling coefficients, which are the input data of the used TSVD imaging algorithm.

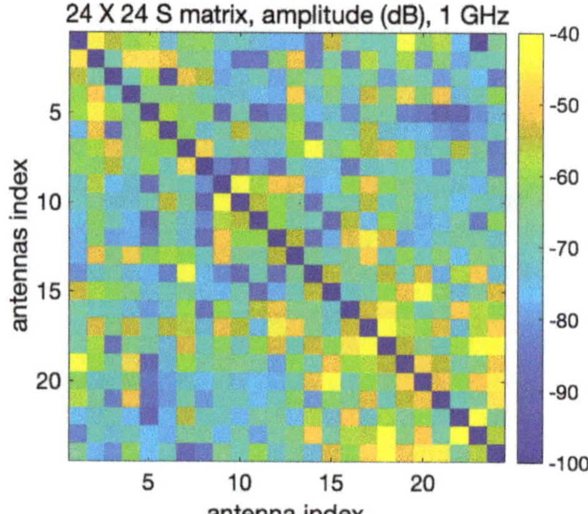

Figure 3. The 24×24 scattering matrix measured in the presence of the head phantom; the self-terms are forced to zero.

It can be noticed that the measured signals are above $-100\,\mathrm{dB}$; considering that the VNA noise floor is $-110\,\mathrm{dBm}$ (at $1\,\mathrm{GHz}$ and with an intermediate frequency (IF) filter equal to $10\,\mathrm{Hz}$), with an input power of $0\,\mathrm{dBm}$ the measured data are then above the VNA noise floor [25]. Moreover, as expected, the 24×24 matrix is approximately symmetric, which confirms the reciprocity of the realized system.

2.2.3. Anthropomorphic Head Phantom

The 3D anthropomorphic head phantom used for the validation and testing of the microwave imaging system was made of polyester casting resin. It was realized by additive manufacturing from a stereo-lithography (STL) file derived from MRI scans. The STL file was obtained with computer-aided design (CAD) software by modifying an original file from the Athinoula A. Martinos Center for Biomedical Imaging at Massachusetts General Hospital [26].

The phantom consisted of a cavity, shown in Figure 4a, whose wall thickness and height were equal to 3 mm and 26 cm, respectively. Its maximum cross section was approximately an ellipse whose minor and major axes were equal to 20 cm and 26 cm, respectively. The cavity was filled with a liquid mixture, made of Triton X-100, water and salt, which mimicked the average value of the dielectric characteristics of different brain tissues (white matter, gray matter) [27]. The measured dielectric characteristics of the mixture are reported in Figure 4b.

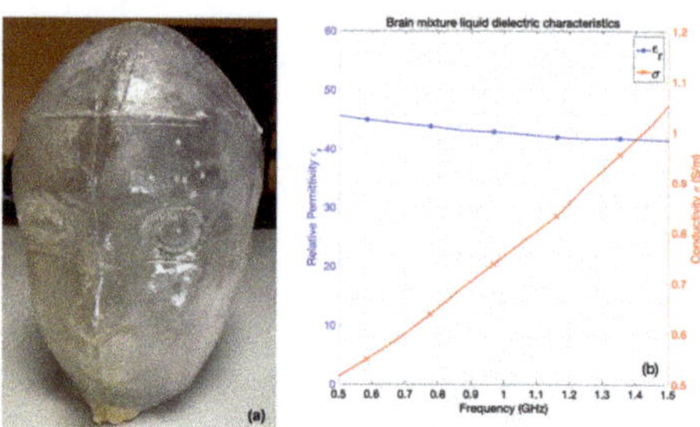

Figure 4. Anthropomorphic head phantom; (**a**) structure, and (**b**) dielectric characteristics of the liquid mixture, mimicking brain tissues as averages, which fills the phantom.

2.2.4. Digital Twin of the Device

The developed device was equipped with a digital twin, namely an accurate full-wave numerical model simulating its behavior, thanks to the use of proper CAD and EM simulation softwares. This tool had a two-fold purpose. First, it allowed us to foresee the expected outcomes of the planned experiments and to analyze them before running the actual experiments. Second, it provided the EM fields E_b needed to build the imaging kernel, as described in Section 2.1.

Figure 5 shows the CAD model of the device which includes both the antenna array and the 3D anthropomorphic head phantom.

Figure 5. Computer-assisted design (CAD) model of the 3D microwave imaging system together with the anthropomorphic head phantom; two printed monopole antennas within the coupling medium bricks are highlighted in yellow.

Each brick, placed on the head, represented one TX/RX antenna together with the dielectric coupling medium; the antenna ports were placed at the end of the coaxial cables leading out from the bricks. The dielectric characteristics of the bricks were the same as the nominal ones used in the realized system (see Section 2.2.2), i.e., $\epsilon_r = 18.5$ and $\sigma = 0.2\,\mathrm{S/m}$. The head phantom was made of a dielectric medium representing the average brain tissues with $\epsilon_r = 42.5$ and $\sigma = 0.75\,\mathrm{S/m}$, according to the properties of the medium adopted in the experimental validation (Section 2.2.3).

To perform the EM simulation, which provided both the S-parameters at the antenna ports and the EM fields in the whole scenario, the CAD model was introduced into an in-house full-wave software, based on the finite element method (FEM) [28]. The implemented software used the standard "curl–curl" formulation for the electric field and Galerkin testing. The whole volume, including the CAD model, was discretized with edge-basis functions, defined over a mesh of tetrahedral cells. The antenna metal parts were modeled as perfect electric conductors (PEC), and all the dielectrics were modeled via sub-volumes with given ϵ_r and σ values. The tetrahedral mesh was terminated at the volume boundaries with appropriate absorbing boundary conditions (ABC). Each antenna port was modeled as a coaxial cable section where, if excited, a tangential electric field distribution was enforced.

In the numerical modeling of the MWI system, the 24 antennas were excited one at a time, setting all the others as receivers in order to generate the 24×24 scattering matrix. Moreover, the field radiated by each antenna was evaluated at different locations within the head to generate the discretized scattering operator $\mathcal{S}$ in (1).

2.2.5. Experiment Set-Up

The microwave imaging system was validated using the 3D anthropomorphic head phantom in which a 1.25 cm-radius plastic sphere was inserted, as shown in Figure 6.

Figure 6. (**a**) Experimental set-up; (**b**) 1.25 cm-radius plastic sphere.

The sphere was fixed via a monofilament polymer line (fishing line) to an external support located above the head and was immersed in the liquid mixture at a location and height that could be varied. The sphere density was higher than the liquid mixture filling the head phantom, and the sphere was therefore not floating on the liquid and could be easily positioned in different locations within the head phantom. Three positions were considered, as shown in Figure 7. However, as their exact location could be affected by inaccuracies, it is obvious that the actual positions may have slightly differed from the expected ones.

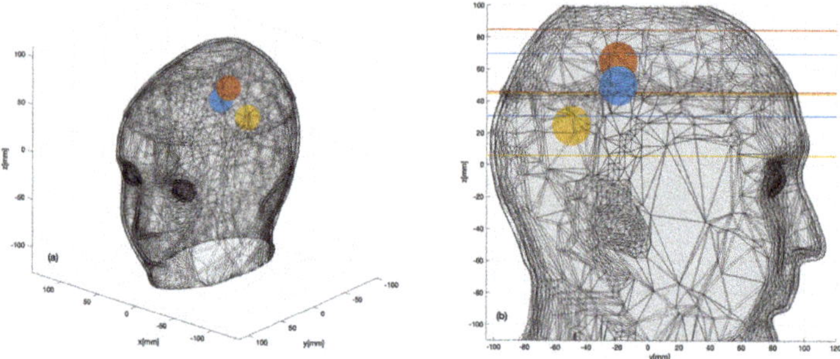

Figure 7. Three sphere locations (color dots) considered within the 3D head phantom: (**a**) 3D view; (**b**) 2D sagittal view.

The aim of these experiments was to identify the 3D shape and location of the sphere that was supposed to represent the region of the brain affected by the stroke. In this respect, it is important to remark that, while the plastic sphere adopted here for the sake of simplicity was not intended to mimic a hemorrhage or a clot, it showed a dielectric contrast with brain tissues which was comparable to a hemorrhage but with the opposite sign. Thus, the experiment was expected to provide a meaningful, though initial, validation of the imaging capabilities of the prototype device.

3. Results

In the following, we describe the validation of the realized 3D microwave imaging system (Section 2.2), whose experimental set-up is detailed above in Section 2.2.5. First, by means of the digital twin, the outcomes of the experiments are foreseen and assessed. Then, the results of the actual experiments are reported. All the imaging results were obtained by using the TSVD algorithm described in Section 2.1. Figure 8 shows the singular values calculated for the relevant scattering operator, which were computed with the digital twin. The truncation index L_t in (2) was set to -20 dB for all the reported cases. To quantitatively assess the quality of the reconstruction images, the root mean square error (RMSE) was evaluated as

$$\text{RMSE} = \sqrt{\frac{\sum_{n=1}^{N_s} (\Delta\hat{\chi} - \Delta\chi)^2}{N_s}}, \tag{3}$$

where N_s is the number of samples of the discretized domain, $\Delta\hat{\chi}$ the retrieved differential contrast and $\Delta\chi$ is the actual contrast.

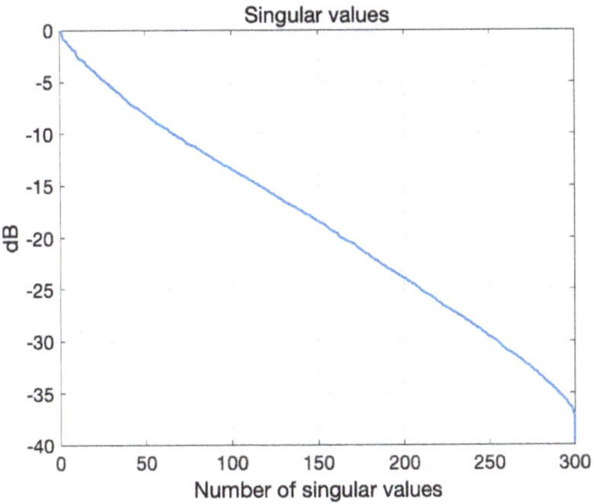

Figure 8. Singular value behavior of the scattering operator.

3.1. Numerical Assessment

To confirm the validity of the above statement, the digital twin of the system described in Section 2.2.4 was exploited. To this end, the experiment for the target corresponding to the blue sphere in Figure 7 was simulated by including a 1.25 cm-radius plastic sphere with $\epsilon_r = 2.1$ in the CAD head phantom. The simulation was repeated for the case of noiseless measurements (to provide an ideal benchmark) and for two SNR levels, namely 65 dB and 55 dB. Noise was modeled as an additive Gaussian white noise, which was added to the simulated data given by the S matrices computed with and without the target, respectively. Figure 9 shows the outcomes of the simulations. In particular, the first row shows the amplitude of the scattering matrices, where, in agreement with the actual experimental situation, the self-contributions have been omitted. It appears that the considered noise level severely affects the data matrix. The middle and bottom rows of Figure 9 show the normalized amplitude of the reconstructed differential contrast $\Delta\chi$ arising from the TSVD algorithm. It is worthwhile to note that the target is properly imaged, even when the SNR is comparable to or higher than the maximum values of the differential S matrix and the corresponding matrices appear very noisy (see Figure 9b,c), as the unavoidable occurrence of some artifacts does not restrict the interpretation of the result. The RMSE values obtained for the reconstructions at the three considered noise levels are 0.04, 0.06 and 0.11, respectively.

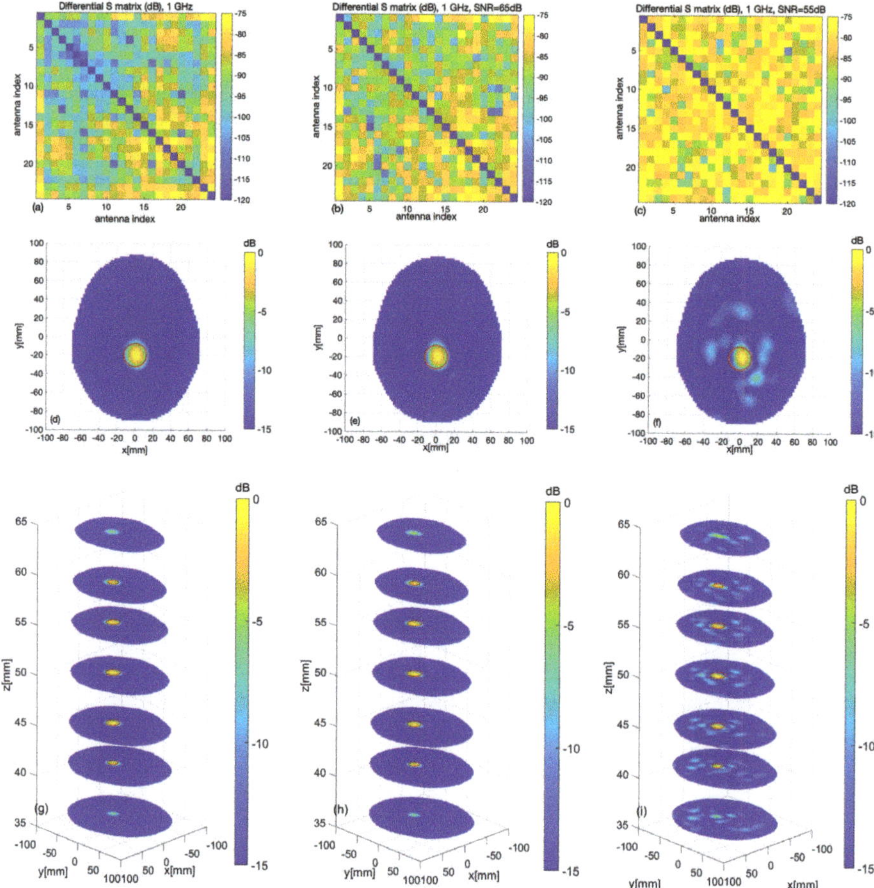

Figure 9. Digital twin: case of a target plastic sphere; the exact sphere location and shape are indicated by red circles; (**a–c**) differential scattering matrices, and (**d–f**) cross-sections of the reconstructed images at the sphere center and (**g–i**) at various levels, for different values of the signal-to-noise ratio (SNR) (left column: no noise, middle column: SNR = 65 dB , right column: SNR = 55 dB).

Once the kinds of results that the device was expected to provide in the experiments had been characterized, the second goal was to show whether—and to what extent—the experiment was relevant for the considered clinical scenario. To this end, the same simulation as before was repeated considering a spherical target with the properties of blood ($\epsilon_r = 63.4$ and $\sigma = 1.6$ S/m) instead of plastic, thus mimicking a hemorrhagic stroke. The results of the simulations are shown in Figure 10. Figure 10d–i show the imaging results, which are comparable with the previous ones but for an expected increased presence of (not meaningful) artifacts. In agreement with this, the RMSE values are essentially the same as for the case of the plastic target, at 0.04, 0.06 and 0.13 for the different considered levels of noise, respectively. According to the above results, we can conclude that the developed prototype will be able to pass the planned experimental validation and that the considered experiments provide a relevant, though initial, test-bed for the designed microwave imaging system.

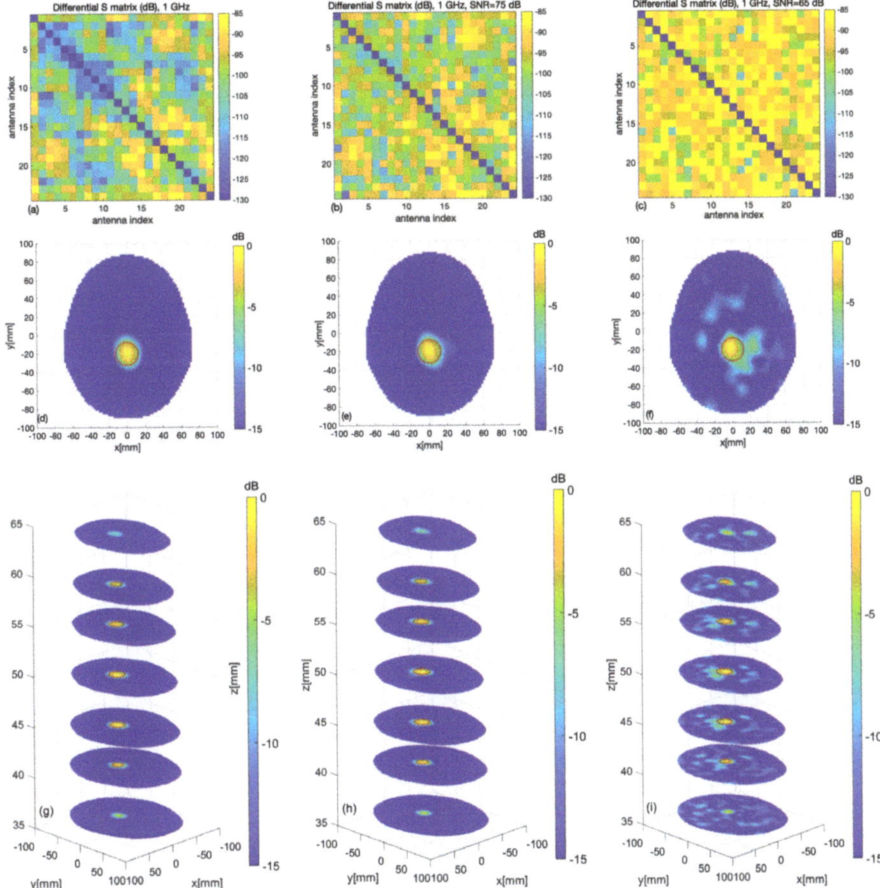

Figure 10. Digital twin: case of a target blood sphere; the exact sphere location and shape are indicated by red circles. (**a–c**) Differential scattering matrices; (**d–f**) cross-sections of the reconstructed images at the sphere center and (**g–i**) at various levels, for different values of the signal-to-noise ratio (left column: no noise, middle column: SNR = 75 dB , right column: SNR = 65 dB).

3.2. Experimental Validation

To perform the experimental validation of the system, for each sphere location, two measurement sets were taken at different times. The first data-set was measured in the absence of the sphere, and the second one was taken when the sphere was positioned inside the phantom. Each pair of measured 24×24 scattering matrices were then differentiated and given as an input to the TSVD algorithm. It can be observed that VNA port calibration was not needed as possible systematic measurement errors were cancelled out via the use of differential data or did not affect the obtained qualitative imaging of the differential contrast function. The time needed to perform the total measurement with the current prototype was less than 4 min for each sphere location, and the elapsed time between the two measurement sets was around 10 min. The kernel simulated with the digital twin (Section 2.2.4) was exploited to generate the images, and the data processing time needed by the TSVD algorithm was negligible (less than 1 s).

The 24×24 differential scattering matrices obtained for the three cases are shown in Figure 11a–c. The second and third rows of Figure 11 display the images obtained for the different sphere locations indicated in Figure 7. The second row corresponds to the horizontal cross-section passing through the sphere center; the expected sphere location and size are highlighted with a red circle. The third row displays horizontal cross-sections at the different levels indicated in Figure 7b. In all cases, the targets are very well retrieved, and the images agree with the simulation results, which confirms our expectations. The RMSE values for these reconstructions are 0.16, 0.19 and 0.18 for the three considered positions of the target, respectively. Moreover, the last row of Figure 11 shows the 3D rendering of the imaged stroke obtained by plotting the values of the normalized differential contrast amplitude, which are above −3 dB. Finally, Figure 12 reports horizontal cross-sections at various levels of the 3D reconstructed image obtained by differentiating two different sets of measurements of the same scenario; all the images are normalized with respect to the maximum value of Figure 11b, which represents a case when the target is present.

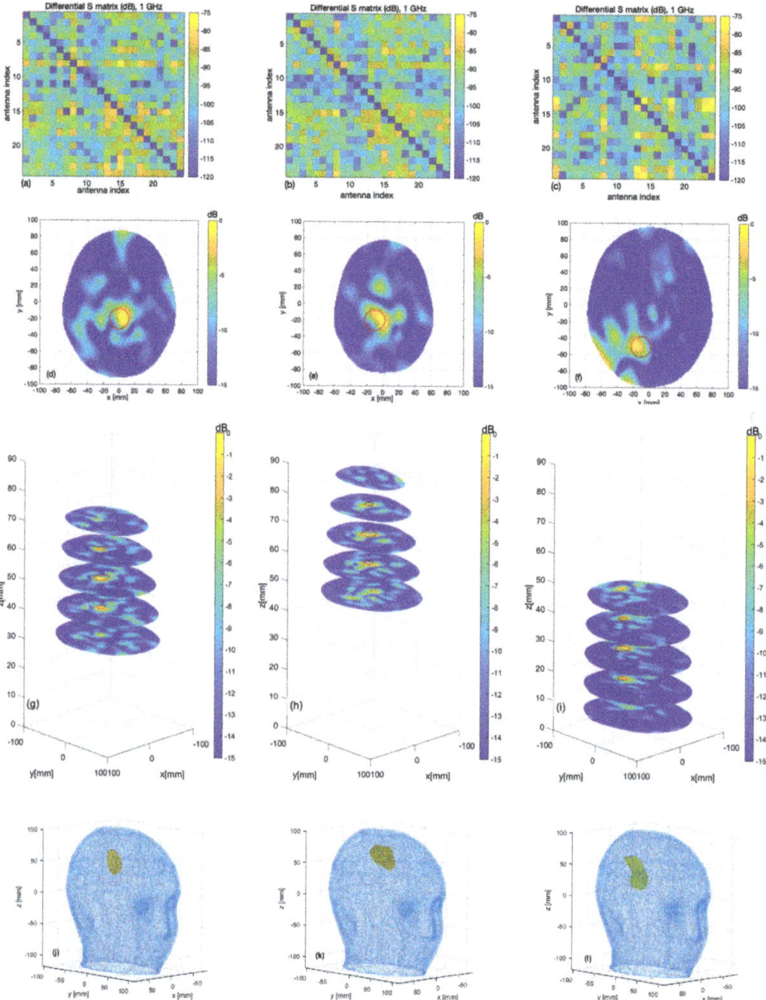

Figure 11. Measurement results: (**a**–**c**) differential scattering matrices; (**d**–**f**) horizontal cross-sections at sphere center and (**g**–**i**) at the various levels as highlighted in Figure 7 (the expected sphere location and shape are indicated by red circles); from left to right, images corresponding to the blue, red and yellow spheres of Figure 7, respectively; (**j**–**l**) the 3D rendering of the imaged stroke.

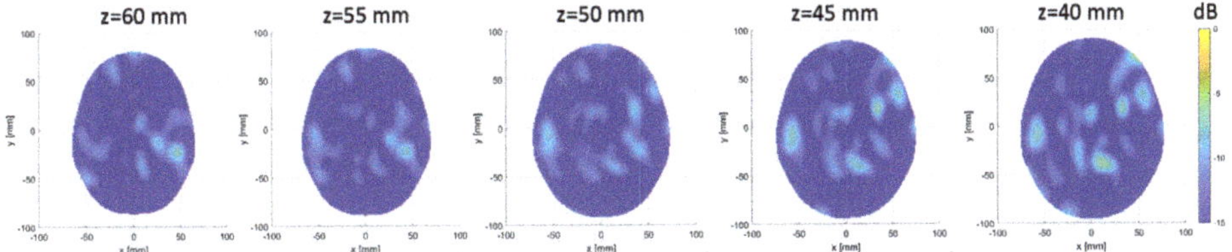

Figure 12. Horizontal cross-sections at various levels of the obtained 3D image, differentiating two sets of different measurements of the same scenario; images normalized with respect to the maximum value of Figure 11b.

4. Discussion

The main goal of the developed device was to image (qualitatively) possible anomalies (clots or hemorrhages) in the head to support clinicians in the evaluation of the effectiveness of the administrated therapies. The results shown in the previous section, although preliminary, confirm the potential of the technology in providing reliable results, as it is capable of imaging a target as small as 1.25 cm in radius.

A second important achievement of the analysis carried out here is represented by the validation of the proposed system through its digital twin, which provides simulated data which are quantitatively consistent with the measured data. As such, the adopted modeling tool provides a reliable representation of the device, making it possible both to build the imaging kernel and to synthetically reproduce laboratory experiments. As a matter of fact, the results from the simulations and measurements appear to be essentially the same, except for a slight deterioration of the images in the case of measured data. In particular, the worse RMSE results in the experimental tests, with respect to the corresponding numerical cases, are due to the possible inaccuracies in the expected positions of the target as well as to model inaccuracies in the digital twin.

A very important feature of the presented device is its robustness against false positives, assessed through a specific experiment, the result of which is shown in Figure 12. We can observe that the reconstructed values, which represent only the overall noise between different sets of measurements, are significantly lower than 0 dB (the maximum value in the 3D reconstructed image is equal to -4.75 dB) and are therefore clearly different from the cases when the target is present (see Figure 11).

To the best of our knowledge, this paper presents the first system based on a low-complexity antenna arrangement conformal to the head which is able to provide full 3D images; other imaging systems available in the literature only provide 2D images and often exploit a large number of antennas.

With this work representing a preliminary validation of the developed hardware, it is worthy of note that, for practical reasons, the target used in the experimental test does not exhibit the same dielectric properties as a stroke. On the other hand, the digital twin can help us to predict what will happen in an experiment dealing with a target mimicking a hemorrhage, for example. As a matter of fact, by comparing results from simulations and measurements, it can be observed that the differential scattering matrices exhibit a similar pattern but are lower in amplitude (by about 10 dB) in the case of simulations due to the lower maximum amplitude of the differential contrast between blood and the average brain with respect to plastic and the average brain (0.49 and -0.95, respectively). This implies that, in an experiment dealing with a target mimicking a hemorrhage, slightly weaker useful signals should be collected. However, this is not a significant limitation, as the amplitudes of the differential scattering matrices are well above the VNA noise floor, which represents the ultimate limitation for accurate measurements [25].

Finally, while the repeatability of the experiment has not been tested in this paper with respect to the possible misalignment of the phantom in the two gathered data sets, from our previous studies, we expect that such uncertainties will produce "structured" artifacts in the final image, which are easily attributable to positioning errors [19].

5. Conclusions and Future Work

In this paper, a prototype of a novel low-complexity device, dedicated to brain stroke monitoring in the post-acute stage, has been presented. The reported experiments aimed to provide an initial validation of the device and confirm that there is an agreement between the design of the system [18] and its performance. This is assessed by the comparison of the achieved results with those obtained from its digital twin. The availability of such a model also allows us to show the kinds of outcomes that are expected to be obtained by the system when operated with more realistic targets.

The next steps of the system validation and development will involve an assessment with the more realistic anthropomorphic head phantom described in [27], which includes an additional cavity modeling a stroke [29]. As this phantom is derived from an STL file, a numerical validation using the digital model of the system will also be carried out in this case. In addition, improvements of the prototype will be performed by considering image reconstruction procedures using multi-frequency data calibration techniques as well as sparsity-promoting regularization schemes; furthermore, there will be a refinement of the antenna array support to make it wearable.

Author Contributions: All authors contributed substantially to the paper. Conceptualization, J.A.T.V., R.S., M.R.C., L.C. and F.V.; Software, J.A.T.V., R.S., G.T. and D.O.R.-D.; Validation, J.A.T.V., R.S., G.B., D.O.R.-D., N.J., B.D. and E.T.; Writing—Original Draft Preparation, J.A.T.V., R.S., G.B., L.C., and F.V.; Writing—Review and Editing, M.R.C., L.C. and F.V.; Supervision, M.R.C., L.C. and F.V.; Project Administration, F.V.; and Funding Acquisition, L.C. and F.V. All authors have read and agreed to the published version of the manuscript.

Funding: This work was funded by the Italian Ministry of University and Research under the PRIN project "MiBraScan—Microwave Brain Scanner for Cerebrovascular Diseases Monitoring", and by the European Union's Horizon 2020 research and innovation program under the EMERALD project, Marie Skłodowska-Curie grant agreement No. 764479.

Acknowledgments: The authors would like to acknowledge the valuable help of Eng. Gianluca Dassano for the realization of the switching matrix.

Conflicts of Interest: The authors declare no conflict of interest. The founding sponsors had no role in the design of the study; in the collection, analyses, or interpretation of data; in the writing of the manuscript, and in the decision to publish the results.

Abbreviations

The following abbreviations are used in this manuscript:

MRI	Magnetic resonance imaging
CT	Computerized tomography
MWI	Microwave imaging
VNA	Vector network analyzer
EM	Electromagnetic
SNR	Signal-to-noise ratio
TSVD	Truncated singular value decomposition
SP4T	Single-pole-four-throw
SPDT	Single-pole-double-throw
IF	Intermediate frequency
TX	Transmitter
RX	Receiver
ABS	Acrylonitrile butadiene styrene
STL	Stereo-lithography
FEM	Finite element method
PEC	Perfect electric conductor
ABC	Absorbing boundary conditions
CAD	Proper computer-aided design
RMSE	Root mean square error

References

1. Benjamin, E.J.; Muntner, P.; Bittencourt, M.S. Heart Disease and Stroke Statistics—2019 Update: A Report from the American Heart Association. *Circulation* **2019**, *139*, e56–e528. [CrossRef] [PubMed]
2. Eurostat. *Healthcare Resource Statistics—Technical Resources and Medical Technology*; Eurostat: Luxembourg, 2019.
3. Walsh, K.B. Non-invasive Sensor Technology for Prehospital Stroke Diagnosis: Current Status and Future Directions. *Int. J. Stroke* **2019**, *14*, 592–602. [CrossRef] [PubMed]
4. Crocco, L.; Karanasiou, I.; James, M.; Conceicao, R.C. *Emerging Electromagnetic Technologies for Brain Diseases Diagnostics, Monitoring and Therapy*; Crocco, L., Karanasiou, I., James, M., Eds.; Springer International Publishing: Cham, Switzerland, 2018.
5. Casu, M.R.; Vacca, M.; Tobon Vasquez, J.A.; Pulimeno, A.; Sarwar, I.; Solimene, R.; Vipiana, F. A COTS-Based Microwave Imaging System for Breast-Cancer Detection. *IEEE Trans. Biomed. Circuits Syst.* **2017**, *11*, 804–814. [CrossRef] [PubMed]
6. Sarwar, I.; Turvani, G.; Casu, M.R.; Tobon Vasquez, J.A.; Vipiana, F.; Scapaticci, R.; Crocco, L. Low-Cost Low-Power Acceleration of a Microwave Imaging Algorithm for Brain Stroke Monitoring. *J. Low Power Electron. Appl.* **2018**, *8*, 43. [CrossRef]
7. Persson, M.; Fhager, A.; Trefnà, H.D.; Yu, Y.; McKelvey, T.; Pegenius, G.; Karlsson, J.E.; Elam, M. Microwave-Based Stroke Diagnosis Making Global Prehospital Thrombolytic Treatment Possible. *IEEE Trans. Biomed. Eng.* **2014**, *61*, 2806–2817. [CrossRef]
8. Candefjord, S.; Winges, J.; Malik, A.; Yu, Y.; Rylander, T.; McKelvey, T.; Fhager, A.; Elam, M.; Persson, M. Microwave Technology for Detecting Traumatic Intracranial Bleedings: Tests on Phantom of Subdural Hematoma and Numerical Simulations. *Med. Biol. Eng. Comput.* **2017**, *55*, 1177–1188. [CrossRef]
9. Fhager, A.; Candefjord, S.; Elam, M.; Persson, M. Microwave Diagnostics Ahead: Saving Time and the Lives of Trauma and Stroke Patients. *IEEE Microw. Mag.* **2018**, *19*, 78–90. [CrossRef]
10. Hopfer, M.; Planas, R.; Hamidipour, A.; Henriksson, T.; Semenov, S. Electromagnetic Tomography for Detection, Differentiation, and Monitoring of Brain Stroke: A Virtual Data and Human Head Phantom Study. *IEEE Antennas Propag. Mag.* **2017**, *59*, 86–97. [CrossRef]
11. Afsari, A.; Abbosh, A.M.; Rahmat-Samii, Y. Modified Born Iterative Method in Medical Electromagnetic Tomography Using Magnetic Field Fluctuation Contrast Source Operator. *IEEE Trans. Microw. Theory Tech.* **2019**, *67*, 454–463. [CrossRef]
12. Mobashsher, A.T.; Abbosh, A.M. On-site Rapid Diagnosis of Intracranial Hematoma using Portable Multi-slice Microwave Imaging System. *Sci. Rep.* **2016**, *6*, 37620. [CrossRef]
13. Alqadami, A.S.M.; Bialkowski, K.S.; Mobashsher, A.T.; Abbosh, A.M. Wearable Electromagnetic Head Imaging System Using Flexible Wideband Antenna Array Based on Polymer Technology for Brain Stroke Diagnosis. *IEEE Trans. Biomed. Circuits Syst.* **2019**, *13*, 124–134. [CrossRef] [PubMed]
14. Maffongelli, M.; Poretti, S.; Salvade, A.; Monleone, R.; Pagnamenta, C.; Fedeli, A.; Pastorino, M.; Randazzo, A. Design and Experimental Test of a Microwave System for Quantitative Biomedical Imaging. In Proceedings of the IEEE International Symposium on Medical Measurements and Applications (MeMeA), Rome, Italy, 11–13 June 2018.
15. Merunka, I.; Massa, A.; Vrba, D.; Fiser, O.; Salucci, M.; Vrba, J. Microwave Tomography System for Methodical Testing of Human Brain Stroke Detection Approaches. *Int. J. Antennas Propag.* **2019**, *2019*, 9. [CrossRef]
16. Karadima, O.; Rahman, M.; Sotiriou, I.; Ghavami, N.; Lu, P.; Ahsan, S.; Kosmas, P. Experimental Validation of Microwave Tomography with the DBIM-TwIST Algorithm for Brain Stroke Detection and Classification. *Sensors* **2020**, *20*, 840. [CrossRef] [PubMed]
17. Bisio, I.; Fedeli, A.; Lavagetto, F.; Pastorino, M.; Sciarrone, A.R.A.; Tavanti, E. A Numerical Study Concerning Brain Stroke Detection by Microwave Imaging Systems. *Multimed. Tools Appl.* **2017**, *77*, 9341–9363. [CrossRef]
18. Scapaticci, R.; Tobon Vasquez, J.A.; Bellizzi, G.; Vipiana, F.; Crocco, L. Design and Numerical Characterization of a Low-Complexity Microwave Device for Brain Stroke Monitoring. *IEEE Trans. Antennas Propag.* **2018**, *66*, 7328–7338. [CrossRef]
19. Scapaticci, R.; Bucci, O.; Catapano, I.; Crocco, L. Differential Microwave Imaging for Brain Stroke Follow up. *Int. J. Antennas Propag.* **2014**, *2014*, 11. [CrossRef]

20. Scapaticci, R.; Donato, L.D.; Catapano, I.; Crocco, L. A feasibility study on Microwave Imaging for Brain Stroke Monitoring. *Prog. Electromagn. Res. B* **2012**, *40*, 305–324. [CrossRef]
21. Bucci, O.M.; Crocco, L.; Scapaticci, R.; Bellizzi, G. On the Design of Phased Arrays for Medical Applications. *Proc. IEEE* **2016**, *104*, 633–648. [CrossRef]
22. Bertero, M.; Boccacci, P. *Introduction to Inverse Problems in Imaging*; CRC Press: Boca Raton, FL, USA, 1998.
23. Tobon Vasquez, J.A.; Scapaticci, R.; Turvani, G.; Bellizzi, G.; Joachimowicz, N.; Duchêne, B.; Tedeschi, E.; Casu, M.R.; Crocco, L.; Vipiana, F. Design and Experimental Assessment of a 2D Microwave Imaging System for Brain Stroke Monitoring. *Int. J. Antennas Propag.* **2019**, *2019*, 12. [CrossRef]
24. Meaney, P.M.; Shubitidze, F.; Fanning, M.W.; Kmiec, M.; Epstein, N.R.; Paulsen, K.D. Surface Wave Multipath Signals in Near-Field Microwave Imaging. *Int. J. Biomed. Imag.* **2012**, *2012*, 11. [CrossRef]
25. Keysight Technologies. Keysight 2-Port and 4-Port PNA Network Analyzer, N5227A 10 MHz to 67 GHz. In *Data Sheet and Technical Specifications*; Keysight Technologies: Santa Rosa, CA, USA, 2019.
26. Graedel, N.N.; Polimeni, J.R.; Guerin, B.; Gagoski, B.; Wald, L.L. An Anatomically Realistic Temperature Phantom for Radio Frequency Heating Measurements. *Magn. Reson. Med.* **2015**, *73*, 442–450. [CrossRef] [PubMed]
27. Joachimowicz, N.; Duchêne, B.; Conessa, C.; Meyer, O. Anthropomorphic Breast and Head Phantoms for Microwave Imaging. *Diagnostics* **2018**, *8*, 85. [CrossRef] [PubMed]
28. Attardo, E.A.; Borsic, A.; Vecchi, G.; Meaney, P.M. Whole-System Electromagnetic Modeling for Microwave Tomography. *IEEE Antennas Wirel. Propag. Lett.* **2012**, *11*, 1618–1621. [CrossRef]
29. Abedi, S.; Joachimowicz, N.; Duchêne, B.; Roussel, H.; Tobon Vasquez, J.A.; Rodriguez-Duarte, D.; Scapaticci, R.; Crocco, L.; Vipiana, F. Benchmark Head Phantoms for Microwave Imaging of Brain Strokes. In Proceedings of the 42nd Photonics & Electromagnetics Research Symposium (PIERS), Xiamen, China, 17–20 December 2019.

Article

Experimental Validation of Microwave Tomography with the DBIM-TwIST Algorithm for Brain Stroke Detection and Classification

Olympia Karadima *, Mohammed Rahman, Ioannis Sotiriou, Navid Ghavami, Pan Lu Syed Ahsan and Panagiotis Kosmas *

Faculty of Natural and Mathematical Sciences, King's College London, Strand, London WC2R 2LS, UK; mohammed.3.rahman@kcl.ac.uk (M.R.); ioannis.sotiriou@kcl.ac.uk (I.S.); navid.ghavami@kcl.ac.uk (N.G); pan.lu@kcl.ac.uk (P.L.); syed.s.ahsan@kcl.ac.uk (S.A.)
* Correspondence: olympia.karadima@kcl.ac.uk (O.K.); panagiotis.kosmas@kcl.ac.uk (P.K.)

Received: 29 November 2019; Accepted: 31 January 2020; Published: 4 February 2020

Abstract: We present an initial experimental validation of a microwave tomography (MWT) prototype for brain stroke detection and classification using the distorted Born iterative method, two-step iterative shrinkage thresholding (DBIM-TwIST) algorithm. The validation study consists of first preparing and characterizing gel phantoms which mimic the structure and the dielectric properties of a simplified brain model with a haemorrhagic or ischemic stroke target. Then, we measure the S-parameters of the phantoms in our experimental prototype and process the scattered signals from 0.5 to 2.5 GHz using the DBIM-TwIST algorithm to estimate the dielectric properties of the reconstruction domain. Our results demonstrate that we are able to detect the stroke target in scenarios where the initial guess of the inverse problem is only an approximation of the true experimental phantom. Moreover, the prototype can differentiate between haemorrhagic and ischemic strokes based on the estimation of their dielectric properties.

Keywords: microwave tomography; stroke detection; DBIM

1. Introduction

Brain stroke is a medical condition that is caused by a blocked (ischaemic) or burst (haemorrhagic) brain vessel, resulting in damage and necrosis of the affected brain tissue. The vast majority of the reported cases, almost 87% of them, are ischaemic. According to the American Heart Association, brain stroke is one of the leading causes of death in the USA [1]. There are different windows of treatment, depending on the type of the stroke and the area of brain that is injured. Determining the stroke type as early as possible is critical, as the wrong or delayed treatment could be lethal [2]. Detection and localization relies on magnetic resonance imaging (MRI) and, more commonly, on computed tomography (CT) scans. While both MRI and CT are accurate and reliable methods, neither of them are truly portable and thus ready to be used widely inside an emergency vehicle for detecting strokes as early as possible. Moreover, MRI is expensive, and CT is associated with health risks due to ionized radiation [3]. These challenges motivate the development of alternative approaches, which aim to be fast, safe, portable and cost-effective.

Microwave imaging (MWI) can potentially offer these advantages, and has thus emerged as a promising alternative for diagnosing cardiovascular diseases which could be used prior to MRI or CT scans [4]. Efforts to develop MWI systems for medical diagnostics go back almost forty years, with several review papers, book chapters, and special issues reporting recent developments [5–7]. Amongst a large body of research in medical microwave imaging, which cannot possibly be fully reviewed here, we note that various experimental systems have been developed for breast cancer detection [8–14], and more recently for stroke monitoring and detection [15–17]. Prototypes based on machine learning have also been developed and clinically tested [18]. With regards to microwave tomography specifically, for brain imaging, a study by Hopfer et al. using a microwave scanner of 177 antennas presented successful experimental reconstruction results for stroke detection and brain monitoring [19]. In addition, the design of a microwave tomography (MWT) scanner for brain imaging has been analysed in References [20] and [21] to determine suitable frequencies and properties of the coupling medium, as well as to optimize the antenna array design.

In a MWT system, an array of antennas transmits electromagnetic (EM) waves which penetrate into the tissue, and receives the scattered field [22]. The MWT algorithm reconstructs a map of the dielectric properties of the imaged region to locate a target with unknown dielectric properties, by solving an inverse and ill-posed EM scattering problem [23,24]. MWI techniques for medical applications rely on the dielectric contrast between healthy and diseased tissues, which depends on their water content [25]. The dielectric properties of human tissues have been extensively studied and reported in References [25,26]. The head's dielectric properties have been measured in References [27] and [28]. For the case of ischaemia, the measurements in [29] reported that the properties of an ischaemic vessel in the brain can vary from -10% to -25% relative to the dielectric properties of a healthy brain.

MWT methods are challenged by the high heterogeneity of the human body [30] as well as the increased computational cost that arises from the non-linearity of the problem and the non-uniqueness of the solution [31]. Based on these considerations, we have developed previously a robust algorithm based on the distorted Born iterative method (DBIM) and the two-step iterative shrinkage/thresholding (TwIST) solver for microwave breast imaging [32,33], which was then incorporated successfully in a prototype tested in experiments with simple cylindrical targets [34–36] and showed advantages in comparison with other DBIM linear solvers like conjugate gradient method for least squares (CGLS) or iterative shrinkage/thresholding (IST) [32,33,36].

Taking into account the challenges of MWT systems and the medical requirements for brain stroke detection and differentiation, the aim of this paper is to validate experimentally DBIM-TwIST and our prototype for stroke imaging using brain tissue and stroke mimicking phantoms. To this end, we present for the first time reconstruction results across a wide frequency range, which demonstrate the potential of differentiating between stroke types based on a quantitative estimation of their dielectric properties. Moreover, we show our method's robustness to differences in the dielectric properties of the background medium assumed in our inverse phantoms relative to the phantom used in the experiment. We therefore believe that these experimental results provide further evidence of MWT's potential to detect strokes based on microwave images produced by a carefully designed system and algorithm.

The remainder of the paper is organized as follows: Section 2 presents the methodology for phantoms preparation and characterization, the experimental data acquisition process, and a summary of our DBIM-TwIST algorithm. Section 3 presents reconstruction results for haemorrhagic and ischaemic mimicking targets, and for phantoms with different dielectric properties than the initial guess used in the inversion. Finally, Section 4 summarises our findings and discusses our future work towards developing a complete MWT system for brain stroke imaging.

2. Materials and Methods

2.1. Phantoms Preparation and Characterization

Experimental testing of MWT systems is critical in order to assess their potential for clinical applications. To this end, phantoms that mimic the dielectric and structure properties of the human head and brain provide efficient, easy to fabricate, and low-cost testbeds for experimental validation prior to clinical trials. Gelatine emulsions made of oil, water and gelatine are popular recipes for preparing such phantoms, as they are cheap and easy to alter, but can suffer from dehydration.

McDermott et al. have proposed an alternative method of constructing solid phantoms made of ceramic and carbon powder, in which solid phantoms were fabricated with polyurethane, graphite, carbon black and isopropanol, to mimic the dielectric properties of the brain and the average properties of the layers surrounding the brain [37]. Solid phantoms are electrically and mechanically stable as well as reusable, but are more difficult to fabricate, time consuming, more expensive than the liquid phantoms, and not easy to modify for additional adjustments in properties or geometry. In Reference [38], an alternative method of 3D printed moulds filled with fluid TX100 salted water mixtures that can mimic the dielectric properties of head tissues was proposed. Those phantoms are cheap, easy to alter, reusable and stable. They require, however, plastic mould materials which can cause disturbances in the scattered signals [39].

As the aim of this work is to validate DBIM-TwIST in more realistic, yet simplified cases for brain stroke detection and classification, we have constructed simple phantoms based on the materials and processes proposed in Reference [40] for breast phantoms. Using different gelatine-oil concentrations compared to those for breast tissues, we fabricated tissue-mimicking materials with specific dielectric properties that mimic average brain tissue, cerebrospinal fluid (CSF), blood and ischaemia. The process is illustrated in Figure 1. First, a water-gelatine mixture is prepared and mixed at 70 °C, until the gelatine particles are fully dissolved into water and the mixture is transparent. Propanol is added to deal with the creation of air bubbles on the surface. Once heated to 70 °C, we add a 50% kerosene-safflower oil solution into the water-gelatine mixture and stir at the same temperature, until the emulsion has an opaque white colour and the oil particles are fully dissolved. We keep stirring the mixture until the temperature drops to 50 °C, when we add the surfactant. Finally, we pour the prepared mixture into the mould when it reaches 35 °C, and let it set overnight before we conduct any measurements.

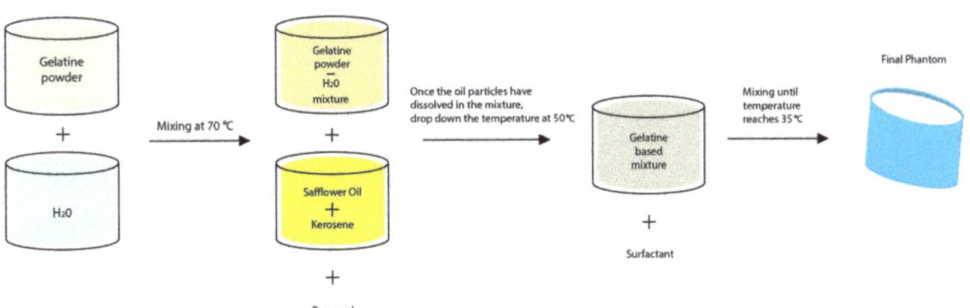

Figure 1. Schematic representation of the phantoms' preparation process.

Table 1 presents the concentrations of the materials used for the phantoms mimicking different brain tissues used in our experiments. We fabricated the four tissue-mimicking phantoms, using as reference the properties reported in Reference [25]. The in-vivo measurements in [29] reported that properties of an ischaemic vessel in the brain can vary from -10% up to -25% relative to the dielectric properties of healthy brain tissue at 1.0 GHz. Therefore we chose to prepare ischaemic phantoms with the maximum reported difference of -25% relative to the properties of the brain phantom. For the experimental testbeds it is assumed that, due to similar dielectric properties of CSF and blood, the same gelatine mixture can be used to mimic the two different tissues. Figure 2 presents the measured dielectric properties of the prepared phantoms for brain, CSF, blood and ischaemia, respectively. The measurements are conducted using Keysight's dielectric spectroscopy kit, over a 0.5–2.5 GHz frequency range at different points of the phantoms. The plots in Figure 2 show very good agreement with reference data for the dielectric constant, but quite lower conductivity values for our phantoms. As MWT relies mostly on contrast in dielectric constant, these discrepancies are not critical for this initial investigation.

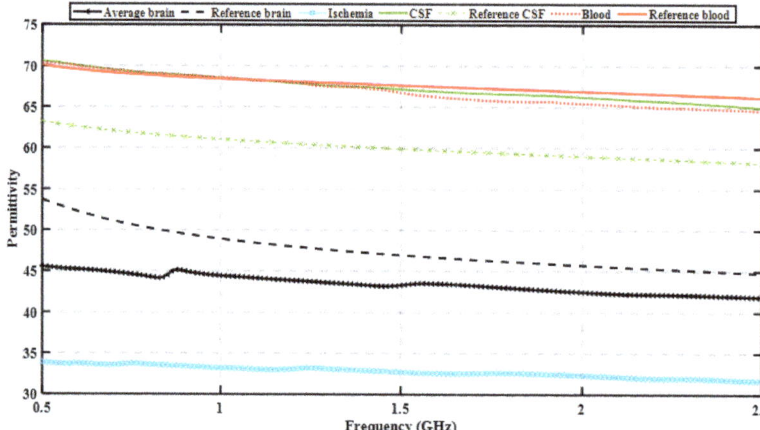

(**a**) Measured permittivity of the of the prepared tissue mimicking phantoms relative to their reference values.

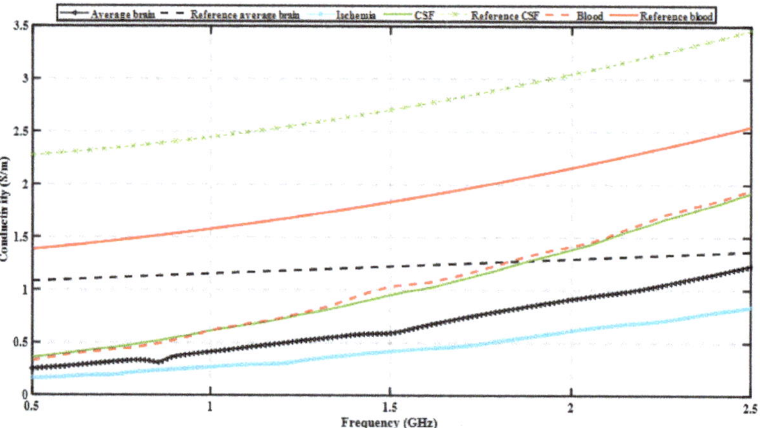

(**b**) Measured conductivity of the of the prepared tissue mimicking phantoms relative to their reference values.

Figure 2. Dielectric properties of the fabricated tissue mimicking phantoms relative to reference values taken from [25] (to the authors' knowledge, there is no literature regarding the dielectric properties of ischaemia in a wide frequency range). Same type of tissues are presented with the same colour.

Table 1. Concentrations of materials used for 100 ml of tissue mimicking phantoms.

Phantom Type	Water	Gelatine Powder	Kerosene	Safflower Oil	Propanol	Surfactant
Average brain	60 mL	11 gr	13 gr	13 gr	2.5 mL	1.5 mL
CSF/Blood	80 mL	16 gr	-	-	4 mL	-
Ischemia	50 mL	8.5 gr	20 gr	20 gr	1.5 mL	1.5 mL

After preparation, phantoms are placed in elliptical plastic acrylonitrile butadiene styrene (ABS) moulds which aim to mimic the brain's shape and multi-layer structure, as shown in Figure 3a,b. We first pour the CSF phantom in the outer layer of 5 mm thickness. After this is solidified, we extract the inner mould and fill the remaining cavity with the brain phantom. For a one-layer model without the

CSF, we simply extract the inner mould and pour the phantom in the outer mould directly. We use an additional cylindrical mould to create a hole (diameter = 30 mm) in the phantom (Figure 3c), which can be filled with the phantom mimicking either blood or ischaemia (Figure 3d), emulating the two cases of brain stroke termed as h-stroke for haemorrhagic and i-stroke for ischaemic, respectively.

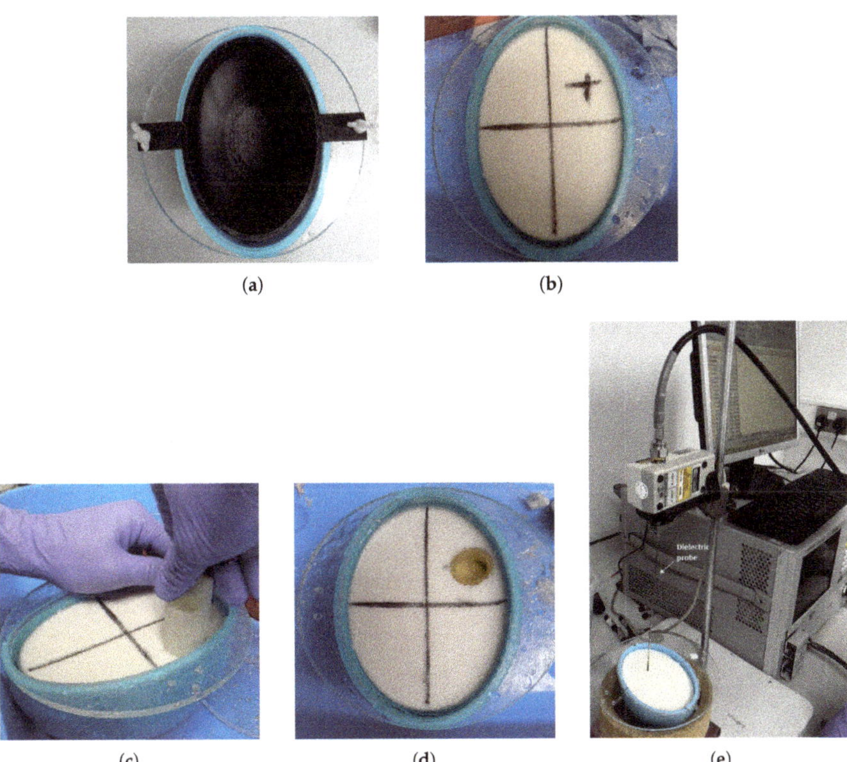

Figure 3. Summary of the head model construction: (**a**) Cylindrical 3-D printed mould to form the cerebrospinal fluid (CSF) and average brain layers; (**b**) The two-layer model after CSF and average brain phantoms are poured into the moulds (the CSF is a thin, transparent layer just inside the blue mould); (**c**) Creating a hole in the phantom to insert the stroke-like target; (**d**) Final two-layer phantom with a target of blood-mimicking phantom; (**e**) Setup with Keysight's slim form probe to measure the dielectric properties of the phantoms.

To further investigate the change in the dielectric properties over time, we prepared two additional brain phantoms and measured their real and imaginary part of permittivity on day 1 and day 6 after their preparation. Table 2 shows that, over a period of 6 days, the dielectric properties increased 10–15% compared to their initial values. This is expected as the water particles have dispersed over time resulting in higher dielectric properties.

Table 2. Dielectric properties of brain phantoms over time at 1 GHz.

Measured Property	Day 1	Day 6
ϵ' sample 1	44.5	51.2
ϵ' sample 2	46.2	51.6
ϵ'' sample 1	0.43	0.51
ϵ'' sample 2	0.42	0.48

2.2. Setup and Data Acquisition Process

Figure 4b presents the experimental setup, which is used for conducting the measurements. It consists of a 300 mm diameter cylindrical tank surrounded by an absorber (ECCOSORB MCS) and enclosed with a metallic shield to decrease surface waves propagating into the perimeter of the tank. The tank is filled with a 90% glycerol-water mixture to improve impedance matching with the phantom and widen the antenna operating frequency range [34]. An eight-antenna elliptical array is immersed inside the tank, and the length of the ellipsoid's axes are 153 mm and 112 mm. From a theoretical point of view, the choice of the number of antennas is defined as:

$$M = 2\beta\alpha, \tag{1}$$

where α is the radius of the reconstruction domain and β is the wave-number [41]. Taking into account our first working frequency of 0.5 GHz, M is approximately equal to 15. However, due to our hardware limitation of the 8-port VNA and our previous experimental work which has showed good results using eight antennas in our MWT prototype, we chose eight antennas to simplify the experiment and the data acquisition process [34]. The experiments presented in this study use spear-shaped patch monopole antennas which operate efficiently in the range of 0.5–2.0 GHz and their design was presented in Reference [42]. This range satisfies the recommendation by the theoretical approach in Reference [20], which suggests that a working frequency up to 1.5 GHz can achieve the optimal trade-off between the required resolution and incident power needed for head imaging applications. Antennas are installed into vertical and horizontal rulers that enable us to adjust their height and the array dimensions.

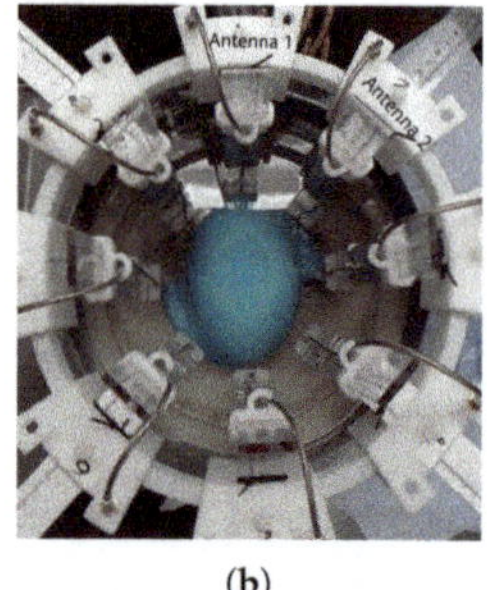

(a)   (b)

Figure 4. (a) Schematic of the simplified "initial guess" model for the inversion (the ellipsoid's axes are 153 mm and 112 mm long); (b) The hardware system prototype.

Phantoms are placed in the bottom of the tank as shown in Figure 4b, and measurements of the S-parameters of the scattered signals are conducted in a frequency range of 0.5–2.5 GHz using an eight-port vector network analyzer (VNA) from Keysight. The eight antennas act both as transmitters and receivers, creating an 8×8 scattering matrix which is fed into DBIM-TwIST. The linear integral equation is discretized iteratively for the entire 8×8 transmit-receive pairs. Overall, two sets of measurements are performed for every imaging scenario. The first set of measurements is performed for the phantom without a target (the "no target" scenario), and then the process is repeated for the phantom which includes the 30 mm diameter h-stroke or i-stroke target ("target" scenario). The target is placed in the upper-right section of the phantom between antennas 1 and 8. Following this process, the S-parameters for "no target" and "target" scenarios are processed by the DBIM-TwIST algorithm [32], which is reviewed in the next subsection.

2.3. Implementation of the DBIM-TwIST Algorithm

To reconstruct the dielectric properties of the phantoms, we apply our formerly developed DBIM-TwIST algorithm, which has been widely validated by our previous studies in different numerical [32,33] and experimental scenarios involving extended cylindrical targets [34–36]. In DBIM, the nonlinear scattering problem which arises from the integral equation of the scattered electric field is discretized under the Born approximation as follows:

$$E_s(r_n, r_m) \approx \iota\omega \int_V E_b(r, r_m) E_b(r, r_n) O(r) dr \quad , \tag{2}$$

where E_s and E_b are the scattered and background fields respectively, and r_n, r_m denote transmitter and receiver positions, respectively. The contrast function $O(r)$ is the difference between the complex permittivity of the known background and the unknown region. The equation is solved iteratively and requires the solution of a linear but ill-posed problem at each DBIM iteration [23]. The background properties are updated at each DBIM iteration with the current solution until the residual error is minimised, and this yields the reconstruction of the dielectric properties of the imaging domain. This is done by discretizing the above integral equation as,

$$Ax = y, \tag{3}$$

where A is an M-by-K propagation matrix, with M the number of transmit-receive pairs in the antenna array and K the number of elements in the discretization inside the reconstruction domain, while y is the M-by-1 vector of the scattered fields recorded at the receivers. The K-by-1 vector x is the unknown dielectric properties contrast function, which is then added to the previous background profile to generate the new background which is used for the next iteration [23]. Our algorithm solves this linear problem with the TwIST method [43], which splits the matrix into a two-step iterative equation.

The background properties are calculated and updated at each DBIM iteration with the TwIST solution until the residual error is minimised, and this yields the reconstruction of the dielectric properties of the imaging domain.

The initial model ("starting guess") for the algorithm is shown in Figure 4a, and consists of a simplified two-dimensional (2D) representation of the experimental setup, which includes the tank filled with the matching medium and an ellipsoid that mimics the average brain tissue with dimensions of 153 × 112 mm. These represent the actual size of the phantom's axial slice at the height where the antennas are located. The algorithm simplifies the system's antennas with line sources, which are placed at the same locations as the experiment. To calibrate the simulated initial model, we use a "no target" reference measurement, as in our previous work (e.g., see Reference [35]). This calibration method relies on the signal difference between the experimental and the simulated "no target" scenarios, to calibrate the signals received from the "target" experiment. It is used to eliminate the errors induced by the differences between the simulated forward model and the experiment.

We must note that an exact "no target" reference measurement will not be available in a realistic clinical scenario such as stroke detection. Aside from an "empty tank" measurement which could always be used as reference, one could also use a reference signal measured from an "average homogeneous head phantom" taken from a set of phantoms with different properties, following the approach used for imaging numerical breast phantoms in Reference [32], for example. This approach, however, would still provide the algorithm with very different signals from the true "no target" signals in stroke detection, as the brain structure is much more complex than any "homogeneous average brain" phantom (see also the Discussion section), and a more complex, in-homogeneous reference phantom may be better suited for data calibration. We also note that our DBIM-TwIST approach has been shown to be capable of reconstructing accurately complex in-homogeneities using simulation data from complex numerical breast phantoms [32], but the problem of detecting the stroke inside the complex yet unknown brain structure may be more challenging. We partially examine this issue in Section 3.2, by using a "no target" scenario from a "day 1 phantom" and a target scenario from a "day 6 phantom" with almost 20% higher dielectric properties. The issue will be fully examined in our future work and presents, arguably, the most significant obstacle that MWT methods need to overcome prior to their clinical use.

To take into account the materials' dispersive dielectric properties, we employ a first-order Debye model for the average brain phantom and the 90% glycerol-water mixture:

$$\epsilon_s(\omega) = \epsilon_\infty + \frac{\Delta\epsilon}{1 + j\omega\tau} + j\frac{\sigma_s}{\omega\epsilon_0}, \tag{4}$$

where ϵ_∞, $\Delta\epsilon$ and σ_s are the parameters of the single pole Debye model provided in Table 3. We use the parameters of Table 3 together with a forward solver based on the finite-difference time-domain (FDTD) method to calculate the scattered and background fields, and then apply DBIM-TwIST to reconstruct the dielectric properties inside the imaging domain with a grid voxel size of 2 mm. The complex permittivity values in (2) are replaced by the frequency-dependent Debye model of (4) and are updated at each DBIM iteration, thus are used to update the contrast function. The reconstructed results present estimated real and imaginary parts of permittivity which are calculated from the updated Debye models at each frequency [23]. Moreover, we perform the same number of 15 DBIM-TwIST iterations at each frequency for both single and multiple frequency reconstructions attempted in this study. The running time for each DBIM iteration was approximate 8 seconds using MATLAB R2019b run on an Intel i7 processor with 16 GB RAM memory, leading to an execution time of 2.5 min for 15 iterations of DBIM at each frequency. By means of an example, the total execution time for the case of the h-stroke with one layer and frequency hopping between 0.5 and 1.5 GHz is 24 min.

Table 3. Debye parameters of materials after curve fitting to measured dielectric properties.

Material Type	$\Delta\epsilon$	ϵ_∞	σ_s
90% glycerol-water	6.56	16.86	0.3232
Average brain	30	10	0.147

3. Results

In our first set of experiments in Section 3.1, our aim is to show our method's potential to determine the type of stroke from estimating its dielectric properties. To this end, we examine two phantoms of average brain tissue in which we inserted a target emulating h- and i-stroke, respectively. In the second set of experiments in Section 3.2, our aim is to test detection performance in cases where the inverse experimental model is slightly different than our forward model. To this end, we examine two additional phantoms with h-stroke targets. The first one is a two-layer phantom including a thin CSF-mimicking layer, and the second one is a brain phantom which we let it set for 6 days before conducting the "target" measurements, so that its properties are different in comparison with the first day when the "no-target" measurements were taken.

3.1. Detection and Classification of Stroke Targets

Figures 5 and 6 present the reconstructed real and imaginary parts of the complex permittivity for the phantom inside the prototype of Figure 4b containing h-stroke and i-stroke targets as well as a target which has 50% lower dielectric constant than the average brain phantom as an extreme case

(50% i-stroke). Images illustrate the whole tank, however, the reconstruction domain contains only the brain ellipsoid inside the antenna array. Representative single-frequency reconstructions in the range 0.7–1.5 GHz are shown in Figure 5, while results from a frequency hopping approach in the same range are shown in Figure 6. These images suggest that the dielectric constant (real part) of 25% i-stroke is detectable at low frequencies albeit with more artefacts in the images, whilst the h-stroke and the 50% i-stroke, is clearly visible in frequencies above 1.1 GHz for both real and imaginary parts. As the dielectric contrast between the constructed 25% i-stroke and brain phantoms is lower than that of h-stroke and 50% i-stroke, detecting the 25% i-stroke target is more challenging than the other cases, with unsuccessful results for the imaginary part of the complex permittivity. The most encouraging observation from these figures, however, is that there is a clear distinction in the estimation of the dielectric properties for all the three targets: the frequency hopping results of Figure 6 estimate the value of real permittivity ϵ' at 1.5 GHz as $\epsilon'=33.8$, $\epsilon'=20.9$ and $\epsilon'=56.25$ for 25% i-stroke, 50% i-stroke and h-stroke, respectively.

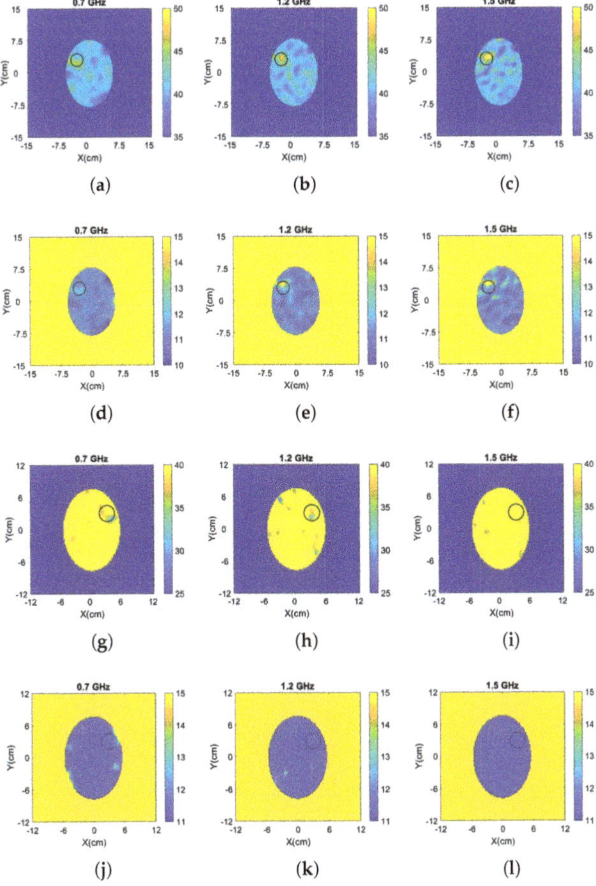

Figure 5. *Cont.*

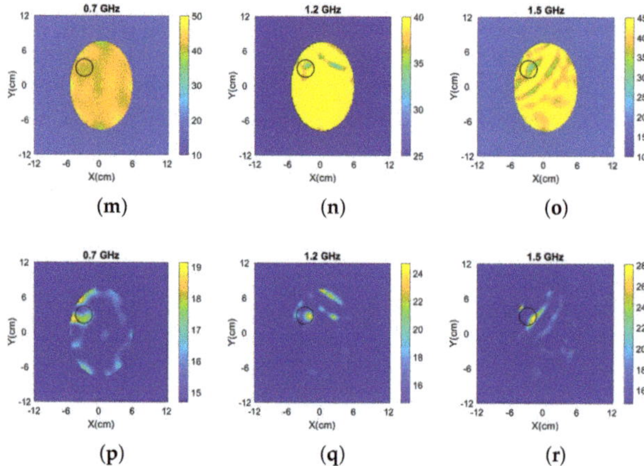

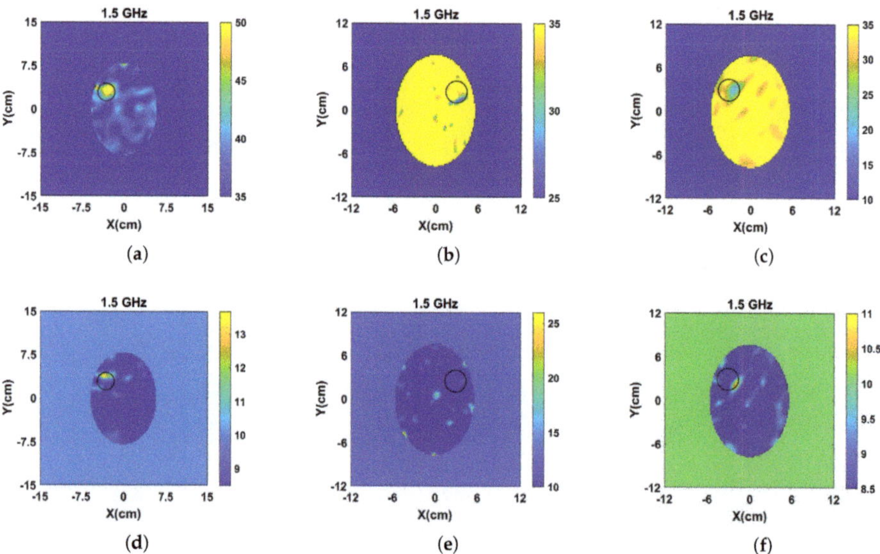

Figure 5. Results from single frequency reconstructions of the real (first, third and fifth line) and imaginary (second, fourth and sixth line) part of the complex permittivity for: (**a–f**) h-stroke, (**g–l**) 25% i-stroke and (**m–r**) 50% i-stroke.

Figure 6. Reconstructed: (**a–c**) real and, (**d–f**) imaginary part of the complex permittivity for h-stroke (left), 25% i-stroke (middle) and 50% i-stroke (right). The complex permittivity was calculated at 1.5 GHz using the frequency hopping approach in a frequency range of 0.7–1.5 GHz.

3.2. Stroke Target Detection for Brain Phantoms with Unknown Properties

To move towards more realistic imaging scenarios where the brain's distribution of tissues will be more complex and their dielectric properties unknown, we have evaluated experimentally the robustness of DBIM-TwIST when the structure and dielectric properties of the brain phantom are slightly different from the "initial guess" of Figure 4a. To examine the effect of differences in the dielectric properties of the average brain phantom where the target is inserted, we prepared a phantom and conducted "no target" measurements after the phantom had set, and then kept it for a week, in which we recorded the change of its dielectric properties over time. Figure 7 shows our measurements for days 4 and 6, where an increase in the permittivity with time is observed possibly due to increase in water dispersion. We then acquired "target" measurements for h-stroke on day 6, to ensure a change in the phantom's dielectric properties from the "no-target" measurements on day 1. In addition to this test, we also performed experiments with a two-layer phantom (CSF and average brain). To this end, we added a CSF layer of 5 mm thickness using the process described in Figure 3, and with dielectric properties plotted in Figure 2.

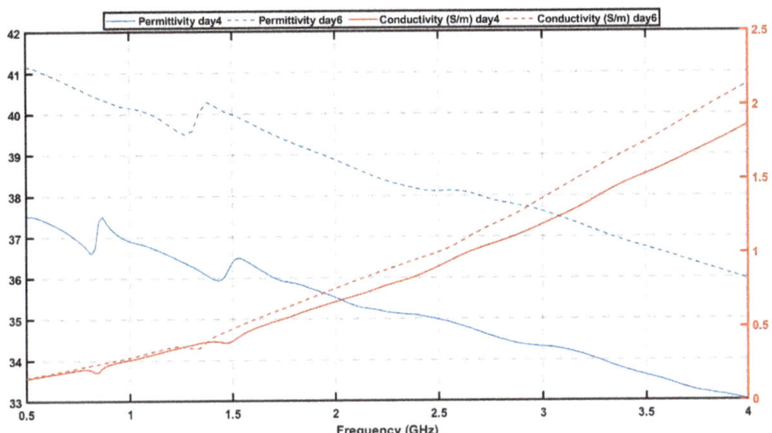

Figure 7. Dielectric properties of brain mimicking phantom at day 4 and day 6 after its preparation.

Figure 8 presents the reconstructed images of the real and imaginary parts of complex permittivity using the data obtained by the measurements at days 1 and 6 after the preparation of the brain phantom, as well as the reconstructed images of the two-layer phantom including the CSF layer. These reconstructions were produced by frequency hopping in the range of 1.1–2.0 GHz. The values of the reconstructed real and imaginary parts for the h-stroke target at 2 GHz are $\epsilon' = 49.28$, $\epsilon'' = 12.19$ for the one-layer phantom, and $\epsilon' = 62.38$, $\epsilon'' = 8.48$ for the two-layer phantom, respectively. With the exception of the imaginary part for the two-layer phantom, these images provide a clear indication of where the h-stroke is located and estimate its dielectric constant with satisfactory accuracy. These reconstructions confirm that the algorithm is sufficiently robust to detect and localize the target successfully in cases with "mild" uncertainties in the true background medium in the experiment and its "initial guess" in the imaging algorithm. We note that we have also performed single frequency reconstructions for both cases (not shown here), which indicate that best results are obtained in the 1.1–1.6 GHz range, which agrees with the operating frequencies of the antennas employed in our measurements [42].

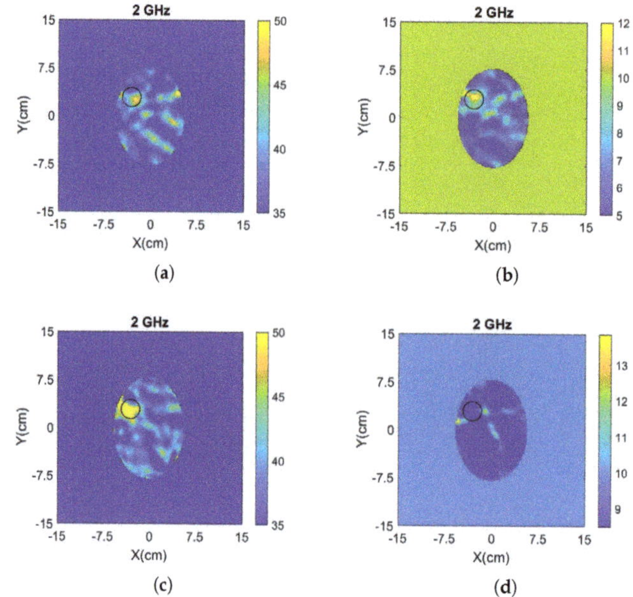

Figure 8. (**a**) Reconstructed real and (**b**) imaginary part of the complex permittivity for the phantom inside the tank of Figure 4b, when "no-target" and "h-stroke target" measurements were conducted at day 1 and day 6 after the preparation of phantom, respectively. (**c**) Reconstructed real and (**d**) maginary part of the complex permittivity for a two-layer phantom with an "h-stroke target" and a CSF layer which is unknown to the imaging algorithm. The complex permittivity was calculated at 2 GHz using a frequency hopping approach in 1.1–2 GHz.

4. Discussion

We presented an initial experimental assessment of a microwave tomography prototype based on the DBIM-TwIST algorithm for brain stroke detection and classification. This MWT system was able to differentiate between targets that mimic haemorrhagic and ischaemic stroke, based on the difference in their estimated dielectric properties. Moreover, the system was able to reconstruct the target inside a brain phantom even when its structure or dielectric properties are different from the "initial guess" used in the inversion. These results benefited from reconstructions in multiple frequencies based on the use of our in-house developed spear-shaped antennas which operate in a wide range from 0.5 GHz to 2.5 GHz. Moreover, the flexible tuning of the DBIM-TwIST parameters allow an easy adaptation of the algorithm to inverse problem at hand. We note that, as with most non-linear inverse methods, the DBIM-TwIST parameters must be tuned based on considerations that relate to both the imaging problem and the experimental prototype. Moreover, experimental measurement errors may result in certain frequencies providing more accurate reconstructions than others. In the past, we have proposed a method to discard data dominated by measurement error, based on a correlation technique [35]. Regardless of whether such a pre-processing step is applied to the measured data, frequency hopping is an excellent way to increase the accuracy and robustness of the algorithm [32].

Moreover, our previous work in Reference [32] has confirmed that the DBIM-TwIST (as any other DBIM implementation) is very sensitive to the initial guess of the inverse model. While this has been partially addressed in this work with our investigation in Section 3.2, it will become a much more challenging problem in a realistic brain stroke detection application, where the structure of the brain is much more complex than a homogeneous average brain phantom.

We also note that our experimental validation required that the antenna array is placed at the same height as the target. We are currently working on a more realistic experimental validation process which involves acquiring data for more than one height and employing a three-dimensional (3D) version of the DBIM-TwIST.

The experimental setup contains a tank filled with matching liquid in which we immersed the antennas and the phantoms. However, our current and future work is focused on designing and proposing a portable helmet which will incorporate the matching medium and the antennas. Algorithm-wise, the image reconstruction can be accelerated by frequency selection [35] and a more accurate initial guess which will also increase the stability of the algorithm. However, knowing a priori information for brain imaging is very challenging. For this reason, we have assumed that the initial model is filled with homogeneous average brain tissue. In Reference [32] a method was proposed to estimate the optimal initial guess based on the residual error of different known cases, which we will explore in our future work.

While there are a few experimental studies on detecting a blood-like target using microwave imaging methods [15–17,44,45], only a few studies exist that demonstrate experimentally the detection of an ischemic-like area. Our results in this respect are encouraging, as they indicate that a classification of the stroke type is possible with a microwave imaging system without the need of training data [18]. This has also been suggested in References [19,46] but with a very different, single-frequency MWT system. This confirms that MWT systems have the potential to be used for early differentiation between i- and h-stroke, which is vital for improving stroke treatment outcomes [47].

We note that the experimental phantoms in this study are oversimplified in relation to the true head and brain anatomy. However, this simplification is necessary to show that our system and algorithm have the potential to differentiate between the two different stroke types based on a quantitative estimation of their dielectric properties. We must acknowledge, however, that the imaging performance of the algorithm and prototype depends strongly on the complexity of the brain, and hence, the experimental phantom that represents it. The skin and skull, for example, introduce strong scattering layers, and the distribution of the CSF can be complex and can obscure the signal from the target. While our previous work has shown that tools such as frequency hopping and optimization of the initial guess can have a significant positive impact, for example in reconstructing breast compositions in the presence of an unknown skin layer [32], application of our approach to a complex brain phantom may require prior information of the phantom's anatomical structure. This in practice could be acquired from average numerical brain models and a careful calibration procedure. These issues are critical to provide a more informed answer to whether MWT can be clinically attractive as a stroke detection method and will be investigated in our future work.

We also note that we are currently on the next version of our head phantom, which will feature a more complex structure with various tissue-mimicking materials resembling different brain tissue layers, as well as outer layers such as the skull and the skin. Moreover, in our future work we will use moulds in realistic head shape that will allow us to include a wig to take into account the impact of hair. As shown in Reference [48], a wig's dielectric properties are very similar to the dielectric properties of natural hair, therefore, a wig can be easily used as an alternative for those experiments. Our successful reconstructions for the phantom with an added thin elliptical CSF layer, although not fully realistic, are a step forward towards our goal of successful detection in complex head phantoms.

Our future work will also focus on validating the proposed MWT approach further towards the development of a portable microwave head scanner designed for stroke detection, which will apply a multiple-frequency 3D DBIM-TwIST algorithm for imaging. This will include assessing the capabilities of our MWT system with phantoms with increased complexity resembling the actual anatomy and dielectric properties of the brain. We will also examine other applications of our MWT approach such as lymph node detection for breast cancer.

Author Contributions: Conceptualization, P.K.; Methodology, O.K., M.R., I.S., N.G. and P.K.; Software, O.K., P.L. and P.K.; Validation, O.K., I.S., and S.A.; Resources, P.K.; Writing–original draft preparation, O.K. and P.K.; Supervision, P.K.; Project Administration, P.K.; Funding Acquisition, P.K. All authors have read and agreed to the published version of the manuscript.

Funding: This work was supported by the EMERALD project funded from the European Union's Horizon 2020 research and innovation programme under the Marie Skodowska-Curie grant agreement No. 764479. The development of the experimental prototype and testbeds has been funded in part by Innovate UK grant number 103920, and in part by the Engineering and Physical Sciences Research Council grant number EP/R013918/1.

Conflicts of Interest: The authors declare no conflict of interest. The founding sponsors had no role in the design of the study; in the collection, analyses, or interpretation of data; in the writing of the manuscript, or in the decision to publish the results.

Abbreviations

The following abbreviations are used in this manuscript:

ABS	Acrylonitrile Butadiene Styrene
CGLS	Conjugate Gradient method for Least Squares
CSF	Cerebrospinal fluid
CT	Computed Tomography
EM	Electromagnetic
DBIM	Distorted Born Iterative Method
FDTD	Finite-Difference Time-Domain
IST	Iterative Shrinkage/Thresholding
MWI	Microwave Imaging
MWT	Microwave Tomography
MRI	Magnetic Resonance Imaging
TwIST	Two-step Iterative Shrinkage Thresholding
VNA	Vector Network Analyzer
2D	Two-dimensional
3D	Three-dimensional

References

1. Mozaffarian, D.; Benjamin, E.J.; Go, A.S.; Arnett, D.K.; Blaha, M.J.; Cushman, M.; Das, S.R.; De Ferranti, S.; Després, J.P.; Fullerton, H.J. Heart disease and stroke statistics-2016 update a report from the American Heart Association. *Circulation* **2016**, *133*, 447–454. [CrossRef]

2. Heros, R.C. Stroke: Early pathophysiology and treatment. Summary of the Fifth Annual Decade of the Brain Symposium. *Stroke* **1994**, *25*, 1877–1881. [CrossRef] [PubMed]

3. Shao, Y.H.; Tsai, K.; Kim, S.; Wu, Y.J.; Demissie, K. Exposure to Tomographic Scans and Cancer Risks. *JNCI Cancer Spectr.* **2019**, pkz072. [CrossRef]

4. Chandra, R.; Zhou, H.; Balasingham, I.; Narayanan, R.M. On the opportunities and challenges in microwave medical sensing and imaging. *IEEE Trans. Biomed. Eng.* **2015**, *62*, 1667–1682. [CrossRef] [PubMed]

5. Nikolova, N.K. *Introduction to Microwave Imaging*; Cambridge University Press: Cambridge, UK, 2017.

6. Bolomey, J.C. Crossed viewpoints on microwave-based imaging for medical diagnosis: From genesis to earliest clinical outcomes. In *The World of Applied Electromagnetics*; Springer International Publishing: Cham, Switzerland, 2018; pp. 369–414.

7. Kosmas, P.; Crocco, L. Introduction to Special Issue on Electromagnetic Technologies for Medical Diagnostics: Fundamental Issues, Clinical Applications and Perspectives. *Diagnostics* **2019**, *9*, 19. [CrossRef]

8. Meaney, P.M.; Fanning, M.W.; Li, D.; Poplack, S.P.; Paulsen, K.D. A clinical prototype for active microwave imaging of the breast. *IEEE Trans. Microw. Theory Tech.* **2000**, *48*, 1841–1853.

9. Meaney, P.M.; Fanning, M.W.; Raynolds, T.; Fox, C.J.; Fang, Q.; Kogel, C.A.; Poplack, S.P.; Paulsen, K.D. Initial clinical experience with microwave breast imaging in women with normal mammography. *Acad. Radiol.* **2007**, *14*, 207–218. [CrossRef]

10. Bahramiabarghouei, H.; Porter, E.; Santorelli, A.; Gosselin, B.; Popović, M.; Rusch, L.A. Flexible 16 antenna array for microwave breast cancer detection. *IEEE Trans. Biomed. Eng.* **2015**, *62*, 2516–2525. [CrossRef]

11. Conceição, R.C.; Mohr, J.J.; O'Halloran, M. (Eds.) *An Introduction to Microwave Imaging for Breast Cancer Detection*; Springer International Publishing: Cham, Switzerland, 2016.

12. O'Loughlin, D.; O'Halloran, M.; Moloney, B.M.; Glavin, M.; Jones, E.; Elahi, M.A. Microwave breast imaging: Clinical advances and remaining challenges. *IEEE Trans. Biomed. Eng.* **2018**, *65*, 2580–2590. [CrossRef]

13. Yu, C.; Yuan, M.; Stang, J.; Bresslour, E.; George, R.T.; Ybarra, G.A.; Joines, W.T.; Liu, Q.H. Active microwave imaging II: 3-D system prototype and image reconstruction from experimental data. *IEEE Trans. Microw. Theory Tech.* **2008**, *56*, 991–1000.

14. Porter, E.; Coates, M.; Popović, M. An early clinical study of time-domain microwave radar for breast health monitoring. *IEEE Trans. Biomed. Eng.* **2015**, *63*, 530–539. [CrossRef] [PubMed]

15. Tobon Vasquez, J.A.; Scapaticci, R.; Turvani, G.; Bellizzi, G.; Joachimowicz, N.; Duchêne, B.; Tedeschi, E.; Casu, M.R.; Crocco, L.; Vipiana, F. Design and Experimental Assessment of a 2D Microwave Imaging System for Brain Stroke Monitoring. *Int. J. Antenn. Propag.* **2019**, *2019*, 8065036. [CrossRef]

16. Mobashsher, A.T.; Bialkowski, K.; Abbosh, A.; Crozier, S. Design and experimental evaluation of a non-invasive microwave head imaging system for intracranial haemorrhage detection. *PLoS ONE* **2016**, *11*, e0152351. [CrossRef] [PubMed]

17. Merunka, I.; Massa, A.; Vrba, D.; Fiser, O.; Salucci, M.; Vrba, J. Microwave Tomography System for Methodical Testing of Human Brain Stroke Detection Approaches. *Int. J. Antenn. Propag.* **2019**, *2019*, 4074862. [CrossRef]

18. Persson, M.; Fhager, A.; Trefná, H.D.; Yu, Y.; McKelvey, T.; Pegenius, G.; Karlsson, J.E.; Elam, M. Microwave-based stroke diagnosis making global prehospital thrombolytic treatment possible. *IEEE Trans. Biomed. Eng.* **2014**, *61*, 2806–2817. [CrossRef]

19. Hopfer, M.; Planas, R.; Hamidipour, A.; Henriksson, T.; Semenov, S. Electromagnetic Tomography for Detection, Differentiation, and Monitoring of Brain Stroke: A Virtual Data and Human Head Phantom Study. *IEEE Trans. Biomed. Eng.* **2017**, *59*, 86–97. [CrossRef]

20. Scapaticci, R.; Di Donato, L.; Catapano, I.; Crocco, L. A feasibility study on microwave imaging for brain stroke monitoring. *Prog. Electromagn. Res.* **2012**, *40*, 305–324. [CrossRef]

21. Scapaticci, R.; Tobon, J.; Bellizzi, G.; Vipiana, F.; Crocco, L. Design and numerical characterization of a low-complexity microwave device for brain stroke monitoring. *IEEE Trans. Antennas Propag.* **2018**, *66*, 7328–7338. [CrossRef]

22. Koutsoupidou, M.; Kosmas, P.; Ahsan, S.; Miao, Z.; Sotiriou, I.; Kallos, T. Towards a microwave imaging prototype based on the DBIM-TwIST algorithm and a custom-made transceiver system. In Proceedings of the 2017 International Conference on Electromagnetics in Advanced Applications (ICEAA), Verona, Italy, 11–15 September 2017; pp. 1004–1007.

23. Shea, J.D.; Kosmas, P.; Hagness, S.C.; Van Veen, B.D. Three-dimensional microwave imaging of realistic numerical breast phantoms via a multiple-frequency inverse scattering technique. *Med. Phys.* **2010**, *37*, 4210–4226. [CrossRef]

24. Azghani, M.; Kosmas, P.; Marvasti, F. Microwave medical imaging based on sparsity and an iterative method with adaptive thresholding. *IEEE Trans. Med. Imaging* **2014**, *34*, 357–365. [CrossRef]

25. Gabriel, S.; Lau, R.; Gabriel, C. The dielectric properties of biological tissues: II. Measurements in the frequency range 10 Hz to 20 GHz. *Phys. Med. Biol.* **1996**, *41*, 2251. [CrossRef] [PubMed]

26. Gabriel, S.; Lau, R.; Gabriel, C. The dielectric properties of biological tissues: III. Parametric models for the dielectric spectrum of tissues. *Phys. Med. Biol.* **1996**, *41*, 2271. [CrossRef] [PubMed]

27. Peyman, A.; Holden, S.; Watts, S.; Perrott, R.; Gabriel, C. Dielectric properties of porcine cerebrospinal tissues at microwave frequencies: in vivo, in vitro and systematic variation with age. *Phys. Med. Biol.* **2007**, *52*, 2229. [CrossRef] [PubMed]

28. Schmid, G.; Neubauer, G.; Mazal, P.R. Dielectric properties of human brain tissue measured less than 10 h postmortem at frequencies from 800 to 2450 MHz. *Bioelectromagnetics* **2003**, *24*, 423–430. [CrossRef]

29. Semenov, S.; Huynh, T.; Williams, T.; Nicholson, B.; Vasilenko, A. Dielectric properties of brain tissue at 1 GHz in acute ischemic stroke: Experimental study on swine. *Bioelectromagnetics* **2017**, *38*, 158–163. [CrossRef]

30. Zhurbenko, V. Challenges in the design of microwave imaging systems for breast cancer detection. *Adv. Electr. Comp. Eng.* **2011**, *11*, 91–96. [CrossRef]

31. Bindu, G.N.; Abraham, S.J.; Lonappan, A.; Thomas, V.; Aanandan, C.K.; Mathew, K. Active microwave imaging for breast cancer detection. *Prog. Electromagn. Res.* **2006**, *58*, 149–169. [CrossRef]

32. Miao, Z.; Kosmas, P. Multiple-frequency DBIM-TwIST algorithm for microwave breast imaging. *IEEE Trans. Antennas Propag.* **2017**, *65*, 2507–2516. [CrossRef]

33. Miao, Z.; Kosmas, P. Microwave breast imaging based on an optimized two-step iterative shrinkage/thresholding method. In Proceedings of the 2015 9th European Conference on Antennas and Propagation (EuCAP), Lisbon, Portugal, 13–17 April 2015; pp. 1–4.

34. Ahsan, S.; Guo, Z.; Miao, Z.; Sotiriou, I.; Koutsoupidou, M.; Kallos, E.; Palikaras, G.; Kosmas, P. Design and experimental validation of a multiple-frequency microwave tomography system employing the dbim-twist algorithm. *Sensors* **2018**, *18*, 3491. [CrossRef]

35. Miao, Z.; Kosmas, P.; Ahsan, S. Impact of information loss on reconstruction quality in microwave tomography for medical imaging. *Diagnostics* **2018**, *8*, 52. [CrossRef]

36. Guo, Z.; Ahsan, S.; Karadima, O.; Sotiriou, I.; Kosmas, P. Resolution Capabilities of the DBIM-TwIST Algorithm in Microwave Imaging. In Proceedings of the 2019 13th European Conference on Antennas and Propagation (EuCAP), Krakow, Poland, 31 March–5 April 2019; pp. 1–4.

37. McDermott, B.; O'Halloran, M.; Porter, E.; Santorelli, A.; Morris, L.; Divilly, B.; McGinley, B.; Jones, M. Anatomically and dielectrically realistic microwave head phantom with circulation and reconfigurable lesions. *Prog. Electromagn. Res. B* **2017**, *78*, 47–60. [CrossRef]

38. Joachimowicz, N.; Duchêne, B.; Conessa, C.; Meyer, O. Anthropomorphic breast and head phantoms for microwave imaging. *Diagnostics* **2018**, *8*, 85. [CrossRef] [PubMed]

39. Rydholm, T.; Fhager, A.; Persson, M.; Geimer, S.D.; Meaney, P.M. Effects of the plastic of the realistic GeePS-L2S-breast phantom. *Diagnostics* **2018**, *8*, 61. [CrossRef] [PubMed]

40. Lazebnik, M.; Madsen, E.L.; Frank, G.R.; Hagness, S.C. Tissue-mimicking phantom materials for narrowband and ultrawideband microwave applications. *Phys. Med. Biol.* **2005**, *50*, 4245. [CrossRef]

41. Bucci, O.; Isernia, T. Electromagnetic inverse scattering: Retrievable information and measurement strategies. *Radio Sci.* **1997**, *32*, 2123–2137. [CrossRef]

42. Guo, W.; Ahsan, S.; He, M.; Koutsoupidou, M.; Kosmas, P. Printed Monopole Antenna Designs for a Microwave Head Scanner. In Proceedings of the 2018 18th Mediterranean Microwave Symposium (MMS), Istanbul, Turkey, 31 October–2 November 2018; pp. 384–386.

43. Bioucas-Dias, J.M.; Figueiredo, M.A. A new TwIST: Two-step iterative shrinkage/thresholding algorithms for image restoration. *IEEE Trans. Image Process.* **2007**, *16*, 2992–3004. [CrossRef]

44. Semenov, S.Y.; Corfield, D.R. Microwave tomography for brain imaging: Feasibility assessment for stroke detection. *Int. J. Antenn. Propag.* **2008**, *2008*, 254830. [CrossRef]

45. Coli, V.L.; Tournier, P.H.; Dolean-Maini, V.; El Kanfoud, I.; Pichot, C.; Migliaccio, C.; Blanc-Féraud, L. Detection of Simulated Brain Strokes Using Microwave Tomography. *IEEE J. Electromagn. RF Microw. Med.* **2019**, *3*, 254–260. [CrossRef]

46. Merunka, I.; Vrba, D.; Fiser, O.; Cumana, J.; Vrba, J. 2D Microwave System for Testing of Brain Stroke Imaging Algorithms. In Proceedings of the 2019 European Microwave Conference in Central Europe (EuMCE), Prague, Czech Republic, 13–15 May 2019; pp. 508–511.

47. Fassbender, K.; Balucani, C.; Walter, S.; Levine, S.R.; Haass, A.; Grotta, J. Streamlining of prehospital stroke management: the golden hour. *Lancet. Neurol.* **2013**, *12*, 585–596. [CrossRef]

48. Mohammed, B.J.; Abbosh, A.M. Realistic head phantom to test microwave systems for brain imaging. *Microw. Opt. Technol. Lett.* **2014**, *56*, 979–982. [CrossRef]

Article

On the Orbital Angular Momentum Incident Fields in Linearized Microwave Imaging

Santi Concetto Pavone, Gino Sorbello and Loreto Di Donato *

Department of Electrical, Electronics and Computer Engineering (DIEEI), University of Catania, viale A. Doria 6, 95125 Catania, Italy; santi.pavone@unict.it (S.C.P.); gino.sorbello@unict.it (G.S.)
* Correspondence: loreto.didonato@unict.it

Received: 15 February 2020; Accepted: 25 March 2020; Published: 30 March 2020

Abstract: Orbital angular momentum (OAM) is gaining great attention in the physics and electromagnetic community owing to an intriguing debate concerning its suitability for widening channel capacity in next-generation wireless communications. While such a debate is still a matter of controversy, we exploit OAM generation for microwave imaging within the classical first order linearized models, i.e., Born and Rytov approximation. Physical insights into different fields carrying ℓ-order OAM are conveniently exploited to propose possible alternative imaging approaches and paradigms in microwave imaging.

Keywords: linearized inverse scattering; microwave imaging; orbital angular momentum; born approximation; rytov approximation

---

1. Introduction

Microwave imaging (MWI) deserves great attention in the electrical engineering community due to its potential applications as disparate as subsurface and planetary exploration (ground penetrating radar), non-destructive testing (NDT), biomedical imaging, and so on and so forth [1–3]. However, while huge efforts have been addressed in recent years towards the development of experimental equipment, which jointly exploits the availability for relatively low cost microwave devices and the ever increasing computing power of modern CPUs [4,5], the underlying inverse scattering problem (ISP) solution still requires great efforts from the methodological and modeling point of view. This is related to two main features of ISPs: non-linearity and ill-posedness. Indeed, if both of them are not properly faced, MWI cannot be used in any practical instance. To face them, a priori information can be exploited in many different ways to obtain reliable inversion or optimization strategies equipped with effective regularization schemes [6–8].

Generally speaking, microwave imaging approaches can be grouped into two main classes: quantitative and qualitative approaches. The first one aims to retrieve both electromagnetic and geometric features, whereas the second one allows only shape characterization of an unknown scattering system in a surveyed region. In general, quantitative methods require the solution of a non-linear problem that inherently results in non-trivial issues, such as local minima [9,10] and regularization of non-convex problems. On the other hand, qualitative methods make a trade-off between the difficulty of solving a non-linear problem and the limited amount of information to be conveyed back from scattered field data. Obviously, a plethora of hybrid methods coexist in the literature, wherein several stepwise optimization strategies, regularization approaches, and approximate models have been proposed to face the inverse scattering problem. Very recently, also machine learning (ML) and deep learning (DL) have been applied to inverse scattering problems [11,12].

Besides the above, with reference to the present paper, it is worth mentioning a recently proposed paradigm for solving inverse scattering problems. Such a paradigm stems from the design of new scattering experiments to recast the original ones into new "virtual" experiments by means of a suitable "design equation". This latter entails a simple linear recombination of scattered field data, which, owing to the linearity of scattering experiments, enforces the total electric field to be tailored with a specific distribution within the imaging domain. Since these scattering experiments are designed without additional measurements (i.e., only via software processing), they have been named virtual experiments (VE) [13–17]. All these approaches have shown an enhanced capability in imaging scatterers not belonging to the weak scattering regime, while facing the problem in a simpler, or more effective/efficient way, than many other imaging approaches currently available in the state-of-the-art.

In this view, other papers have investigated the use of different kinds of probing fields in microwave imaging. For example, in [18], orbital angular momentum (OAM) incident fields of order higher than $\ell = 0$, generated through a planar array, were exploited to perform 3D imaging with an observed improvement in the resolution beyond the Rayleigh limit. In [19], sub-wavelength focused near-field (NF) beams and Bessel beams [20–22] were proposed to perform imaging in scenarios when possible undesired scatterers are present, and in [23], OAM incident fields were used for accurate recovery of sparse objects through mask-constrained sparse reconstruction.

In order to investigate the capability of OAM antennas in microwave tomographic imaging, possibly exploiting additional degrees of freedom carried by the topological charges of such fields, in this paper, we consider the use of OAM incident fields generated by properly feeding a cylindrical array of filamentary currents. Specifically, the "view diversity" conventionally exploited in the scattering experiments is traded with the "mode diversity" carried by OAM incident fields of different order. In doing this, we adopt linearized imaging procedures valid under weak scattering regimes both for small and large scatterers, namely Born and Rytov approximations. Although the problem is strongly simplified under this assumption, the study of linearized inverse scattering problems allows understanding, often by analytical findings, the role of fundamental parameters in the reconstruction capabilities of an imaging method, under adopted measurement configurations, such as the frequency, the number and configuration of probes, etc. Moreover, linearized approaches can be practically useful also when the working hypotheses underlying the weak scattering regime are not fully satisfied, that is when one wants to pursue only a qualitative characterization through microwave imaging.

The paper is structured as follows. In Section 2, the mathematical formulation of the scattering problem is given with respect to the scalar transverse magnetic (TM) 2D case. In Section 3, the linearized imaging procedure through OAM probing fields and Born and Rytov approximations are introduced. In Section 4, the proposed imaging strategy is validated against numerical examples. The conclusions end the paper.

2. Mathematical Formulation of the Inverse Scattering Problem

We consider the two-dimensional inverse scattering problem dealing with the TM polarization (scalar formulation) wherein nonmagnetic scatterers ($\mu_s = \mu_0$) are embedded into a homogeneous background medium. The location and the electromagnetic properties inside the domain are unknowns, and the vector $\underline{r}(x,y)$ denotes the position inside the investigation domain Ω. Time-harmonic dependence $e^{j\omega t}$, with angular frequency $\omega = 2\pi f$, is assumed dropped, and bold text notation for the electric fields is used hereafter.

To reconstruct the geometrical and dielectric properties of the scatterers, the investigated domain is probed with a set of incident fields $\mathbf{E}_i(\underline{r}_v, \underline{r}) = \mathbf{E}_i^v(\underline{r})$, where $\underline{r}_v$ denotes the position of the primary sources (filamentary currents) placed outside the investigated area and v the indexing of the source. The interaction between the incident waves and the scatterers gives rise to a secondary field that is measured by receivers located at $\underline{r}_m \in \Gamma$, still outside the investigation domain. A sketch of the adopted measurement configuration is reported in Figure 1.

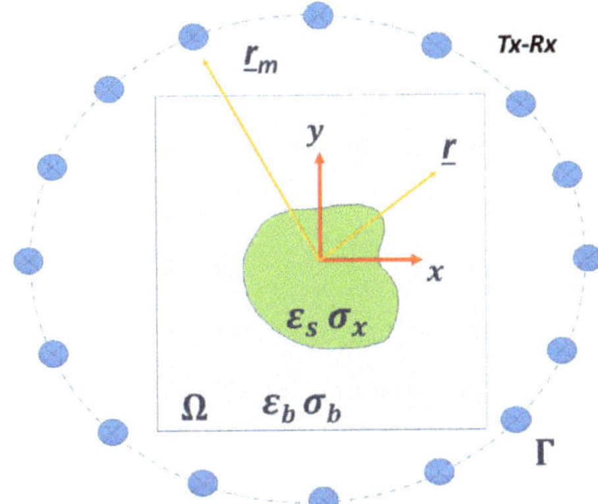

Figure 1. A sketch of the adopted multiview-multistatic measurement configuration to probe the region of interest by means of Tx-Rx primary sources (filamentary currents) placed on a circumference Γ of radius r_m.

As is well known, the total field $E_t^v(\underline{r})$ and the incident field $E_i^v(\underline{r})$ must satisfy the following Helmholtz Equations:

$$[\nabla^2 + k^2(\underline{r})]E_t^v(\underline{r}) = 0 \tag{1}$$

$$[\nabla^2 + k_b^2]E_i^v(\underline{r}) = 0 \tag{2}$$

where $k(\underline{r})$ is the wavenumber in Ω and $k_b = \omega\sqrt{\varepsilon_0\varepsilon_b\mu_0}$ is the wavenumber of the homogeneous embedding background medium having complex permittivity $\varepsilon_b = \varepsilon_b' - j\frac{\sigma_b}{\omega\varepsilon_0}$. On the other hand, the scattered field, defined as $E_s^v(\underline{r}) = E_t^v(\underline{r}) - E_i^v(\underline{r})$, satisfies the following Helmholtz Equation:

$$[\nabla^2 + k_b^2(\underline{r})]E_s^v(\underline{r}) = -k_b^2\chi(\underline{r})E_t^v(\underline{r}) \tag{3}$$

where the contrast function χ, which relates, at a given frequency ω, the properties of the unknown anomalies to those of the background medium, is defined as:

$$\chi(\underline{r}) = \frac{\varepsilon_s(\underline{r})}{\varepsilon_b} - 1 \tag{4}$$

with $\varepsilon_s = \varepsilon_s' - j\frac{\sigma_s}{\omega\varepsilon_0}$ the complex dielectric permittivity of the scatterer. By means of the vector potential theory, the equations governing the scattering phenomenon can be conveniently expressed through a couple of integral Equations:

$$E_s^v(\underline{r}_m) = k_b^2 \iint_\Omega g(\underline{r}_m,\underline{r}')\chi(\underline{r}')E_t^v(\underline{r}')d\underline{r}' \qquad \underline{r}_m \in \Gamma, \quad v = 1,...,V \tag{5}$$

$$E_t^v(\underline{r}) - E_i^v(\underline{r}) = k_b^2 \iint_\Omega g(\underline{r},\underline{r}')\chi(\underline{r}')E_t^v(\underline{r}')d\underline{r}' \qquad \underline{r} \in \Omega, \quad v = 1,...,V \tag{6}$$

where $g(\underline{r},\underline{r}') = -\frac{j}{4}H_0^{(2)}(k_b|\underline{r} - \underline{r}'|)$ is the scalar Green function of the homogeneous background, in which $\underline{r}'$ and $\underline{r}$ denote the generic source point in Ω and the observation point in Γ or Ω, respectively. Finally, $H_0^{(2)}(\cdot)$ is the Hankel function of zero order and second kind. The Green function is the kernel of the radiation operators $\mathcal{A}_e[\cdot] : L^2(\Omega) \rightarrow L^2(\Gamma)$ and $\mathcal{A}_i[\cdot] : L^2(\Omega) \rightarrow L^2(\Omega)$, which relate the induced contrast source $J^v(\underline{r}) = \chi(\underline{r})E_t^v(\underline{r})$ to the field scattered in Γ and in Ω, respectively.

Equations (5) and (6) are first and second kind Fredholm equations and are also known as data and state equations. According to the above, the inverse scattering problem is cast as the retrieval of the unknown contrast χ ($\underline{r} \in \Omega$) from measured scattered field $E_s^v(\underline{r}_m \in \Gamma)$ and known incident fields $E_i^v(\underline{r} \in \Omega)$.

3. Linear Imaging with OAM Incident Fields

As stated above, the solution of the problems Equations (5) and (6) entails facing a non-linear problem, since the total electric field has to be also retrieved for each transmitting antenna. To overcome this drawback, the first order linearized problem has been proposed in the past both for penetrable and impenetrable media. In particular, the Born and Rytov approximations allow tackling imaging of small and large weak scattering systems, provided that the deviation of the dielectric properties of anomalies, with respect to those of the background medium, keeps very small [24].

The linearized approach under Born approximation entails the solution of the following integral Equation:

$$E_s^v(\underline{r}_m) \quad = \quad k_b^2 \iint_\Omega g(\underline{r}_m, \underline{r}')\chi(\underline{r}')E_{inc}^v(\underline{r}')d\underline{r}' \qquad \underline{r}_m \in \Gamma, \quad v = 1, ..., V \tag{7}$$

where the total field is substituted by the incident one, neglecting the effect of the scattering system on the total field. Similar considerations can be applied for the Rytov approximation, where the Equation to be solved turns out to be:

$$\Phi_s^v(\underline{r}_m) \quad = \quad \frac{k_b^2}{E_{inc}^v(\underline{r}_m)} \iint_\Omega g(\underline{r}_m, \underline{r}')E_{inc}^v(\underline{r}')\chi(\underline{r}')d\underline{r}' \qquad \underline{r}_m \in \Gamma, \quad v = 1, ..., V \tag{8}$$

with $\Phi_s^v(r_m)$ the complex scattered phase [24], and $E_{inc}^v(\underline{r}_m)$ the value of the incident fields at the measurement points. In both Equations (7) and (8), the incident field is commonly given by a sequential illumination of single filamentary currents placed in the near- or far-field of the imaging domain. The scattered field is collected by all the other antennas working as receivers when only one is acting as a transmitter. This is the standard scattering experiment procedure in a tomographic microwave imaging system. Hereafter, we refer to such a scheme of data gathering as "sequential illumination".

On the other hand, we want to exploit OAM incident fields properly generated by a progressive phase change in circular array elements, that is:

$$E_{inc}^\ell(\underline{r}) = -\frac{j}{4} \sum_{v=0}^{V-1} H_0^{(2)}(k_b|\underline{r}_v - \underline{r}|)e^{j\ell\varphi_v}, \quad \varphi_v = \frac{2v\pi}{V}, \quad \ell = 0, ..., \ell_{max} \tag{9}$$

According to Equation (9), the investigation domain is illuminated through a set of different ℓ-order OAM fields rather than each single filamentary current placed at different angular positions. In a first approximation, without prior information on the scattering system, the maximum OAM order is related to the electrical dimension of the investigation domain, being $|\ell| \simeq \beta a$, with $\beta = \mathcal{Re}[k_b]$ and a the radius of the minimum convex hull enclosing the investigation domain. Incident fields arising from Equation (9) for three different orders are shown in Figure 2.

The data Equation we are going to consider hereafter is formally the same of Equations (7) and (8), with the corresponding scattered field collected under simultaneous illumination given by the incident fields Equation (9). In such a way, the role of the vth illuminations is exchanged with the role of the ℓth OAM order used to probe the investigation domain Ω. Therefore, as commonly done, the linearized tomography problem can still be solved in a regularized fashion. In this respect, we exploit the standard truncated singular value decomposition (TSVD) method [13], wherein the number N_T of the relevant singular values to be used in the reconstruction formula is simply chosen by the cutoff of the singular values below the threshold of 20 dB lower than the maximum one.

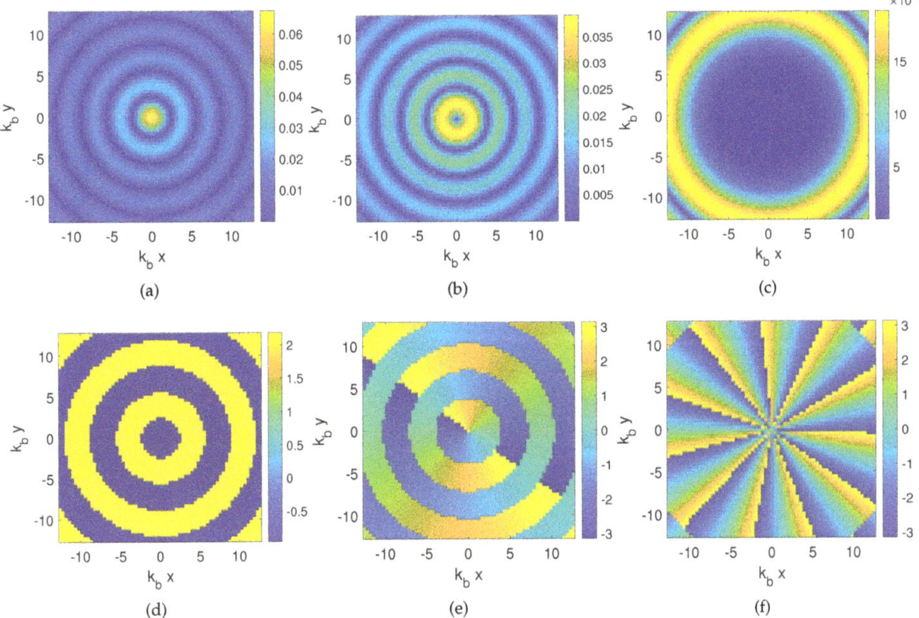

Figure 2. Amplitude (not normalized) and phase distribution of the OAM incident fields used to probe the imaging domain with a circular array of V = 36 filamentary currents. (a),(d) $\ell = 0$; (b),(e) $\ell = 1$; (c),(f) $\ell = 11$.

4. Numerical Benchmarks

In order to investigate the opportunity to perform microwave imaging by means of OAM incident fields, we perform some proof-of-concept examples, under the weak scattering regime underlying both the Born and Rytov approximations.

The first example is concerned with a circular scatterer with radius $0.3\lambda_b$ and relative permittivity $\varepsilon_s = 1.2 - j0.06$, located at $r = (0,0)$ in an investigation domain of $4\lambda_b \times 4\lambda_b$. The background medium is the vacuum, and $V = 2\beta a + 1 = 37$ antennas are used to illuminate the scenario [25] at a distance $r_v = 40\lambda_b$. Accordingly, the same number M of measurement points, at the same distance ($r_m = r_v$), are used to probe the scattered field, as commonly done in any experimental microwave imaging apparatus. Using Equation (9), we generate OAM incident fields up to the order $|\pm \ell_{max}| = 18$. Accordingly, the forward problem is solved by means of a method-of-moments (MoM) based solver by properly discretizing the investigation domain into $N_c = 65 \times 65$ cells, according to the Richmond rule [26]. Finally, the scattered field data are corrupted with an additive white Gaussian noise (AWGN) of SNR=20dB. The reconstruction results are evaluated by the standard metric based on the least squares mean error, which is defined as $err = \frac{||\chi_{act} - \chi_{rec}||^2}{||\chi_{act}||^2}$, wherein χ_{act} and χ_{rec} stem for the actual and the reconstructed contrast profile, respectively.

The reconstruction results using orders $\ell = 0, \pm[1 - 18]$ are shown in Figure 3b–f. As can be seen, the method is able to retrieve the contrast correctly both in its real and imaginary part. After that, we consider only the first four lowest order ($\ell = 0, \pm[1 - 3]$) in solving the Born equation, and as shown in Figure 3c,g, the result is still good in terms of the reconstruction capability, since the reconstruction error is slightly larger than the previous one and mainly related to background reconstruction artifacts. On the other hand, it is worth noting that the dimension of the scattering matrix operator is $N_c \times (7 \times M)$ for $\ell = [-3, +3]$ and $N_c \times (37 \times M)$ for $\ell = [-18, +18]$, with $N_c = (65)^2$ in both cases. If we take into account that also for the multiview-multistatic configuration, the matrix has dimension $N_c \times (V \times M)$

($V = M = 37$) too, the computational advantage in the SVD numerical evaluation is not negligible in the case of few OAM modes. For the sake of completeness, the reconstruction performed with the standard multiview-multistatic field data acquisition is shown Figure 3d–h, and as expected, it is fully comparable with those achieved by means of the OAM incident fields.

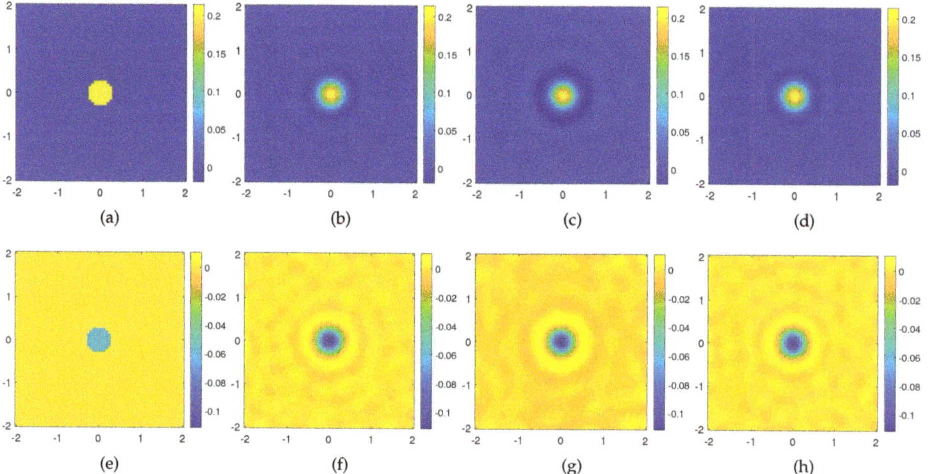

Figure 3. Reconstruction of a small weak circular scatterer through the Born approximation. (**a**) Real and (**e**) imaginary part of the actual contrast profile; (**b**) real and (**f**) imaginary part of the retrieved contrast profile for $\ell = 0, \pm[1 - 18]$, $err = 0.2028$ with a cutoff value in the TSVD equal to $N_T = 252$; (**c**) real and (**g**) imaginary part of the retrieved contrast profile for $\ell = 0, \pm[1 - 3]$, $err = 0.2384$ with a cutoff value in the TSVD equal to N_T=142; (**d**) real and (**h**) imaginary part of the retrieved contrast profile using $V = M = 37$ equispaced filamentary currents $err = 0.2020$ with a cutoff value in the TSVD equal to $N_T = 252$. The axes of the imaging domain are expressed in background wavelengths.

In the second example, we consider a scattering system made of two off-centered targets embedded in the same investigation domain described for the previous example. The targets are shaped as a circle and a square scatterers, with permittivity $\varepsilon_s = 1.2 - j0.06$ and leading dimension $0.6\lambda_b$; see Figure 4a,e. We looks for the solution of the Born equation in three different cases. In the first case, we exploit all the OAM incident fields ($\ell = 0, \pm[1 - 18]$) needed to probe the entire domain, whereas in the second case, only the lowest order OAM incident fields ($\ell = 0, \pm[1 - 4]$), and finally, only the highest order ones ($\ell = \pm[6 - 11]$). As can be seen in Figure 4, the reconstruction accounts for the whole scattering system in the first case (see Figure 4b,f), only the innermost scatterer in the second case (see Figure 4c,g), and only the square target in the third case (see Figure 4d,h). From these results, it is evident as the topological properties of the different OAM orders used to probe the domain are able to image targets whose support is mainly illuminated by the OAM rings (cores) of given order.

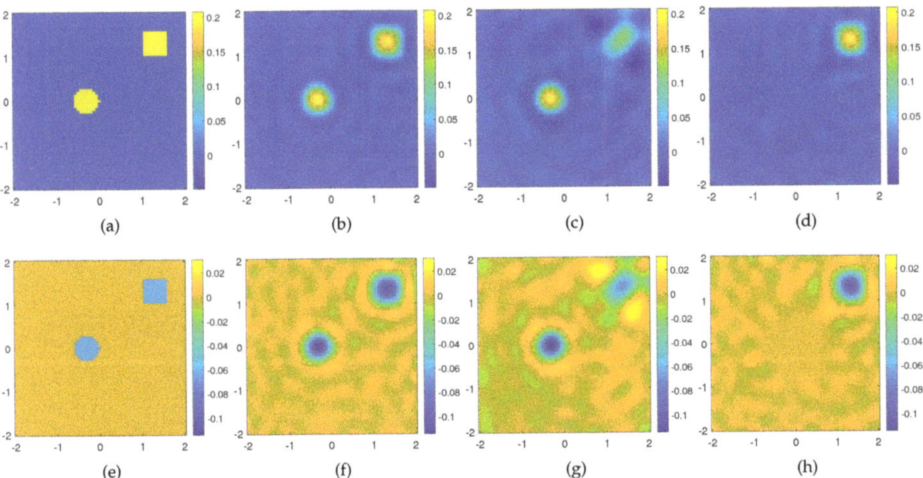

Figure 4. Reconstruction of a two small weak scatterers through the Born approximation. (**a**) Real and (**e**) imaginary part of the actual contrast profile; (**b**) real and (**f**) imaginary part of the recovered contrast profile for $\ell = 0, \pm[1 - 18]$, *err* $= 0.2241$ with a cutoff value in the TSVD equal to $N_T = 252$; (**c**) real and (**g**) imaginary part of the recovered contrast profile for $\ell = 0, \pm[1 - 4]$, *err* $= 0.4017$ with a cutoff value in the TSVD equal to $N_T = 166$; (**d**) real and (**h**) imaginary part of the recovered contrast profile for $\ell = \pm[6 - 11]$ (without the lowest order modes), *err* $= 0.60$ with a cutoff value in the TSVD equal to $N_T = 187$. The axes of the imaging domain are expressed in background wavelengths.

Finally, the third example is concerned with a large lossless elliptically shaped target with semi-axes $5\lambda_b$ and $4\lambda_b$, respectively, embedded in an imaging domain of $12\lambda_b \times 12\lambda_b$ discretized into $N_c = 129 \times 129$ cells. The background is the vacuum, and the permittivity of the target is changed 2% with respect to the permittivity of the vacuum, while $\mathcal{I}m\{\varepsilon_s\} = -0.006$. For such kinds of objects, the Rytov approximation holds true. According to the "rule of thumb" suggested by the electromagnetic field degrees of freedom [25], the domain needs to be probed by means of $V = 107$ equispaced filamentary currents placed in the far-field of the imaging domain ($r_m = 120\lambda_b$), and the corresponding scattered fields are collected through $M = 107$ measurement points. We consider the solution of the forward problem with a set of OAM incident fields up to $|\pm \ell_{max}| = 30$ and solve the underlying inverse problem by means of the Rytov Equation (8). In this case also, the scattered field is corrupted with AWGN of SNR = 20 dB. The reconstruction results are shown in Figure 5b–e and allow appraising the dielectric features of the target. It is worth noting that, if the SVD of the multiview-multistatic scattering matrix, of dimensions $(N_c) \times (M \times V)$, is computed via MATLAB on a standard CPU Intel Core i7 8GB RAM, it results in an "out-of-memory" warning. Instead, by using the proposed OAM based inversion, a meaningful result is found without the need for higher performance computers. Furthermore, for this example, we consider also reconstruction by processing the scattered field gathered in the near-field (though non-reactive zone) of the imaging domain by setting the distance of the Tx-Rx probes at $r_m = 9\lambda$, namely the minimum circle enclosing the surveyed area. As we can appraise from Figure 5c–f, the reconstruction results are fully comparable as the reconstruction errors attain the same values in both considered cases.

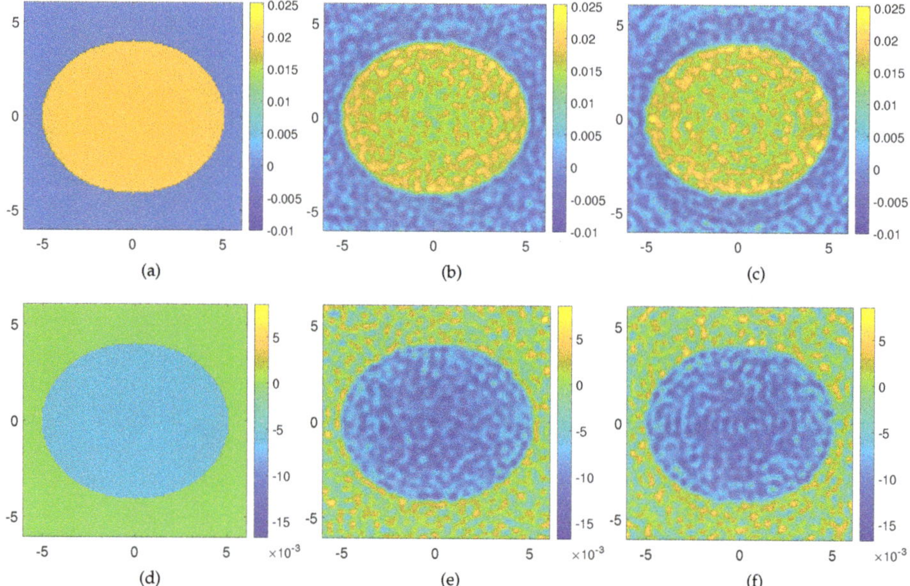

Figure 5. Reconstruction of a large circular scatterer through the Rytov approximation exploiting OAM fields generated in the far-field ($\underline{r}_m = 100$) m and near-field ($\underline{r}_m = 9$) m of the imaging domain. The OAM orders used are $\ell = 0, \pm[1 - 35]$. (**a**) Real and (**d**) imaginary part of the actual contrast profile; (**b**) real and (**e**) imaginary part of the retrieved contrast profile in the far-field measurement configuration, reconstruction $err = 0.1108$ with a cutoff value in the TSVD reconstruction $N_T = 1856$; (**c**) real and (**f**) imaginary part of the retrieved contrast profile in the near-field measurement configuration, reconstruction $err = 0.1267$ with a cutoff value in the TSVD reconstruction $N_T = 1557$. The axes of the imaging domain are expressed in background wavelengths.

5. Conclusions

The use of OAM incident fields in linearized diffraction tomography has been investigated and analyzed in this paper. The main conclusions are concerned with the possibility to adopt alternative measurement setup that give additional flexibility in performing tomographic imaging.

As an example, this is the case when the investigated domain contains undesired scatterers that should be neglected in the reconstruction process, such as, for example, in those approaches wherein part of the electromagnetic features of the scene is known, such as non-destructive testing (NDT) for the detection of faults in an "undesired" background. Indeed, this can be done via hardware, without resorting to computationally heavy differential (or distorted) imaging procedures, wherein the knowledge of the Green function is also required. Another possible context of interest is in subsurface imaging where the goal may be to address the inversion strategy at a given depth and/or location, on the base of a priori available information.

On the other hand, in all those applications concerned with the detection of small scatterers, OAM probing fields can allow to significantly reduce gathering time and computational burden, as only few modes have to be exploited to achieve satisfactory reconstructions. Indeed, the size of the multiorder-multistatic scattering operator is often smaller than the multiview-mutistatic counterpart, depending on the dimension of the scattering system, and hence on the $|\ell_{max}|$ order used to probe the scenario. This may be of particular interest for the development of fast 3D tomographic approaches with reduced computational burden.

As main drawback, OAMs imply a complication of the feeding network (phase shifters and possibly amplifiers) that can be traded, on the other hand, with a lower acquisition time through digital beamforming networks that avoid sequentially transmitting antennas and switching circuitry. Last, but not least, it is worth noticing that, when no a priori information is available about the scattering system, it is even possible to scan the OAM cores along some directions over the investigation domain, according to the well known phased array theory.

Author Contributions: Conceptualization and methodology, S.C.P., G.S., and L.D.D.; numerical validation S.C.P. and L.D.D.; writing, original draft preparation, L.D.D.; writing, review and editing, S.C.P., G.S., and L.D.D. All authors have read and agreed to the published version of the manuscript.

Funding: This research received no external funding.

Conflicts of Interest: The authors declare no conflict of interest.

References

1. Pastorino, M. *Microwave Imaging*; John Wiley & Sons: Hoboken, NJ, USA, 2010.
2. Daniels, D.J. Ground penetrating radar. In *Encyclopedia of RF and Microwave Engineering*; Institute of Electrical Engineers: London, UK, 2005.
3. Amin, M.G. *Through-The-Wall Radar Imaging*; CRC Press: Boca Raton, FL, USA, 2017.
4. Salucci, M.; Poli, L.; Oliveri, G. Full-Vectorial 3D Microwave Imaging of Sparse Scatterers through a Multi-Task Bayesian Compressive Sensing Approach. *J. Imaging* **2019**, *5*, 19. [CrossRef]
5. Fedeli, A.; Maffongelli, M.; Monleone, R.; Pagnamenta, C.; Pastorino, M.; Poretti, S.; Randazzo, A.; Salvadè, A. A Tomograph Prototype for Quantitative Microwave Imaging: Preliminary Experimental Results. *J. Imaging* **2018**, *4*, 139. [CrossRef]
6. Golnabi, A.H.; Meaney, P.M.; Geimer, S.D.; Paulsen, K.D. 3-D microwave tomography using the soft prior regularization technique: Evaluation in anatomically realistic MRI-derived numerical breast phantoms. *IEEE Trans. Biomed. Eng.* **2019**, *66*, 2566–2575. [CrossRef] [PubMed]
7. Bayat, N.; Mojabi, P. Incorporating Spatial Priors in Microwave Imaging via Multiplicative Regularization. *IEEE Trans. Antennas Propag.* **2019**, *68*, 1107–1118. [CrossRef]
8. Bevacqua, M.T.; Scapaticci, R.; Bellizzi, G.G.; Isernia, T.; Crocco, L. Permittivity and Conductivity Estimation of Biological Scenarios via 3D Microwave Tomography. In Proceedings of the 13th European Conference on Antennas and Propagation (EuCAP), Kraków, Poland, 31 March–4 April 2019; pp. 1–3.
9. Isernia, T.; Pascazio, V.; Pierri, R. A nonlinear estimation method in tomographic imaging. *IEEE Trans. Geosci. Remote Sens.* **1997**, *35*, 910–923. [CrossRef]
10. Isernia, T.; Pascazio, V.; Pierri, R. On the local minima in a tomographic imaging technique. *IEEE Trans. Geosci. Remote Sens.* **2001**, *39*, 1596–1607. [CrossRef]
11. Wei, Z.; Chen, X. Deep-learning schemes for full-wave nonlinear inverse scattering problems. *IEEE Trans. Geosci. Remote. Sens.* **2018**, *57*, 1849–1860. [CrossRef]
12. Li, L.; Wang, L.G.; Teixeira, F.L.; Liu, C.; Nehorai, A.; Cui, T.J. DeepNIS: Deep neural network for nonlinear electromagnetic inverse scattering. *IEEE Trans. Antennas Propag.* **2018**, *67*, 1819–1825. [CrossRef]
13. Crocco, L.; Catapano, I.; Di Donato, L.; Isernia, T. The Linear Sampling Method as a way for quantitative inverse scattering. *IEEE Trans. Antennas Propag.* **2012**, *4*, 1844–1853. [CrossRef]
14. Crocco, L.; Di Donato, L.; Catapano, I.; Isernia, T. The factorization method for virtual experiments based quantitative inverse scattering. *Prog. Electromagn. Res.* **2016**, *157*, 121–131. [CrossRef]
15. Bevacqua, M.; Crocco, L.; Di Donato, L.; Isernia, T. An Algebraic Solution Method for Nonlinear Inverse Scattering. *IEEE Trans. Antennas Propag.* **2015**, *63*, 601–610. [CrossRef]
16. Di Donato, L.; Bevacqua, M.T.; Crocco, L.; Isernia, T. Inverse scattering via virtual experiments and contrast source regularization. *IEEE Trans. Antennas Propag.* **2015**, *63*, 1669–1677. [CrossRef]
17. Bevacqua, M.; Crocco, L.; Donato, L.D.; Isernia, T.; Palmeri, R. Exploiting sparsity and field conditioning in subsurface microwave imaging of nonweak buried targets. *Radio Sci.* **2016**, *51*, 301–310. [CrossRef]
18. Li, L.; Li, F. Beating the Rayleigh limit: Orbital-angular-momentum-based super-resolution diffraction tomography. *Phys. Rev.* **2013**, *88*, 033205. [CrossRef] [PubMed]

Sensors **2020**, *20*, 1905

19. Bayat, N.; Mojabi, P. On the Use of Focused Incident Near-Field Beams in Microwave Imaging. *Sensors* **2018**, *18*, 3127. [CrossRef] [PubMed]

20. Pavone, S.C.; Mazzinghi, A.; Freni, A.; Albani, M. Comparison between broadband Bessel beam launchers based on either Bessel or Hankel aperture distribution for millimeter wave short pulse generation. *Opt. Express* **2017**, *25*, 19548–19560. [CrossRef]

21. Comite, D.; Fuscaldo, W.; Pavone, S.; Valerio, G.; Ettorre, M.; Albani, M.; Galli, A. Propagation of nondiffracting pulses carrying orbital angular momentum at microwave frequencies. *Appl. Phys. Lett.* **2017**, *110*, 114102. [CrossRef]

22. Pavone, S.C.; Martini, E.; Albani, M.; Maci, S.; Renard, C.; Chazelas, J. A novel approach to low profile scanning antenna design using reconfigurable metasurfaces. In Proceedings of the International Radar Conference, Lille, France, 13–17 October 2014; pp. 1–4.

23. Fan, Q.; Yin, C.; Liu, H. Accurate Recovery of Sparse Objects With Perfect Mask Based on Joint Sparse Reconstruction. *IEEE Access* **2019**, *7*, 73504–73515. [CrossRef]

24. Slaney, M.; Kak, A.C.; Larsen, L.E. Limitations of imaging with first-order diffraction tomography. *IEEE Trans. Microw. Theory Tech.* **1984**, *32*, 860–874. [CrossRef]

25. Bucci, O.M.; Isernia, T. Electromagnetic inverse scattering: Retrievable information and measurement strategies. *Radio Sci.* **1997**, *32*, 2123–2138. [CrossRef]

26. Richmond, J. Scattering by a dielectric cylinder of arbitrary cross section shape. *IEEE Trans. Antennas Propag.* **1965**, *13*, 334–341. [CrossRef]

Article

An Imaging Plane Calibration Method for MIMO Radar Imaging

Yuanyue Guo *, Bo Yuan, Zhaohui Wang and Rui Xia

Key Laboratory of Electromagnetic Space Information, Chinese Academy of Sciences,
University of Science and Technology of China, Hefei 230026, China; yuanb@mail.ustc.edu.cn (B.Y.);
wzh01@mail.ustc.edu.cn (Z.W.); xrke928@mail.ustc.edu.cn (R.X.)
* Correspondence: yuanyueg@ustc.edu.cn

Received: 10 November 2019; Accepted: 27 November 2019; Published: 29 November 2019

Abstract: In two dimensional cross-range multiple-input multiple-output radar imaging for aerial targets, due to the non-cooperative movement of the targets, the estimated imaging plane parameters, namely the center and the posture angles of the imaging plane, may have deviations from true values, which defocus the final image. This problem is called imaging plane mismatch in this paper. Focusing on this problem, firstly the deviations of spatial spectrum fulfilling region caused by imaging plane mismatch is analyzed, as well as the errors of the corresponding spatial spectral values. Thereupon, the calibration operation is deduced when the imaging plane parameters are accurately obtained. Afterwards, an imaging plane calibration algorithm is proposed to utilize particle swarm optimization to search out the imaging plane parameters. Finally, it is demonstrated through simulations that the proposed algorithm can accurately estimate the imaging plane parameters and achieve good image focusing performance.

Keywords: two-dimensional radar imaging; multiple-input multiple-output (MIMO) radar; Particle Swarm Optimization (PSO); imaging plane calibration algorithm (IPCA)

---

1. Introduction

Aerial targets imaging is an important research direction in the field of radar imaging technology. Especially, it plays a crucial role in the military field, such as, aerial defense [1] and anti-missile defense [2] and so on. Multiple-input multiple-output (MIMO) radar is a new radar technique, which adopts multiple transmitters and receivers. By transmitting orthogonal space–time block codes [3] or frequency diversity signals [4], a MIMO radar with M transmitters and N receivers can eventually form a virtual array of the aperture length up to M times that of the receive array, which greatly saves the hardware cost. In addition, MIMO radar can obtain the images of the aerial targets with only one snapshot, and thus has enormous superiority in radar image acquisition time [5,6].

Since MIMO radar technique was proposed, it has been desired to build high performance imaging algorithms. The researches of MIMO radar imaging algorithms mainly focus on two aspects: The first is the wave-number domain imaging methods, and their imaging performances mainly depend on the spatial spectrum fulfilling region, which is generally required to be uniformly fulfilled, so that the fast Fourier transform (FFT) can be applied [7,8]. In this regard, Prof. Yarovoy et al. have conducted a large number of studies and verified the feasibility of the proposed algorithms in security check [9], wall penetrating [10] and ground penetrating imaging applications [7]. In addition to the traditional wave-number domain methods, the iterative optimization methods have also been applied in the MIMO radar imaging field. For example, Prof. Li's team of University of Florida proposed several iterative imaging algorithms, such as iterative adaptive approaches (IAA) algorithm [11] and the sparse learning via iterative minimization (SLIM) [12] algorithm, which have been well applied and verified in MIMO radar scenarios.

In MIMO radar imaging, the model mismatch caused by system errors or array spatial position errors will degrade the imaging quality. Therefore, the study of model error calibration algorithms is an important research direction of high-quality MIMO radar imaging. In this regard, a large number of studies have been used to calibrate the phase error [13,14], carrier frequency deviation [15], array position error [16], off-grid problem [17,18], and so on. The degradation of MIMO radar resolution under the condition of phase error from the perspective of point spread function (PSF) is analyzed in [13], and the sparse imaging via expectation maximization (SIEM) algorithm is proposed, which alternately estimates the phase errors and the target image, and obtains better imaging quality. Subsequently, the degradation of MIMO radar resolution under the condition of carrier frequency deviation from the perspective of PSF is analyzed in [15] as well, and an iterative algorithm employing iteration strategy similar to reference [13] is proposed, and good imaging results are achieved. MIMO imaging with array position errors is studied in reference [16], while in reference [17,18], the off-grid problem of MIMO radar imaging is studied. The algorithms proposed in [16–18] all employ sparse optimization by alternately estimating the target image and the errors during iterations, hence clear images are finally obtained.

Most of the above methods are proposed for two dimensional (2D) imaging in range and cross-range directions. However, for the target plane parallel to cross-range direction, these methods cannot be directly applied. In fact, it is necessary to set the origin of the coordinates at the center of the imaging plane [5,19] for the 2D cross-range imaging methods based on the spatial spectrum. In practical applications, as the target is non-cooperative, it is required to estimate the scene center and posture angles of the target plane, so as to make the final image focus on the image scene center and the target plane. However, there are always some deviations between the estimated imaging plane parameters and the real situations, which resulting in unfocused image and poor imaging quality. This problem is called the imaging plane mismatch problem in this paper.

To solve this problem, firstly, the deviations of spatial spectral fulfilling region caused by imaging plane mismatch are analyzed, and the location errors between estimated spatial spectral point and real spatial spectral point under the condition of imaging plane mismatch are deduced, as well as the errors of the corresponding spatial spectral values. Subsequently, in order to estimate imaging plane parameters and to be able to calibrate the locations and values of spatial spectral points, an imaging plane calibration algorithm (IPCA) is proposed. Aiming at minimizing the image entropy as well as promoting target sparsity, IPCA utilizes a particle swarm optimization (PSO) [20,21] algorithm to search out the parameters of imaging plane center deviation and pose angles deviations, and then calibrates the locations of spatial spectral points according to these parameters, so as to obtain images with better quality.

This paper is organized as follows. Section 2 introduces the spatial spectral imaging model of MIMO radar, and analyzes the problem of imaging plane mismatch, and the deviations between the estimated spatial spectral point and the real spatial spectral point position under the imaging plane mismatch are deduced. Section 3 provides the design and the detailed flow of IPCA. In Section 4, the validity of the proposed algorithm, robustness to noise, and tolerance to mismatching parameters are verified by simulations. Section 5 is the conclusion of this paper.

2. Problem Formulation of Imaging Plane Mismatch

In this section, the spatial spectral imaging model of MIMO radar is reviewed firstly, and then the model mismatch problem caused by the imaging plane mismatch is analyzed. Under the condition of imaging plane mismatch, the deviations between the positions of the obtained spatial spectral points and the positions of the real spectral points are analyzed, as well as the values of the corresponding spatial spectral points. Afterwards, the calibration operation is deduced to obtain the focused image after obtaining the imaging plane parameters. Notice that the radar system discussed here is the frequency diversity MIMO (f-MIMO) radar [4,22].

2.1. Space Spectral Imaging Model of Multiple-Input Multiple-Output (MIMO) Radar

In this subsection, the spatial spectral imaging model of MIMO radar is reviewed firstly.

Figure 1 illustrates the general 2D cross-range imaging scenario of MIMO radar. Let (x, y, z) be Cartesian coordinates with the origin O located at the center of the imaging plane and the 2D target is supposed to be on the imaging plane. The location of the p-th transmitting antenna and the q-th receiving antenna are denoted as $\boldsymbol{r}_p = (r_p, \theta_p, \varphi_p)$ and $\boldsymbol{r}_q = (r_q, \theta_q, \varphi_q)$ in spherical coordinate respectively.

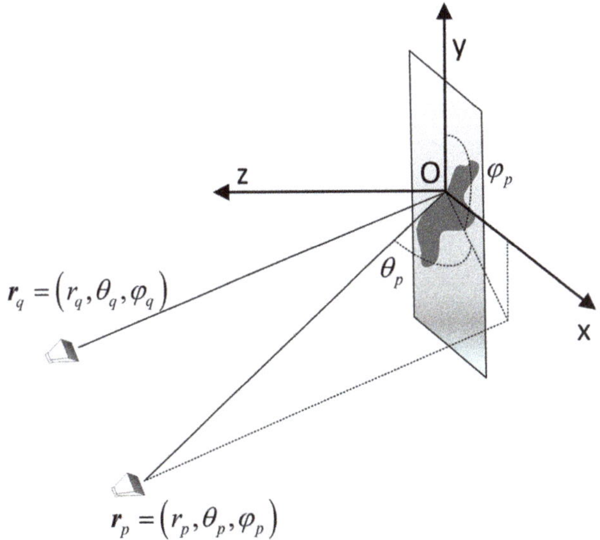

Figure 1. Space spectral imaging model of multiple-input multiple-output (MIMO) radar.

Without loss of generality, regardless of the loss associated with the free-space propagation, the echo signal at the q-th receiver by the p-th transmitter is given by [5]:

$$s_{p,q}(t) = \iint \sigma(x_T, y_T) \exp\left(j2\pi \left[f_p\left(t - \frac{R_{p,T}}{c} - \frac{R_{q,T}}{c}\right)\right]\right) dx_T dy_T \tag{1}$$

where $\sigma(x_T, y_T)$ denotes the reflectivity of the scatterer at (x_T, y_T) on the imaging plane, f_p is the transmitting frequency of the p-th transmitter and c is the speed of light. $R_{p,T} = |\boldsymbol{r}_p - \boldsymbol{r}_T|$, $R_{q,T} = |\boldsymbol{r}_q - \boldsymbol{r}_T|$, $\boldsymbol{r}_T = (x_T, y_T, 0)$.

Then, down conversion is applied to the received signal, which can be achieved by multiplying it by the following reference signal:

$$s_{ref}(t) = \exp\left(-j2\pi \left[f_p\left(t - \frac{|\boldsymbol{r}_p|}{c} - \frac{|\boldsymbol{r}_q|}{c}\right)\right]\right) \tag{2}$$

In the case of the far field, approximate conditions can be used:

$$R_{p,T} = |\boldsymbol{r}_p - \boldsymbol{r}_T| = |\boldsymbol{r}_p| - \boldsymbol{r}_T \cdot \hat{\boldsymbol{e}}_p$$
$$R_{q,T} = |\boldsymbol{r}_q - \boldsymbol{r}_T| = |\boldsymbol{r}_q| - \boldsymbol{r}_T \cdot \hat{\boldsymbol{e}}_q \tag{3}$$

where $\hat{\boldsymbol{e}}_p = \boldsymbol{r}_p / |\boldsymbol{r}_p|$ and $\hat{\boldsymbol{e}}_q = \boldsymbol{r}_q / |\boldsymbol{r}_q|$.

Thus it can get:

$$s_{p,q}(t) \cdot s_{ref}(t) = \iint \sigma(x_T, y_T) \exp\left(j2\pi \left[f_p \left(t - \frac{R_{p,T}}{c} - \frac{R_{q,T}}{c} \right) \right] \right)$$

$$\times \exp\left(-j2\pi \left[f_p \left(t - \frac{|r_p|}{c} - \frac{|r_q|}{c} \right) \right] \right) dx_T dy_T \tag{4}$$

$$= \iint \sigma(x_T, y_T) \exp\left(j2\pi K_{p,q} \cdot r_T \right) dx_T dy_T$$

where $K_{p,q} = \left(k_{p,q}^x, k_{p,q}^y \right)$. $k_{p,q}^x$ and $k_{p,q}^y$ are:

$$k_{p,q}^x = \frac{f_p}{c} \left(\cos\theta_p \sin\varphi_p + \cos\theta_q \sin\varphi_q \right)$$

$$k_{p,q}^y = \frac{f_p}{c} \left(\cos\theta_p \cos\varphi_p + \cos\theta_q \cos\varphi_q \right) \tag{5}$$

Therefore, the value of the 2D spatial spectral point $(k_{p,q}^x, k_{p,q}^y)$ of the imaging plane is obtained:

$$G\left(k_{p,q}^x, k_{p,q}^y \right) = \iint \sigma(x_T, y_T) \exp\left(j2\pi \left(x_T k_{p,q}^x + y_T k_{p,q}^y \right) \right) dx_T dy_T \tag{6}$$

Finally, the common algorithms can be applied on spatial spectral point value to obtain the target image, such as the inverse fast Fourier transform (IFFT) algorithm, the back-projection (BP) algorithm [23], and the non-uniform fast Fourier transform [8] and so on.

2.2. Analysis of Model Mismatch Problem Caused by Imaging Plane Mismatch

In the actual MIMO radar imaging applications, especially in aerial target imaging, the center and the posture angles of the imaging plane are uncertain due to the target's non-cooperative movement state. When the parameters of the imaging plane have deviations, the fulfilling region of the obtained spatial spectrum will deviate from the real region, which will cause images to be unfocused.

Figure 2 shows the imaging geometry of the scenario with imaging plane mismatch. Let α denote the estimated target plane, while let β denote the real target plane. Besides, set up the coordinate $O - x_\alpha y_\alpha z_\alpha$ with the origin O located at the center of the estimated target plane and plane $x_\alpha O y_\alpha$ coincides with the estimated target plane. The location vectors of p-th transmitting antenna and q-th receiving antenna are $r_p = (r_p, \theta_p, \varphi_p)$ and $r_q = (r_q, \theta_q, \varphi_q)$ respectively in Coordinate $O - x_\alpha y_\alpha z_\alpha$. Likewise, set up the coordinate $O' - x_\beta y_\beta z_\beta$ with the origin O' located at the center of the real target plane and plane $x_\beta O y_\beta$ coincides with the real target plane. Meanwhile, the location vectors of p-th transmitting antenna and q-th receiving antenna are $r_p' = \left(r_p', \theta_p', \varphi_p' \right)$ and $r_q' = \left(r_q', \theta_q', \varphi_q' \right)$ respectively in coordinate $O' - x_\beta y_\beta z_\beta$. In addition, the changes between plane β and plane α consist of the translation change and the posture angles' change. The direction vector $d = (d_\beta, \theta_\beta, \varphi_\beta)$ represents the translation of the origin of coordinate $O' - x_\beta y_\beta z_\beta$ relative to the origin of coordinate $O - x_\alpha y_\alpha z_\alpha$. Define (δ, μ, ζ) as the posture angle of plane β in Coordinate $O - x_\alpha y_\alpha z_\alpha$, where δ denotes the angle between the axis x_β of coordinate $O' - x_\beta y_\beta z_\beta$ and the plane $x_\alpha O y_\alpha$, μ denotes the angle between the projection of the axis x_β on the plane $x_\alpha O z_\alpha$ and the axis x_α, and ζ denotes the angle between the plane $y_\beta O' z_\beta$ and the plane $y_\alpha O z_\alpha$.

What is more, $\theta_p', \theta_q', \varphi_p', \varphi_q'$ and $\theta_p, \theta_q, \varphi_p, \varphi_q$ have relations:

$$\theta_p' = \theta_p + \delta \sin\varphi_p + \mu \cos\varphi_p, \qquad \varphi_p' = \varphi_p + \zeta + \delta \cos\varphi_p + \mu \sin\varphi_p$$

$$\theta_q' = \theta_q + \delta \sin\varphi_q + \mu \cos\varphi_q, \qquad \varphi_q' = \varphi_q + \zeta + \delta \cos\varphi_q + \mu \sin\varphi_q \tag{7}$$

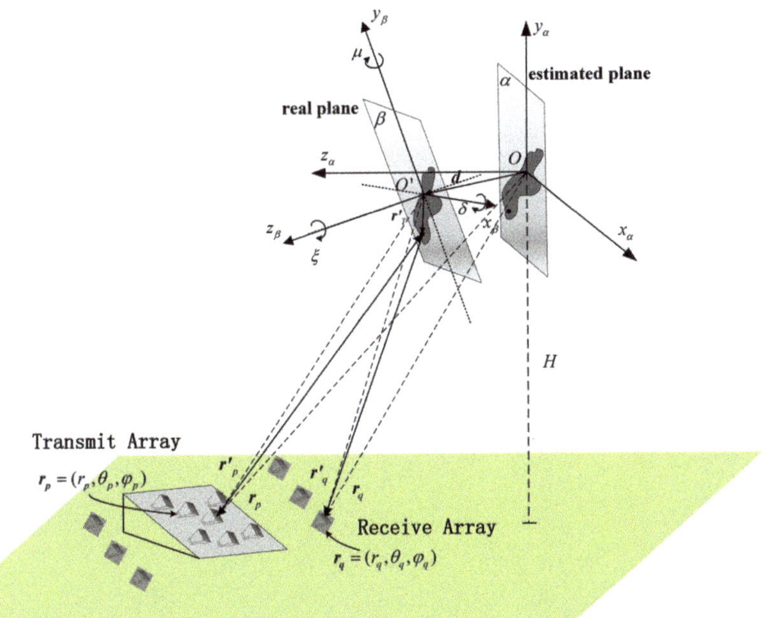

Figure 2. Imaging geometry of MIMO radar with imaging plane mismatch.

Hence, r'_p and r'_q can be computed by r_p, r_q and d, namely

$$\left| r'_p \right| = \left| r_p - d \right| = \sqrt{\left| r_p \right|^2 + d_\beta^2 - 2 \left| r_p \right| \cdot d_\beta \left(\cos\theta_p \cos\theta_\beta \cos\left(\varphi_p - \varphi_\beta\right) + \sin\theta_p \sin\theta_\beta \right)}$$
$$\left| r'_q \right| = \left| r_q - d \right| = \sqrt{\left| r_q \right|^2 + d_\beta^2 - 2 \left| r_q \right| \cdot d_\beta \left(\cos\theta_q \cos\theta_\beta \cos\left(\varphi_q - \varphi_\beta\right) + \sin\theta_q \sin\theta_\beta \right)} \tag{8}$$

Substitute Equation (8) into Equation (3), so that it can get:

$$
\begin{aligned}
\tau_{p,T} &= \frac{R_{p,T}}{c} = \frac{\left| r'_p \right| + x_T \cos\theta'_p \sin\varphi'_p + y_T \cos\theta'_p \cos\varphi'_p}{c} \\
&= \frac{\sqrt{\left| r_p \right|^2 + d_\beta^2 - 2 \left| r_p \right| d_\beta \left(\cos\theta_p \cos\theta_\beta \cos\left(\varphi_p - \varphi_\beta\right) + \sin\theta_p \sin\theta_\beta \right)}}{c} \\
&\quad + \frac{x_T \cos\theta'_p \sin\varphi'_p + y_T \cos\theta'_p \cos\varphi'_p}{c} \\
\tau_{q,T} &= \frac{R_{q,T}}{c} = \frac{\left| r'_q \right| + x_T \cos\theta'_q \sin\varphi'_q + y_T \cos\theta'_q \cos\varphi'_q}{c} \\
&= \frac{\sqrt{\left| r_q \right|^2 + d_\beta^2 - 2 \left| r_q \right| d_\beta \left(\cos\theta_q \cos\theta_\beta \cos\left(\varphi_q - \varphi_\beta\right) + \sin\theta_q \sin\theta_\beta \right)}}{c} \\
&\quad + \frac{x_T \cos\theta'_q \sin\varphi'_q + y_T \cos\theta'_q \cos\varphi'_q}{c}
\end{aligned}
\tag{9}
$$

Since the real imaging plane parameters are not accurately known, the reference signal $s_{ref}(t)$ is still the same as Equation (2). Substitute Equation (9) into Equation (4) and using far field approximation:

$$
\begin{aligned}
s_{p,q}(t) \cdot s_{ref}(t) = \iint \sigma\left(x_T, y_T\right) \exp \Big\{ j2\pi \frac{f_p}{c} \Big(&- d_\beta(\cos\theta_p \cos\theta_\beta \cos\left(\varphi_p - \varphi_\beta\right) + \sin\theta_p \sin\theta_\beta) \\
&- d_\beta(\cos\theta_q \cos\theta_\beta \cos\left(\varphi_q - \varphi_\beta\right) + \sin\theta_q \sin\theta_\beta) \\
&+ x_T \big[\cos(\theta_p + \delta\sin\varphi_p + \mu\cos\varphi_p)\sin(\varphi_p + \xi + \delta\cos\varphi_p + \mu\sin\varphi_p) \\
&\quad + \cos(\theta_q + \delta\sin\varphi_q + \mu\cos\varphi_q)\sin(\varphi_q + \xi + \delta\cos\varphi_q + \mu\sin\varphi_q) \big] \\
&+ y_T \big[\cos(\theta_p + \delta\sin\varphi_p + \mu\cos\varphi_p)\cos(\varphi_p + \xi + \delta\cos\varphi_p + \mu\sin\varphi_p) \\
&\quad + \cos\left(\theta_q + \delta\sin\varphi_q + \mu\cos\varphi_q\right)\cos(\varphi_q + \xi + \delta\cos\varphi_q + \mu\sin\varphi_q) \big] \Big) \Big\} dx_T dy_T
\end{aligned}
\tag{10}
$$

Therefore Equation (6) actually gets:

$$
\begin{aligned}
G\left(k^x_{p,q}, k^y_{p,q}\right) = \iint \sigma\left(x_T, y_T\right) \exp \Big\{ j2\pi \left(x_T k^x_{p,q} + y_T k^y_{p,q}\right) + j2\pi \frac{f_p}{c} \Big(\\
- d_\beta(\cos\theta_p \cos\theta_\beta \cos\left(\varphi_p - \varphi_\beta\right) + \sin\theta_p \sin\theta_\beta) \\
- d_\beta(\cos\theta_q \cos\theta_\beta \cos\left(\varphi_q - \varphi_\beta\right) + \sin\theta_q \sin\theta_\beta) \\
+ x_T \big[\sin\left(\delta\sin\varphi_p + \mu\cos\varphi_p\right)\left(\sin\theta_p \sin\varphi_p + \sin\theta_q \sin\varphi_q\right) \\
\quad + \sin\left(\xi + \delta\cos\varphi_q + \mu\sin\varphi_q\right)\left(\cos\theta_p \cos\varphi_p + \cos\theta_q \cos\varphi_q\right) \big] \\
+ y_T \big[\sin\left(\delta\sin\varphi_p + \mu\cos\varphi_p\right)\left(\sin\theta_p \cos\varphi_p + \sin\theta_q \cos\varphi_q\right) \\
\quad + \sin\left(\xi + \delta\cos\varphi_q + \mu\sin\varphi_q\right)\left(\cos\theta_p \sin\varphi_p + \cos\theta_q \sin\varphi_q\right) \big] \Big) \Big\} dx_T dy_T
\end{aligned}
\tag{11}
$$

The location of space spectral points obtained by the above processing is not completely matched with the real spatial spectral fulfilling region. As a result, the entire image is not focused on the real target plane.

2.3. The Calibration Operation

In order to obtain more accurate imaging results, it is necessary to search out the parameters of the imaging plane, so as to eliminate the mismatch between the estimated spatial spectral fulfilling region and the real spatial spectral fulfilling region, and the according spatial spectral values. After the parameters of the imaging plane are obtained, the following calibration operation can be applied to achieve focused image.

When the parameters $d = \left(d_\beta, \theta_\beta, \varphi_\beta\right)$ and (δ, μ, ξ) have been obtained, the reference signal can be calibrated as:

$$
s'_{ref}\left(t, r'_p, r'_q\right) = \exp\left\{ -j2\pi f_p \left(t - \frac{\left|r'_p\right| + \left|r'_q\right|}{c}\right)\right\}
\tag{12}
$$

After the coherent processing in Equation (10), the final spatial spectral value is:

$$
G'\left(k'^x_{p,q}, k'^y_{p,q}\right) = \iint_T \sigma\left(x_T, y_T\right) e^{j2\pi\left(x_T k'^x_{p,q} + y_T k'^y_{p,q}\right)} dx_T dy_T
\tag{13}
$$

where T denotes the imaging area and $k_{p,q}^{\prime x}$, $k_{p,q}^{\prime y}$ are defined in Equation (14).

$$
\begin{aligned}
k_{p,q}^{\prime x} &= \frac{f_p \left(\cos \theta_p' \sin \varphi_p' + \cos \theta_q' \sin \varphi_q' \right)}{c} \\
&= \frac{f_p}{c} \left(\left(\cos \left(\theta_p + \delta \sin \varphi_p + \mu \cos \varphi_p \right) \sin \left(\varphi_p + \xi + \delta \cos \varphi_p + \mu \sin \varphi_p \right) \right. \right. \\
&\quad \left. + \cos \left(\theta_q + \delta \sin \varphi_q + \mu \cos \varphi_q \right) \sin \left(\varphi_q + \xi + \delta \cos \varphi_q + \mu \sin \varphi_q \right) \right) \\
k_{p,q}^{\prime y} &= \frac{f_p \left(\cos \theta_p' \cos \varphi_p' + \cos \theta_q' \cos \varphi_q' \right)}{c} \\
&= \frac{f_p}{c} \left(\cos \left(\theta_p + \delta \sin \varphi_p + \mu \cos \varphi_p \right) \cos \left(\varphi_p + \xi + \delta \cos \varphi_p + \mu \sin \varphi_p \right) \right. \\
&\quad \left. + \cos \left(\theta_q + \delta \sin \varphi_q + \mu \cos \varphi_q \right) \cos \left(\varphi_q + \xi + \delta \cos \varphi_q + \mu \sin \varphi_q \right) \right)
\end{aligned}
\tag{14}
$$

Ultimately, the focused images can be obtained using $k_{p,q}^{\prime x}$, $k_{p,q}^{\prime y}$ and the corresponding spatial spectral value $G' \left(k_{p,q}^{\prime x}, k_{p,q}^{\prime y} \right)$.

3. The Proposed Imaging Plane Calibration Algorithm (IPCA)

When there are deviations between the estimated values of the imaging plane parameters $(d_\beta, \theta_\beta, \varphi_\beta, \delta, \mu, \xi)$ and the real values, it will lead to the problem of spatial spectrum mismatch. The mismatch of spatial spectrum results in the image not focusing at the real imaging plane and the whole image quality is deteriorated seriously.

In order to achieve better imaging performance, it is necessary to search out the real imaging plane parameters, namely the six parameters $(d_\beta, \theta_\beta, \varphi_\beta, \delta, \mu, \xi)$.

However, searching directions cannot be clear at beginning and always changed in the processing. Commonly, a good objective function will make the searching process in desirable directions.

Generally, image entropy can be used to measure the focusing performance of an image. In radar imaging, it is defined as follows:

$$
E = - \sum_{k=0}^{M-1} \sum_{n=0}^{N-1} \frac{|\sigma(x_k, y_n)|^2}{S} \ln \frac{|\sigma(x_k, y_n)|^2}{S}
\tag{15}
$$

where, $S = \sum_{k=0}^{M-1} \sum_{n=0}^{N-1} |\sigma(x_k, y_n)|^2$, and the image is discretized into $M \times N$ grids. $\sigma(x_k, y_n)$ denotes the scattering coefficient of the grid in k-th row at n-th column. At the same time, in the aerial imaging applications, the targets usually have sparse characteristics.

Therefore, both the focusing performance and the sparsity of the targets are considered, hence the objective function is to minimize the following function:

$$
f(d, \delta, \mu, \xi) = - \sum_{k=0}^{M-1} \sum_{n=0}^{N-1} \frac{|\sigma(x_k, y_n)|^2}{S} \ln \frac{|\sigma(x_k, y_n)|^2}{S} + \gamma \sum_{k=0}^{M-1} \sum_{n=0}^{N-1} |\sigma(x_k, y_n)|_1
\tag{16}
$$

where $\sigma(x_k, y_n) = IFFT \left(G_{p,q} \left(k_x, k_y, d, \delta, \mu, \xi \right) \right)$ and γ is a tunable parameter to adjust the weights of the image entropy and the sparsity.

In such a way, the estimation problem for (d, δ, μ, ξ) can be transformed into the following optimization problem:

$$
\begin{aligned}
(d, \delta, \mu, \xi) &= \arg \min_{d, \delta, \mu, \xi} f(d, \delta, \mu, \xi) \\
\text{s.t.} \quad \sigma(x_k, y_n) &= IFFT \left(G_{p,q} \left(k_x, k_y, d, \delta, \mu, \xi \right) \right)
\end{aligned}
\tag{17}
$$

However, the above optimization problem is not a convex problem. Commonly, among non-convex optimization algorithms, metaheuristic optimization [24] is becoming increasingly popular. Particle swarm optimization (PSO) algorithm is one of the most popular metaheuristic optimization algorithms, and has been successfully applied in radar imaging problems [25,26]. Hence, the imaging plane calibration algorithm (IPCA) is proposed to utilize PSO algorithm for solving the optimization problem in Equation (17). In IPCA, each particle $p_i = (d_{\beta i}, \theta_{\beta i}, \varphi_{\beta i}, \delta_i, \mu_i, \xi_i)$ denotes one estimation of $(d_\beta, \theta_\beta, \varphi_\beta, \delta, \mu, \xi)$, while the velocity v_i of each particle denotes one search direction. In PSO, each particle remembers its personal best position. Meanwhile, global best position is also recorded. In each iteration, the velocity of each particle is updated combining the individual movement states and the group movement states. Particles acquire their new positions by constantly updating their speed, and eventually all particles will converge to the global optimal value [20,21].

The detailed algorithm of IPCA is described in Algorithm 1 and the flow chart is illustrated in Figure 3.

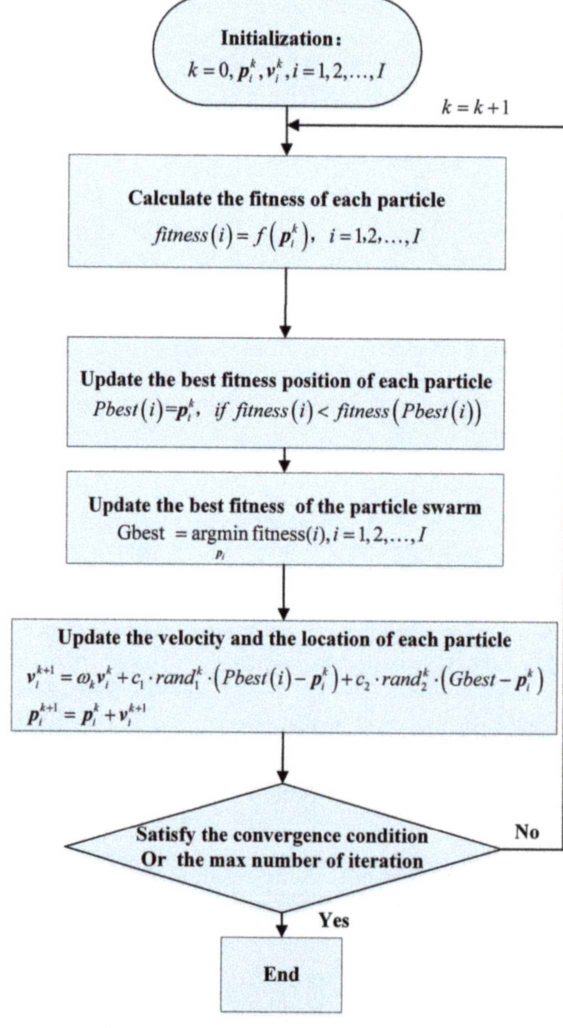

Figure 3. Flow chart of imaging plane calibration algorithm (IPCA).

Algorithm 1: IPCA

    Input: the maximum number of iterations K_{max} , Convergence condition ϵ, the number of particle I

1   Initialization: $d = (d_\beta, \theta_\beta, \varphi_\beta) = (0,0,0)$, $\delta = 0$, $\mu = 0$, $\xi = 0$; the location of the i-th particle p_i^0 and the velocity v_i. Let $Pbest(i)$ denotes the best fitness position of the i-th particle and $Gbest$ denotes the best position of all the particles.

2   **for** $k = 1 : K_{max}$ **do**

3   **for** $i = 1 : I$ **do**

4   compute the fitness of each particle: $fitness(i) = f(p_i^k)$;

5   **if** $fitness(i) < fitness(Pbest(i))$ **then**

6             $Pbest(i) = p_i^k$

7   **end**

8   **end**

9   Update $Gbest$ with $Gbest = \underset{p_i}{argmin}\, fitness(i), i = 1, 2, \ldots, I$;

10   Update the location and velocity of each particle using the following equations where w_k, c_1, c_2 are parameters controlling the iteration step and $rand_1^k$ and $rand_2^k$ are random numbers in $(0, 1)$.

$$v_i^{k+1} = w_k v_i^k + c_1 \cdot rand_1^k \cdot \left(Pbest(i) - p_i^k\right) + c_2 \cdot rand_2^k \cdot \left(Gbest - p_i^k\right)$$

$$p_i^{k+1} = p_i^k + v_i^{k+1}$$

11   **if** *Convergence condition reached* **then**

12   break

13   **end**

14 **end**

    Output: $(d_\beta, \theta_\beta, \varphi_\beta, \delta, \mu, \xi)$ and the image σ

---

4. Simulations

In this section, several simulations are carried out to verify the effectiveness of IPCA.

The imaging distance is set to 1 km in the simulations and the target is a B727 airplane with its point scattering model shown in Figure 4a. The size of the imaging plane, approximately parallel to the radar antenna array, is 60×60 m. The radar consists of 31×31 transmitters and 4 receivers and all of the antennas are located on the same transceiver plane, as is illustrated in Figure 4b. Besides, the size of the radar array is 60×60 m and the radar works at C-band. The detailed simulation parameters are given in Table 1.

Table 1. Simulation parameters.

Parameter	Value
Imaging distance	1 km
The size of the imaging plane	60 m × 60 m
The size of the radar antenna array	60 m × 60 m
Number of transmitters	31 × 31
Number of receivers	4
Carrier frequency	5 GHz
Bandwidth	200 MHz

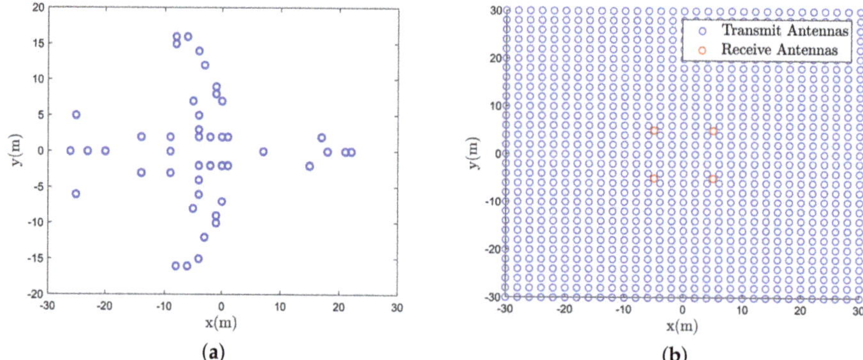

Figure 4. Target and antenna array (**a**) Scattering points distribution of the B727 airplane, (**b**) Locations of transmit antennas and receive antennas.

4.1. Imaging Simulation

To verify the effectiveness of the proposed method, the following simulation was carried out. System parameters in the simulations are shown in Table 1. The parameters of the center and pose angles of the imaging plane both have errors. The real parameters are given in Table 2.

Table 2. Imaging plane parameters.

Parameter	Value
Deviation of the center of the imaging plane	$d = (d_\beta, \theta_\beta, \varphi_\beta) = (1.5\,\text{m}, -0.785\,\text{rad}, 0.340\,\text{rad})$
Deviation of the posture angle of the imaging plane	$(\delta, \mu, \xi) = (\frac{1}{180}\pi, -\frac{1}{180}\pi, -\frac{1}{180}\pi)\,\text{rad}$

According to the derivation in Section 2 and the deviation parameters of the imaging plane, the estimated and real fulfilling region of the spatial spectrum are illustrated in Figure 5a,b respectively. It can be seen that when there are deviations of the imaging plane parameters, the estimated fulfilling region of the spatial spectrum is quite different from the fulfilling region of the real spatial spectrum. The recovered image is illustrated in Figure 5c when deviations of the imaging plane parameters are not known, and the adopted algorithm is the IFFT algorithm. It can been seen that even when there are slight deviations of the imaging plane parameters, e.g., the deviations of the posture angles of the imaging plane are within 1°, the image is badly defocused, and the target's contour is not clear.

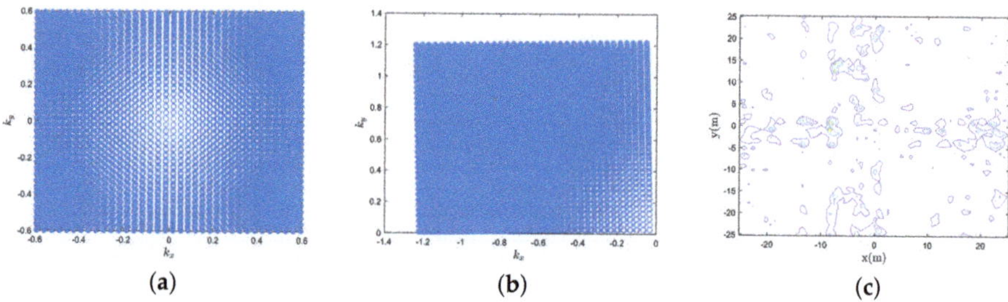

Figure 5. Spatial spectrum and reconstructed image. (**a**) The estimated fulfilled region of the spatial spectrum, (**b**) The real fulfilled region of the spatial spectrum, (**c**) Image reconstructed using estimated spatial spectrum values.

In the following, the proposed IPCA is taken to search out the six parameters, namely $(d_\beta, \theta_\beta, \varphi_\beta, \delta, \mu, \xi)$, and then the calibration operation is taken to obtain the final image. The number of the particles, I, in IPCA is set to 100 and the maximum number of iterations, K_{max}, is set to 100 too. Meanwhile some other parameters in PSO are set: $\omega_k = 0.8 - 0.5 \times \frac{k}{K_{max}}$, $c_1 = c_2 = 1$.

According to Figure 6, the image has clearer target outlines and better focusing performance with fewer noisy points around the target. In the final image, each scattering point of the target can be clearly identified. Meanwhile, the image entropy is smaller using IPCA than that using IFFT, as is shown in Table 3, which means better focusing performance.

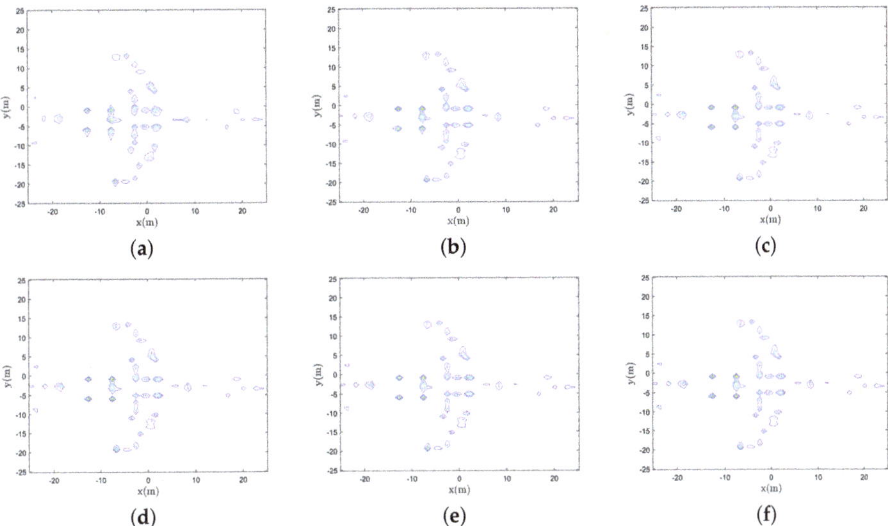

Figure 6. Reconstructed images of different iterations, (a–f) images of the 10th, 20th, 30th, 50th, 80th, and 100th iteration respectively.

Table 3. Image entropy.

Adopted Algorithm	IFFT	The Proposed IPCA
Image entropy	5.66	3.729

Figure 7 shows the value of the objective function and the image entropy during the IPCA iterations. It can been seen that when the number of iterations exceeds 80, the value of the objective function tends to be stable, that is, the search results gradually converge.

From what has been discussed above, the method proposed in this paper can effectively converge. Meanwhile, the entropy of the inversion image is lower and the image quality is higher.

Figure 8a shows the estimated $d = (d_x, d_y, d_z)$ during the iterations. It can be seen that, with the increase of the number of iterations, d_z gradually approaches the real value, and finally converges to the real value. The values of d_x d_y are still deviated from the real values. This is because the deviations d_x, d_y of the image will only make the image translation in the imaging plane, which have no effect on the image focus effect and target sparsity. Therefore, d_x, d_y do not affect the imaging quality and IPCA cannot guarantee d_x, d_y convergence to the real value. Likewise, Figure 8b shows the estimated imaging plane posture angles (δ, μ, ζ) with the number of iterations. It can be seen that as the number of iterations increases, the value of δ, μ converges to the real value while ζ not. This is because the estimation error of the rotation angle ζ, whose rotating axis parallel to the line of sight, will make the image rotate in the imaging plane, which has no influence on the image entropy and target sparsity. Therefore ζ does not affect the imaging quality and IPCA cannot guarantee ζ convergence to the real value.

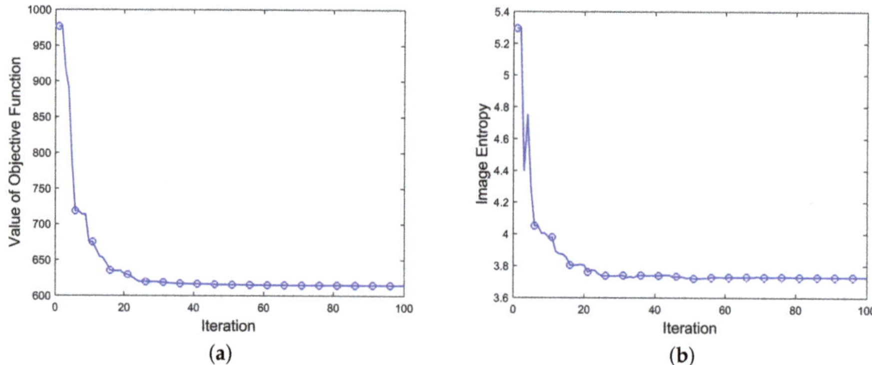

(a)   (b)

Figure 7. Value of objective function and image entropy during iterations. (**a**) Value of objective function during iterations (**b**) Image entropy during iterations.

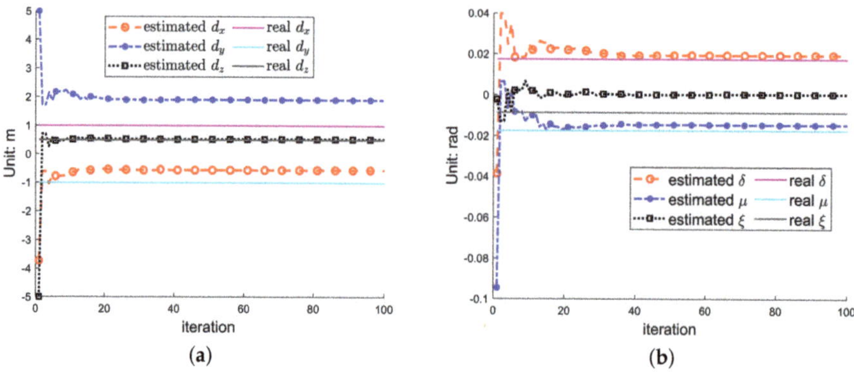

(a)   (b)

Figure 8. Values of $\boldsymbol{d} = (d_x, d_y, d_z)$ and (δ, μ, ξ) during iterations. (**a**) $\boldsymbol{d} = (d_x, d_y, d_z)$, (**b**) (δ, μ, ξ).

Meanwhile, since the PSO algorithm is a metaheuristic optimization algorithm, the computation time of IPCA is hard to predict. Therefore, 10 Monte Carlo trials are taken to count the computation time. The computation times of 10 Monte Carlo trials vary from 954 s to 2017 s with average computation time is 1617 s.

To sum up, the algorithm proposed in this paper can obtain more accurate imaging plane parameters, resulting in better image focusing performance and better imaging quality.

4.2. Simulations with Different Tunable Parameter γ

In Equation (16), a tunable parameter γ is defined to adjust the weights of the image entropy and the sparsity. In order to analysis the influence of γ, the below simulations are taken.

The simulation parameters are set the same with that in Tables 1 and 2. γ is set to [0.1, 0.5, 1, 3, 5, 10, 50]. The imaging results are illustrated in Figure 9, meanwhile the image entropy are given in Table 4.

It can be seen from Figure 9 that the reconstructed images are all well focused and the image entropy are all lower than 4 when γ is chosen with different values. So it can conclude that γ has seldom influence on the final imaging results. Therefore, γ can be selected from a wide range in IPCA.

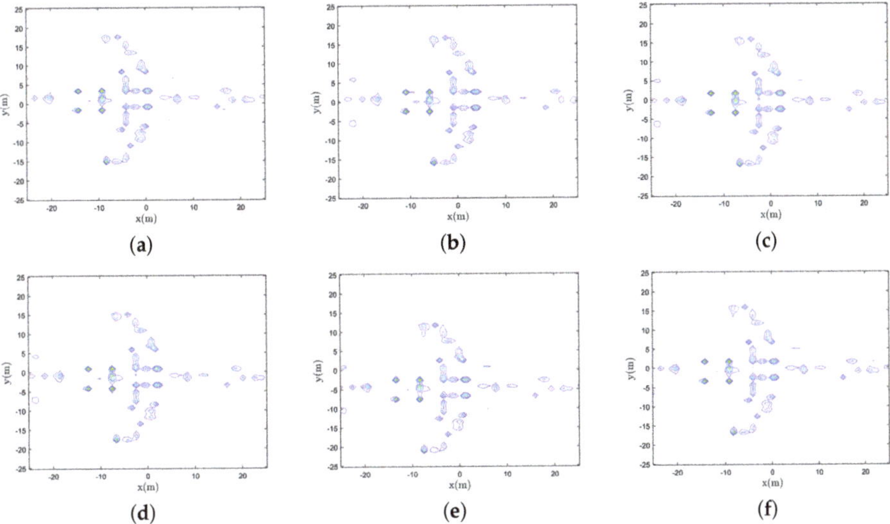

Figure 9. Reconstructed images with different γ, (**a**) $\gamma = 0.1$, (**b**) $\gamma = 1$, (**c**) $\gamma = 3$, (**d**) $\gamma = 5$, (**e**) $\gamma = 10$, (**f**) $\gamma = 50$.

Table 4. Image entropy with different γ.

γ	0.1	0.5	1	3	5	10	50
Image entropy	3.76	3.70	3.63	3.73	3.73	3.74	3.75

4.3. Simulations under Different Signal-to-Noise Ratios (SNRs)

In order to verify the robustness to the noise of IPCA, the following simulation is carried out in different echo signal-to-noise ratio. The echo SNRs are set from 5 dB to 30 dB. The simulation parameters are the same with that in Tables 1 and 2. The simulation results are in Figure 10.

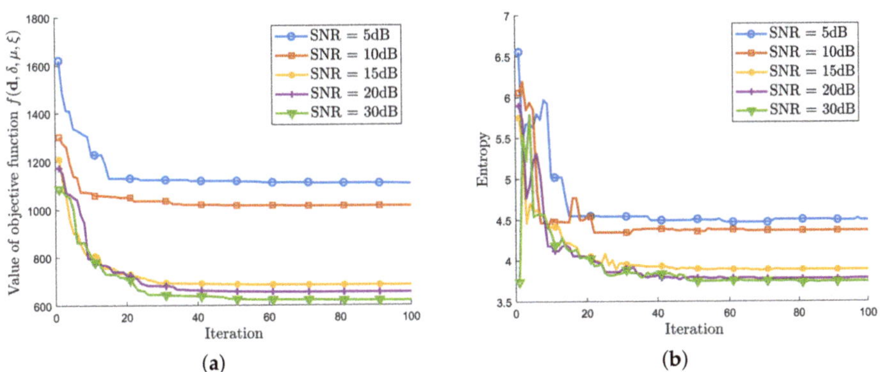

Figure 10. (**a**) Value of the objective function $f(\boldsymbol{d}, \delta, \mu, \xi)$ during iterations under different SNRs, (**b**) Image entropy during iterations under different SNRs.

The value of the objective function $f(\boldsymbol{d}, \delta, \mu, \xi)$ during the iterations is illustrated in Figure 10a. As it is shown, the value of the objective function $f(\boldsymbol{d}, \delta, \mu, \xi)$ decreases iteratively and finally converges. When the SNR is above 15 dB, the final values of $f(\boldsymbol{d}, \delta, \mu, \xi)$ are closed to each other.

The image entropy during the iterations is illustrated in Figure 10b. It can be seen that image entropy decreases iteratively. When the SNR is equal to or above 15 dB, the final image entropy is lower than 4 which means good image focusing performance.

4.4. Simulations under Different Parameter Ranges

In order to verify that the proposed method has a certain degree of tolerance for the deviations of the imaging parameters, the following simulations are conducted. According to the first subsection in this section, the center deviation parameter d_x, d_y of the imaging plane only cause the image to shift in the imaging plane, and has no influence on the imaging quality and focusing performance. In addition, the posture angle ζ makes the image rotate in the imaging plane, while the focusing performance is not affected. So in the following simulations, only d_z, δ and μ are considered.

In the simulations, d_z, δ and μ are uniform random selected in $[-\frac{\Delta d_z}{2}, \frac{\Delta d_z}{2}]$, $[-\frac{\Delta \delta}{2}, \frac{\Delta \delta}{2}]$ and $[-\frac{\Delta \mu}{2}, \frac{\Delta \mu}{2}]$. Δd_z are set to $[0.1, 0.2, 0.3, 0.5, 0.8, 1.0]$, while $\Delta \delta$ are set to $[1, 2, 3, 5, 8, 10]/180\pi$ whereas $\Delta \mu$ are set to $[1, 2, 3, 5, 8, 10]/180\pi$. For each Δd_z, $\Delta \delta$ and $\Delta \mu$, 10 Monte Carlo trials are taken. Meanwhile, when the value d_z, δ and μ are changing, d_x, d_y and ζ are set to be zeros.

The errors between the estimated d_z, δ, μ and the real values and the image entropy obtained by 10 Monte Carlo trials with different Δd_z, $\Delta \delta$, $\Delta \mu$ are illustrated using boxplot in Figure 11. On each box, the central mark indicates the median, and the bottom and top edges of the box indicate the 25th and 75th percentiles, respectively. The whiskers extend to the most extreme data points not considered outliers, and the outliers are plotted individually using the red '+' symbol. According to Figure 11a–c, the errors between the final estimated d_z and the real values are within ± 0.01 m, and the errors between the final estimated δ and the real value are within $\pm 0.35°$ when $\Delta \delta$ is chosen little than $3°$. Moreover the errors between the final estimated μ and the real values are within $\pm 0.35°$ when $\Delta \mu$ is chosen little than $5°$. In addition, according to Figure 11d–f, the image entropy are all lower than 4 when Δd_z, $\Delta \delta$, $\Delta \mu$ are chosen within 1 m, $3°$, $5°$ respectively.

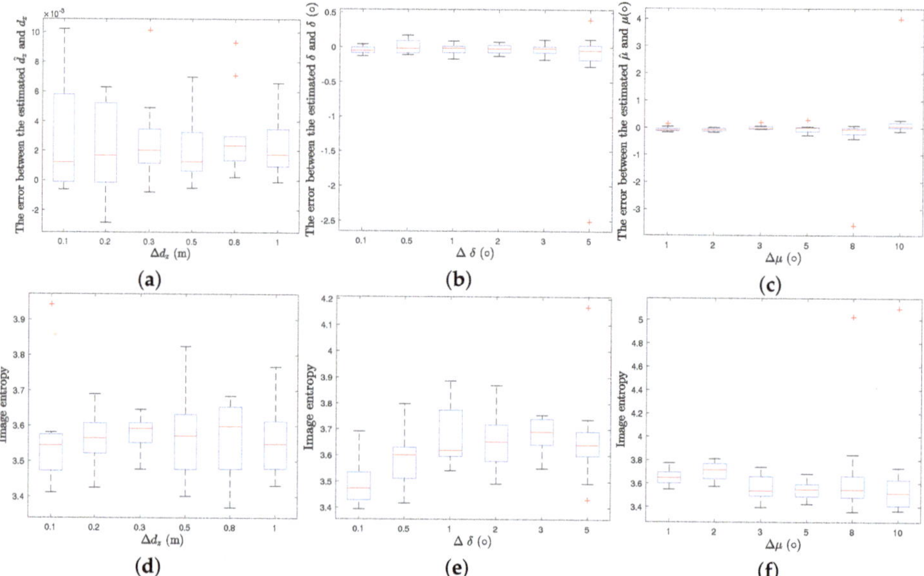

Figure 11. Errors between the estimated d_z, δ, μ and real values and Image entropy of 10 Monte Carlo trials, (**a**–**c**). Errors between the estimated d_z, δ, μ and real values respectively when different Δd_z, $\Delta \delta$ and $\Delta \mu$ are chosen; (**d**–**f**) image entropy with different Δd_z, $\Delta \delta$ and $\Delta \mu$.

In short, IPCA can obtain the final estimated d_z, δ and μ with errors at most ± 0.01 m, $\pm 0.35°$ and $\pm 0.35°$ when Δd_z, $\Delta \delta$, $\Delta \mu$ are chosen within 1 m, $3°$, $5°$ respectively, and the image entropy is smaller than 4.

5. Conclusions

In this paper, the imaging plane mismatch problem of 2D cross-range MIMO radar imaging is analyzed, and the deviations between the estimated spatial spectral point location and the real spatial spectral point location are deduced as well as the corresponding spatial spectral values. To solve this problem, IPCA is proposed in this paper. Aiming at minimizing the image entropy and sparsity of the image, PSO is utilized in IPCA to obtain the image with better focusing performance. Simulation results verify the effectiveness of the proposed algorithm and the robustness to noise. At the same time, when the parameters of the imaging plane are different, the proposed algorithm can obtain the imaging plane parameters of the real values, and then carry out the imaging plane calibration operation to obtain the focused image. The method proposed in this paper can solve the imaging plane mismatch problem and obtain high quality MIMO images.

Author Contributions: All authors contributed extensively to the work presented in this paper. Y.G. proposed the original idea and designed the study; B.Y. and Z.W. performed the simulations and wrote the paper; Y.G. supervised the analysis and edited the manuscript; R.X. provided his valuable suggestions to improve this study.

Funding: This research was funded by the National Natural Science Foundation of China under contact No. 61771446 and No. 61431016.

Conflicts of Interest: The authors declare no conflict of interest.

References

1. Constancias, L.; Cattenoz, M.; Brouard, P.; Brun, A. Coherent collocated MIMO radar demonstration for air defence applications. In Proceedings of the 2013 IEEE Radar Conference (RadarCon13), Ottawa, ON, Canada, 29 April–3 May 2013 ; pp. 1–6.
2. Chen, X.P.; Zeng, X.N. Superiority Analysis of MIMO Radar in Aerial Defence and Anti-missile Battle. *Telecommun. Eng.* **2009**, *10*.
3. Fishler, E.; Haimovich, A.; Blum, R.; Chizhik, D.; Cimini, L.; Valenzuela, R. MIMO radar: An idea whose time has come. In Proceedings of the IEEE Radar Conference, Philadelphia, PA, USA, 26–29 April 2004; pp. 71–78.
4. Li, X.R.; Zhang, Z.; Mao, W.X.; Wang, X.M.; Lu, J.; Wang, W.S. A study of frequency diversity MIMO radar beamforming. In Proceedings of the IEEE International Conference on Signal Processing, Beijing, China, 24–28 October 2010.
5. Wang, D.W.; Ma, X.Y.; Su, Y. Two-dimensional imaging via a narrowband MIMO radar system with two perpendicular linear arrays. *IEEE Trans. Image Process.* **2009**, *19*, 1269–1279. [CrossRef] [PubMed]
6. Hu, X.; Tong, N.; Song, B.; Ding, S.; Zhao, X. Joint sparsity-driven three-dimensional imaging method for multiple-input multiple-output radar with sparse antenna array. *IET Radar Sonar Navig.* **2016**, *11*, 709–720. [CrossRef]
7. Sakamoto, T.; Sato, T.; Aubry, P.J.; Yarovoy, A.G. Ultra-wideband radar imaging using a hybrid of Kirchhoff migration and Stolt FK migration with an inverse boundary scattering transform. *IEEE Trans. Antennas Propag.* **2015**, *63*, 3502–3512. [CrossRef]
8. Wang, J.; Cetinkaya, H.; Yarovoy, A. NUFFT based frequency-wavenumber domain focusing under MIMO array configurations. In Proceedings of the 2014 IEEE Radar Conference, Cincinnati, OH, USA, 19–23 May 2014; pp. 1–5.
9. Savelyev, T.; Yarovoy, A. 3D imaging by fast deconvolution algorithm in short-range UWB radar for concealed weapon detection. *Int. J. Microw. Wirel. Technol.* **2013**, *5*, 381–389. [CrossRef]
10. Zhuge, X.; Yarovoy, A.G.; Savelyev, T.; Ligthart, L. Modified Kirchhoff migration for UWB MIMO array-based radar imaging. *IEEE Trans. Geosci. Remote Sens.* **2010**, *48*, 2692–2703. [CrossRef]
11. Roberts, W.; Stoica, P.; Li, J.; Yardibi, T.; Sadjadi, F.A. Iterative adaptive approaches to MIMO radar imaging. *IEEE J. Sel. Top. Signal Process.* **2010**, *4*, 5–20. [CrossRef]

12. Tan, X.; Roberts, W.; Li, J.; Stoica, P. Sparse learning via iterative minimization with application to MIMO radar imaging. *IEEE Trans. Signal Process.* **2010**, *59*, 1088–1101. [CrossRef]

13. Ding, L.; Chen, W. MIMO radar sparse imaging with phase mismatch. *IEEE Geosci. Remote Sens. Lett.* **2014**, *12*, 816–820. [CrossRef]

14. Yun, L.; Zhao, H.; Du, M. A MIMO radar quadrature and multi-channel amplitude-phase error combined correction method based on cross-correlation. In Proceedings of the Ninth International Conference on Graphic and Image Processing (ICGIP 2017), Qingdao, China, 13–15 October 2017; International Society for Optics and Photonics: Bellingham, DC, USA, 2018; Volume 10615, p. 1061555.

15. Ding, L.; Chen, W.; Zhang, W.; Poor, H.V. MIMO radar imaging with imperfect carrier synchronization: A point spread function analysis. *IEEE Trans. Aerosp. Electron. Syst.* **2015**, *51*, 2236–2247. [CrossRef]

16. Liu, C.; Yan, J.; Chen, W. Sparse self-calibration by MAP method for MIMO radar imaging. In Proceedings of the 2012 IEEE International Conference on Acoustics, Speech and Signal Processing (ICASSP), Kyoto, Japan, 25–30 March 2012; pp. 2469–2472.

17. Tan, Z.; Nehorai, A. Sparse direction of arrival estimation using co-prime arrays with off-grid targets. *IEEE Signal Process. Lett.* **2014**, *21*, 26–29. [CrossRef]

18. He, X.; Liu, C.; Liu, B.; Wang, D. Sparse frequency diverse MIMO radar imaging for off-grid target based on adaptive iterative MAP. *Remote Sens.* **2013**, *5*, 631–647. [CrossRef]

19. Duan, G.Q.; Dang, W.W.; Xiao, Y.M.; Yi, S. Three-Dimensional Imaging via Wideband MIMO Radar System. *IEEE Geosci. Remote Sens. Lett.* **2010**, *7*, 445–449. [CrossRef]

20. Eberhart, R.; Kennedy, J. A new optimizer using particle swarm theory. In Proceedings of the Sixth International Symposium on Micro Machine and Human Science, Nagoya, Japan, 4–6 October 1995; pp. 39–43.

21. Kennedy, J. Particle swarm optimization. In *Encyclopedia of Machine Learning*; Springer: Boston, MA, USA 2010; pp. 760–766.

22. Gao, C.; Teh, K.C.; Liu, A. Orthogonal Frequency Diversity Waveform with Range-Doppler Optimization for MIMO Radar. *IEEE Signal Process. Lett.* **2014**, *21*, 1201–1205. [CrossRef]

23. Chen, A.L.; Wang, D.W.; Ma, X.Y. An improved BP algorithm for high-resolution MIMO imaging radar. In Proceedings of the 2010 International Conference on Audio, Language and Image Processing, Shanghai, China, 23–25 November 2010; pp. 1663–1667.

24. Yang, X.S. Metaheuristic optimization: Algorithm analysis and open problems. In Proceedings of the 10th International Conference on Experimental Algorithms, Crete, Greece, 5–7 May 2011, pp. 21–32.

25. Liu, L.; Zhou, F.; Tao, M.; Zhang, Z. A Novel Method for Multi-Targets ISAR Imaging Based on Particle Swarm Optimization and Modified CLEAN Technique. *IEEE Sens. J.* **2015**, *16*, 97–108. [CrossRef]

26. Luo, C.; Wang, G.; Lu, G.; Wang, D. Recovery of moving targets for a novel super-resolution imaging radar with PSO-SRC. In Proceedings of the 2016 CIE International Conference on Radar (RADAR), Guangzhou, China, 10–13 Octorber 2016.

Article

Field Representation Microwave Thermography Utilizing Lossy Microwave Design Materials

Christoph Baer *, Kerstin Orend, Birk Hattenhorst and Thomas Musch

Institute of Electronic Circuits, Ruhr University Bochum, Universitaetsstr. 150, 44801 Bochum, Germany; Kerstin.Orend@rub.de (K.O.); Birk.Hattenhorst@rub.de (B.H.); Thomas.Musch@est.rub.de (T.M.)
* Correspondence: christoph.baer@rub.de; Tel.: +49-234-3227606

Abstract: In this contribution, we are investigating a technique for the representation of electromagnetic fields by recording their thermal footprints on an indicator material using a thermal camera. Fundamentals regarding the interaction of electromagnetic heating, thermodynamics, and fluid dynamics are derived which allow for a precise design of the field illustration method. The synthesis and description of high-loss dielectric materials is discussed and a technique for a simple estimation of the broadband material's imaginary permittivity part is introduced. Finally, exemplifying investigations, comparing simulations and measurements on the fundamental TE10-mode in an X-band waveguide are presented, which prove the above introduced sensing theory.

Keywords: microwave; thermography; field illustration; permittivity

Citation: Baer, C.; Orend, K.; Hattenhorst, B.; Musch, T. Field Representation Microwave Thermography Utilizing Lossy Microwave Design Materials. *Sensors* **2021**, *21*, 4830. https://doi.org/10.3390/s21144830

Academic Editors: Andrea Randazzo, Cristina Ponti and Alessandro Fedeli

Received: 18 June 2021
Accepted: 12 July 2021
Published: 15 July 2021

Publisher's Note: MDPI stays neutral with regard to jurisdictional claims in published maps and institutional affiliations.

1. Introduction

The measurement and visualization of electrical and magnetic fields has been an important task in the big scope of instrumentation, measurement, and education for many years as it enables a better understanding of physical phenomena. Here, several differentiations can be made for classifying methods and techniques. In terms of static and quasi-static field observation up to 100 MHz, optical sensors utilizing the Kerr effect, showing a two-dimensional field distribution, have been already reported in the 1970s [1]. Further instrumentation approaches utilize optical sensors [2,3]. However, when increasing the alternating frequency of the applied field up to wave propagation, sensor and probe design is often reduced to single point measurements that record a scalar value. Therefore, the illustration of complete field distributions is only possible by the time consuming scanning of the area of interest. Corresponding sensor and measurement techniques have been reported in [4–6] and include dielectric tips and dipole probes.

Anyway, the fast recording of microwave field distributions remains an interesting approach for several scientific and engineering fields including the investigation of field radiation in terms of antenna design, EMC testing and of course educational purposes. However, large scale and multi-dimensional visualization is challenging, because the desired resolution of the illustrated field image demands for a large amount of sensor array elements, which are used in near field sensor devices. Consequently, when applied to bigger investigation areas, these devices are costly, power consuming, and disturb the incident electromagnetic wave. Hence, far field measurements are often performed for antenna and electromagnetic compatibility (EMC) testing, which however give only information on radiated field components. Finally, nowadays simulation based illustrations are a key technology in antenna design and education. Yet, conformity between virtual and real world results must be verified. Another method for the illustration of electromagnetic fields and their interaction with matter is the microwave thermography (MWT) approach. Here, the thermal footprint of an incident electromagnetic wave on lossy dielectrics or metallic objects is recorded by means of a thermal camera and subsequently processed. While, materials for the illustration of two-dimensional laser beams of non-optical wavelengths,

147

e.g., infrared lasers, are very well known and have been already commercialized and are distributed by Macken Instruments or Cascadelasers, the illustration of microwave fields is challenging because field strenghts' are significantly lower. To date, MWT has been mostly used for non-destructive testing as described in [7]. Further recording techniques rely on thermo-elastic optical indicators [8,9], or solid state quantum sensing [10]. In [11–13], promising approaches for the microwave field illustration are presented that make use of films with magnetic losses in combination with thermo-fluoresecent indicators. However, due to the magnetic loss mechanism, the frequency range of the illustrated field is limited. Moreover, the presented materials demand for high feeding powers up to several watts, which result in comparably low heating and therefore low measurement effects.

Taking previous research results into consideration reveals that a frequency gap between several 100 MHz up to optical frequencies is present for field representation thermography, which we want to partly close in this paper. Next to a holistic modeling of the multi-physical background, we propose novel approaches on the description of lossy dielectric material mixing and further realizations for the accurate representation of electro-magnetic field in the microwave region between 1 GHz and 100 GHz. Obviously, for Field-Representation-MWT (FRMWT), an adequate indicator material must be found, which has to meet several requirements: its parameters regarding dielectric and thermal properties must be chosen wisely in order to increase the overall efficiency and resolution of the thermal footprint. This is desirable as it allows reduced excitation powers and the applicability of low-cost thermal cameras. Moreover, the chosen size of the indicator material plays an important role for its transparency in terms of field interference. Finally, special test benches for FRMWT must be constructed to reduce disturbing environmental influences, such as infrared reflections as well as forced and additional natural convection, which can tamper the field recording.

This manuscript is organized as follows: Section 2 explains necessary fundamentals regarding "wave-matter interaction", "dielectric heating", "thermal loss theory", and "dielectric design materials and mixing". In Section 3, we discuss and set up a novel transient FRMWT model, while Section 4 presents self-made design materials for an optimized FRMWT. Finally, Section 5 shows microwave thermography results in simulation and measurements. In this contribution, the exemplary field distribution under investigation is the fundamental TE10-mode of a rectangular X-band waveguide. Section 6 concludes the manuscript and gives an outlook on future modifications and investigations.

2. Fundamentals

Since the description of FRMWT demands for the combination of several physical phenomena from electromagnetics, thermodynamics, and flow dynamics, the following section gives a brief overview on the dependencies between the different disciplines.

2.1. Wave Interaction and Dielectric Heating

The description of the interaction between an electromagnetic wave with the indicator material is very fundamental for FRMWT. In this context we need to observe two properties: the dielectric power dissipation, and the total reflection coefficient. The dielectric power dissipation is essential for the heating process of the indicator material and therefore important for the later field illustration. The total reflection coefficient, however, is of great interest as it gives information on field interferences, which can lead to distorted field illustration. In the simplest case, a plane wave interacts with a dielectric material, which is oriented perpendicular to the Poynting-vector of the incident wave. At the interface between free space and material, reflection coefficients occur that depend on wave impedances of the free space (Z_0) and inside of the material (Z_1).

$$\Gamma = \frac{Z_1 - Z_0}{Z_1 + Z_0} \tag{1}$$

At this point, we need to distinguish between the wave's entrance and exit reflection of the material as their coefficients differ. Furthermore, it is obvious that a part of the incident wave's power P_i remains inside of the material and is continuously attenuated by the material as illustrated in Figure 1. The power P_{di}, which remains inside of the dielectric material can be calculated as follows:

$$P_{di} = P_i \cdot (1 - \Gamma_1^2) \cdot \alpha(f)^2 \cdot \frac{1}{1 - (\Gamma_2 \alpha(f))^2}. \tag{2}$$

In (2), the factors Γ_1 and Γ_2 stand for the reflection coefficient at the interfaces free-space-to-material and vice versa. The attenuation α depends on the material's thickness d and its dielectric properties, which are frequency dependent:

$$\alpha(f) = e^{-\frac{2\pi f}{c_0} \text{Im}\left\{\sqrt{\varepsilon_r(f)}\right\} \cdot d}. \tag{3}$$

The total reflection coefficient, which is important for the interference with the incident field, is the square root of the relation of the reflected power P_{refl} to the incident power P_i and can be described as the superposition of the reflections at the interfaces free space-material and material-free-space. Because of the material's thickness d, the contribution of the second reflection performs an additional phase shift, which must be considered for the superposition. By considering the refractive indices $n_0 = 1$ and $n_1 = \sqrt{\varepsilon_{r,1} \mu_{r,1}}$ for free-space and a material, respectively. The total reflection Γ_{tot} is calculated by:

$$\Gamma_{tot} = \sqrt{\frac{P_{refl}}{P_i}} = \frac{j(n_0^2 - n_1^2) \sin(kd)}{2n_0 n_1 \cos(kd) - j(n_0^2 + n_1^2) \sin(kd)}. \tag{4}$$

In (4), j is the imaginary unit, while k represents the corresponding wave number. As we consider the dielectric losses of the material, the heating process of the material is directly connected with the magnitude of the electrical field strength $|E|$ of the incident wave. By considering the wave impedance Z and the length of the electric field vector l_1, it can be described by:

$$|E| = \sqrt{P_{di} \cdot Z / l_1} \tag{5}$$

Finally, the dissipated power depends on the imaginary part of the permittivity ε_r'' [14], the frequency f and the material's volume V:

$$P_{loss} = 2\pi f \varepsilon_0 \cdot \varepsilon_r'' \cdot |E|^2 \cdot V. \tag{6}$$

The actual heating, which can be regarded as the power transfer from electromagnetic to thermal power, is a transient process and depends on the material's mass m and its specific thermal capacity c_w. It can be described by:

$$\Delta T = \frac{P_{loss}}{c_w \cdot m} \cdot t. \tag{7}$$

Equations (5)–(7) show that the heating of an indicator material is directly proportional to the microwave excitation power. However, for calculating the electrical microwave field distribution we need to consider the square root of the recorded thermal footprint.

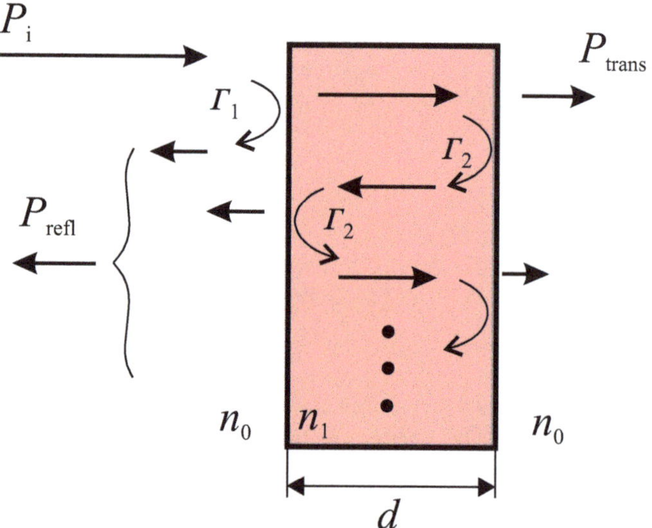

Figure 1. Schematical presentation of the wave interaction with a thin dielectric material sheet.

2.2. Thermal Losses

As we imbalance the thermodynamic equilibrium by the dielectric heating, the consideration of thermal losses is mandatory at this point. Otherwise, the material's temperature would rise indefinitely, because the heating Equation (7) is a linear function with time. Consequently, different thermal loss mechanisms affect a cooling, which limits the heating process to a final temperature. Here, three main thermodynamic transfer processes must be taken into consideration: thermal conduction, radiation, and convection. In a first approximation, conduction effects can be neglected as we observe pure dielectric materials. Consequently, the main contribution to thermal losses are radiation (rad) and free convection (fc). Therefore, the total thermal power loss can be described by:

$$P_{\text{th,tot}} = \Delta P_{\text{rad}} + P_{\text{fc}}. \tag{8}$$

Regarding radiation losses ΔP_{rad}, we only need to consider the additional losses caused by the temperature difference to the surrounding. In good approximation, the phenomenon can be described using the Stefan–Boltzmann law for black bodied radiators, because for the kind of measurements carried out in this research, where temperature differences are measured, the choice of an emissivity is not fundamental. By using the Stefan–Boltzmann constant $\sigma_{\text{SB}} = 5.67 \cdot 10^{-8}$ Wm^{-2}K^{-4}, we receive:

$$\Delta P_{\text{rad}} = P_{\text{rad,Tm}} - P_{\text{rad,T}\infty} \tag{9}$$

$$= \sigma_{\text{SB}} \cdot A_{\text{rad}} \cdot \left(T_{\text{m}}^4 - T_{\infty}^4 \right). \tag{10}$$

In (9) and (10), $P_{\text{rad,Tm}}$ and $P_{\text{rad,T}\infty}$ represent the thermal radiation powers caused by the temperatures T_{m} and T_{∞} on the material and in the environment, respectively. The parameter A_{rad} indicates the heated area of the indicator material. Free convection refers to a mechanism of fluid mechanics and is caused by a spatially non-uniform distribution of density. In this case, this non-uniformity is caused by the temperature difference between indicator material and environment [15]. In a first approximation, the transported heat in steady state can be regarded as:

$$P_{\text{fc}}(t \to \infty) = A_{\text{rad}} \cdot \alpha_{\text{th}} \cdot \Delta T_{\text{env}}. \tag{11}$$

In (11), A_{rad} is the heated area that affects free convection while $\Delta T_{\text{env}} = T_{\text{m}} - T_{\infty}$ describes the temperature difference between the actual indicator material's temperature and the environmental temperature. In case of FRMWT indicator foils, the heated area must be taken twice into consideration as free convection occurs in front and behind of the foil. The parameter α_{th} is the heat transfer coefficient. Its determination leads to a problem of flow dynamics and therefore to the determination of the functional relationship between the non-dimensional group of Rayleigh (Ra), Prandtl (Pr), and Nusselt (Nu) numbers [16,17]. Unfortunately, this approach bases on fitted measurement models and the further investigation is not very precise. Anyway, for vertical material orientation, the literature approximates the following equations, which have been constructed from measurement series [18,19]:

$$\alpha_{\text{th}} = \frac{\text{Nu} \cdot \lambda_{\text{th}}}{l_1}, \tag{12}$$

$$\text{Nu} = \left(0.825 + 0.387 \cdot (\text{Ra} \cdot \text{Pr}_{\text{eff}})^{1/6}\right)^2, \tag{13}$$

$$\text{Pr}_{\text{eff}} = \left(1 + \left(\frac{0.492}{Pr}\right)^{9/16}\right)^{-16/9}, \tag{14}$$

$$\text{Ra} = \frac{g \cdot l_2^3 \cdot \Delta T_{\text{env}} \cdot \text{Pr}}{T_{\infty} \cdot \nu^2}. \tag{15}$$

In (12)–(15), the parameters λ_{th} and ν represent the surrounding gas' thermal conductivity and viscosity, respectively. Further, l_1 and l_2 represent geometrical lengths for thermal imbalance alongside the direction of convection as well as the characteristic length of the flow. Finally, g is the gravitational acceleration. Thermal losses are highly non-linear regarding time and geometrical setup. Therefore, the presented equations must be regarded as approximation but they facilitate setting up a model for the transient behavior, which is interesting in terms of FRMWT.

2.3. Design Materials and Mixing

As shown before, the successful FRMWT strictly depends on the right selection of the indicator material. Here, we prefer materials that on the one hand easily heat up and store the additional temperature, while the total reflection coefficient should be kept low in order to minimize field interferences on the other hand. Moreover, thermal conduction should be kept low as well, because it would negatively affect the field illustration's resolution. Consequently, several compromises must be met to serve these requirements. In order to meet as many requirements as possible, an indicator material was designed by mixing several raw ingredients and their corresponding physical parameters. In order to make accurate predictions, so-called mixing equations can be utilized in order to estimate final parameters. In terms of the thermal capacity c_{w} and mass density ρ, the resulting quantity is directly dependent on the volume fractions ζ of the corresponding materials. In contrast to this, the description of the resulting, complex valued permittivity for dielectric mixing is not straightforward. However, the manufacturing of phantom materials for the substitution of, e.g., medical substances and surrogate explosives utilizing loss-free materials has been already published in [20,21], respectively. A general overview on dielectric mixing and corresponding models is given in [22]. In terms of FRMWT-materials, the most relevant material parameter are the losses in microwave range. Losses in dielectric materials can be caused by a small remaining conductivity, which leads to small conduction currents, or by replacement currents, which describe dipole orientation etc. Usually, for the general description of a material's complex valued permittivity those effects are tantamount.

However, referring to the Wiedemann–Franz law [23], electrical and thermal conductivity are directly proportional so that an increased electrical conductivity will deteriorate the resolution of the FRMWT. Consequently, it is important to tune the material's losses by increasing replacement currents but keeping conductivity as low as possible. Since the electrical polarization strongly depends on the time-variation of the excitation, the permittivity depends on the frequency of those field variations. This effect is called dispersion and can be seen in both real and imaginary part of the permittivity. Part of a material's dispersion is caused by the different polarization effects like orientation polarization as well as atomic and electronic polarization, which cause a resonance behavior in the permittivity's real part associated with a peak in the imaginary part [22]. In any real material, several polarization mechanisms take place and superimpose, which contribute to different dispersion effects within the entire electromagnetic spectrum. Obviously, a holistic model is difficult to formulate. That is why in the literature several polarization response models can be found that describe the aforementioned dispersion and loss effects for limited frequency ranges. Because the proposed FRMWT approach in this manuscript handles the microwave frequency range, we choose the Debye model henceforth. The Debye model is an easy to understand model that considers single relaxations processes very well. As we only handle solid materials in this work, we do not expect various relaxation processes in the microwave region, but continuously decreasing permittivity's real part values. Therefore, the Debye model is a good choice for modeling our indicator materials. Because of the induced torque of the dipole moments, the polarization requires time to reach equilibrium. This relaxation time τ is essential for the description of the complex valued permittivity. Moreover, Debye model defines the edges of the observed spectrum. Therefore, $\varepsilon_{r,\infty}$ and $\varepsilon_{r,s}$ describe the permittivities at optical and low frequencies, respectively. The description of a mixed material with Debye-type inclusions in a dispersion and loss-free background with permittivity $\varepsilon_{r,e}$ can be aligned to the basic Maxwell–Garnett mixing rule. The resulting material is also of Debye-type and can be formulated as follows:

$$\varepsilon_{r,\text{eff}} = \varepsilon_{r,\infty,\text{eff}} + \frac{\varepsilon_{r,s,\text{eff}} - \varepsilon_{r,\infty,\text{eff}}}{1 + j\omega\tau_{\text{eff}}}. \tag{16}$$

Consequently, the parameters of the mixed material are:

$$\varepsilon_{r,\infty,\text{eff}} = \varepsilon_{r,e} + \frac{3\zeta\varepsilon_{r,e}(\varepsilon_{r,\infty} - \varepsilon_{r,e})}{\varepsilon_{r,\infty} + 2\varepsilon_{r,e} - \zeta(\varepsilon_{r,\infty} - \varepsilon_{r,e})}, \tag{17}$$

$$\varepsilon_{r,s,\text{eff}} = \varepsilon_{r,e} + \frac{3\zeta\varepsilon_{r,e}(\varepsilon_{r,s} - \varepsilon_{r,e})}{\varepsilon_{r,s} + 2\varepsilon_{r,e} - \zeta(\varepsilon_{r,s} - \varepsilon_{r,e})}, \tag{18}$$

$$\tau_{\text{eff}} = \tau\frac{(1 - \zeta)\varepsilon_{r,\infty} + (2 + \zeta)\varepsilon_e}{(1 - \zeta)\varepsilon_{r,s} + (2 + \zeta)\varepsilon_e}. \tag{19}$$

Next to the Maxwell–Garnett averaging of the static and optical permittivity in (17) and (18), respectively, (19) reveals that the relaxation time of the including particles is also altered [22]. This is of great interest for the FRMWT as the material mixing allows for an optimization of the losses for a predefined operation frequency.

Since the measurement of lossy materials results in extremely low signal-to-noise ratios, especially the determination of the the loss-describing part, i.e., the imaginary part of the permittivity, can be crucial. Moreover, measurement probes are prone to delivering wrong loss values when the probe-material connection is not properly realized and generates radiation. However, obviously the connection between frequency behavior of the real and imaginary part can be formulated. Starting from the basic physical principle of causality, which means that the polarization response cannot lead the cause, the Kramers–Kronig relation must be fulfilled for the real and imaginary part of the permittivity [24]. As consequence, this very fundamental connection allows for an estimation of the imaginary part when the broadband real part permittivity is known. The estimation itself can be done in various ways, such as solving the Kramers–Kronig integrals, applying Hilbert transformation [25], or simply using the complex valued completion (CVC) which is described further on. The CVC allows for the estimation of the permittivity's imaginary part, when the corresponding permittivity's real part is known. It makes use of the dependencies of complex valued time signals and their spectra, which are subdivided in their even and odd signal and spectral parts. Figure 2 illustrates the well known dependencies. Within the figure, the sign ○——● indicates the Fourier transformation from time (○) to frequency domain (●). Moreover, the indices E and O represent even and odd signal and spectral parts, while the superscripts R and I indicate real and imaginary parts. As initial point, we consider a measured broadband, real part permittivity. As this quantity depends on frequency, we can regard it as a spectrum. Moreover, any commercial measuring probe will deliver measurement values for positive frequencies only. Therefore, the measured permittivity is the combination of its even and odd part:

$$\varepsilon_r^R = 1/2(\varepsilon_{r,E}^R + \varepsilon_{r,O}^R). \tag{20}$$

By means of the following CVC-steps, illustrated in Figure 3, we can calculate the right-sided, complex permittivity. As initial step we form the evenly-shaped spectrum of the measured real part permittivity. Therefore, the spectrum ε_r^R is multiplied by the heaviside step function Θ and divided by 2. Additionally, the resulting data vector is mirrored to the negative frequency range in order to achieve an axis-symmetric spectrum, which is described as the evenly-shaped spectral part of the real part permittivity $\varepsilon_{r,E}^R$. According to Figure 2, the application of the inverse Fourier transformation delivers the evenly-shaped impulse response of the real part permittivity $E_{r,E}^R(t)$. In [22], several interpretations of this pulse response signal are given, however in our case the time domain signal is a pure mathematical intermediate step and further interpretation is not relevant. As the next step, we transform the evenly-shaped impulse response to be oddly-shaped. Therefore, we once more apply the heaviside function and add the point symmetric signal for $t < 0$. By applying the Fourier transformation to the oddly-shaped time signal, we receive the oddly-shaped spectrum of the permittivity's imaginary part $\varepsilon_{r,O}^I(\omega)$. By doubling this spectrum, applying another heaviside step function, and adding the right-sided spectrum of the real part permittivity, we finally receive the right sided complex valued permittivity $\underline{\varepsilon}_r^+(\omega)$.

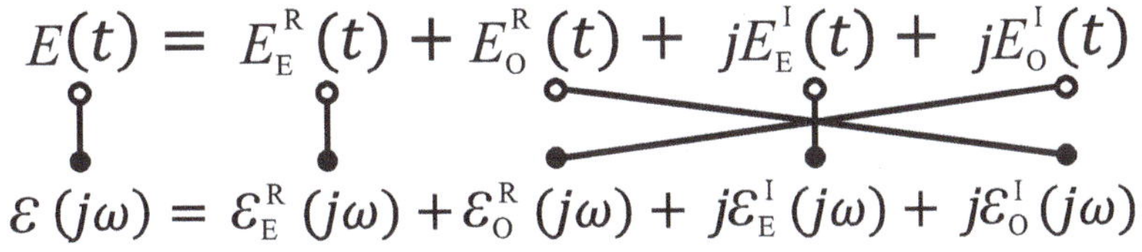

$$E(t) = E_E^R(t) + E_O^R(t) + jE_E^I(t) + jE_O^I(t)$$
$$\mathcal{E}(j\omega) = \mathcal{E}_E^R(j\omega) + \mathcal{E}_O^R(j\omega) + j\mathcal{E}_E^I(j\omega) + j\mathcal{E}_O^I(j\omega)$$

Figure 2. Dependencies of complex-valued time signal and the corresponding spectrum sub-divided into even (E) and odd (O) valued functions. The handle bar shows the intercorrespondencies.

It needs to be mentioned that the CVC is only applicable if the measured permittivity data is sufficiently broadband and covers all relevant polarization behaviors. In terms of FRMWT and the corresponding design material manufacturing, CVC plays an important role for the precise description of the high loss indicator materials as well as their corresponding mixing equations.

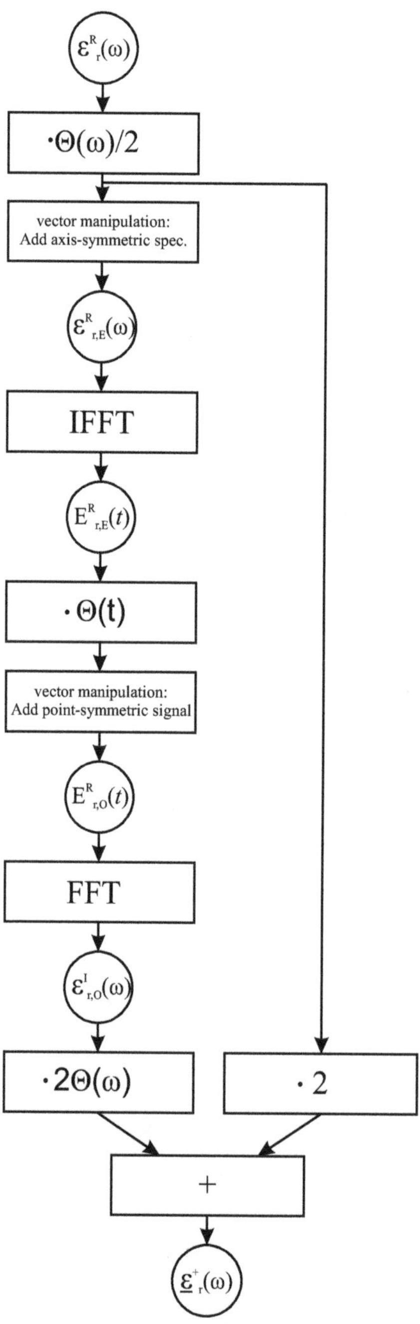

Figure 3. Processing steps of the complex valued completion (CVC). While rectangular boxes describe signal operation, circles indicate the current signal form. The applied function Θ indicates the heaviside function in frequency-domain (ω) and time-domain (t).

3. Transient FRMWT Model

In the following, we will observe the special case of FRMWT for mode characterization of a rectangular waveguide in X-band, as illustrated in Figure 4. The corresponding geometrical dimensions for the investigated WR90 rectangular waveguide and IEC 60154-2:2016 flange are: $l_1 = 10.16\,\text{mm}$, $l_2 = 41.4\,\text{mm}$, and $w = 22.86\,\text{mm}$. Moreover, the indicator material's thickness is considered to be $d = 0.9\,\text{mm}$ and it is positioned in close proximity to the waveguide's flange.

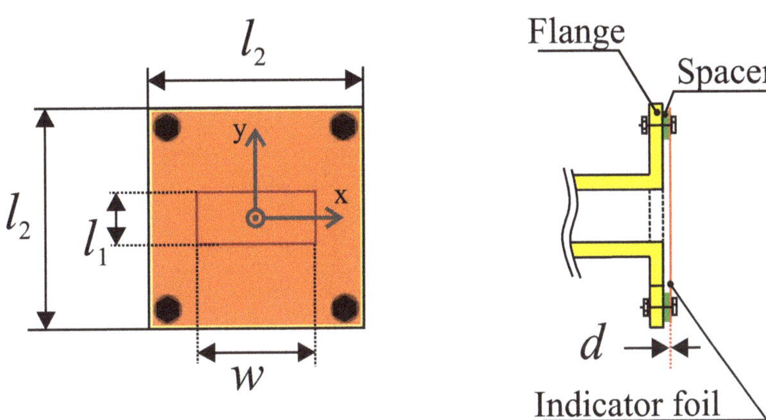

Figure 4. Sketch of the investigated waveguide flange.

However, it does not touch its metal boundary in order to prevent additional thermal conduction. After reaching steady state, i.e., when the microwave power which is transferred to heat equals the thermal losses, the heating-up process stops and the indicator material's end-temperature T_end is reached. Consequently, by equating the dissipated microwave power with the thermal loss power, we can calculate T_end and therefore the final temperature change ΔT_end. Figure 5 combines the thermal heating model with the microwave scenario by showing the theoretical temperature change ΔT_end as a function of the thermal loss power for the described setup and different surrounding gases. The determination of the temperature change is of great interest as it has direct influence on the sensitivity of the thermal sensor and the applied microwave power. Therefore, for a given thermal sensor sensitivity, the required microwave power can be reduced, which is beneficial in many scenarios. Apparently, a higher temperature change is reached when choosing surrounding gases like Helium or Xenon. However, the highest temperature change is achieved when the setup is kept in vacuum, because free convection vanishes and only thermal radiation contributes to the thermal loss. Anyway, this insight shows that the construction of an optimized FRMWT test stand that guarantees for a highly defined environment scenario is desirable. Yet, low cost setups also yield sufficient results in many applications.

Since the heating process of the indicator material can be interpreted as charging process, the transient description of microwave induced heating in FRMWT can be expressed as:

$$\Delta T(t) = \Delta T_\text{end} \cdot \left(1 - e^{-\frac{\beta_\text{th}}{\Delta T_\text{end}} \cdot \frac{P_\text{loss}}{c_\text{w} \cdot m} \cdot t}\right) = \Delta T_\text{end} \cdot \left(1 - e^{-\frac{t}{\tau_\text{th}}}\right). \tag{21}$$

In (21), the parameter β_th is called thermal retardation, a form factor that compensates measurement setup geometries etc. Once the thermal retardation is evaluated the time constant τ_th can be obtained. It combines all the previous theoretical considerations from electromagnetics in Section 2.1 and thermodynamics in Section 2.2. This time constant is of great interest, because it gives indication for the compromise that needs to be found for FRMWT-measurements: On the one hand, the induced temperature differences of the indicator should be as high as possible in order to meet the thermal resolution of the utilized recording device, such as a thermal camera. On the other hand, the footprint recording should be performed as fast as possible after initial excitation in order to prevent image blurring and further non-linear behavior such as conduction effects. Therefore, we propose to take the thermal snapshot for field illustration purposes within the linear part of the transient model described in (21), i.e., for

$$\tau_{th} < t_{rec} < 3 \cdot \tau_{th},\tag{22}$$

So that the heating process already reached between 68% and 95% of the final temperature.

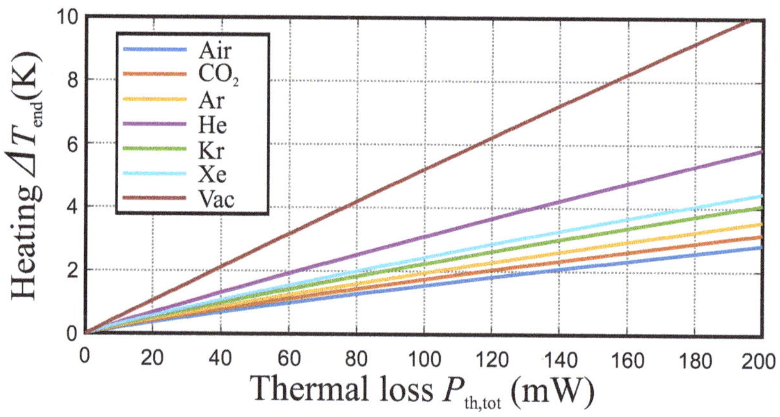

Figure 5. Theoretical temperature change for maximum thermal losses in steady state and different gaseous environments.

4. Design Materials

For the manufacturing of adequate indicator material foils, several material mixtures have been investigated. In order to meet the materials requirements, i.e., providing high dielectric losses by keeping the conductivity low and enabling the possibility of forming robust foils, we chose epoxy resin as matrix material and carbon black as additive. The utilized epoxy resin Breddepox E300 from Breddemann is a low viscosity, two component casting resin, providing a high UV resistance. As carbon black, we used the DUREX 0-Powder from manufacturer Orioncarbons. This amorphous carbon black provides low conductivity, offers very good processability, exhibits low compression set, and has good dynamic properties by keeping high elasticity. For tuning and fitting the mixing equation, we produced several test mixtures with different carbon black volume fractions ζ_{cb} between 0.01 and 0.09, which than were characterized by means of a DAK1.2E dielectric probe kit from SPEAG. As mentioned earlier, the measurement of high loss materials can lead to inaccuracies, especially for the permittivity's imaginary part. Figure 6a–c show the broadband measured real and imaginary part as well as the loss tangent, respectively, for a volume fraction of $\zeta_{CB} = 0.07$. In (a) we added a fitting curve, which is the result of a least-square-fit of the Debye models permittivity from (16). While in (a) this curve perfectly balances the measured values, it absolutely mismatches the measured imaginary part in (b). Therefore, we added CVC-processed curves from both measurement and fitting data. Obviously, as the fitting curves and CVC-processed curve are nearly identical, the directly measured imaginary part is highly erroneous and will be ignored further on.

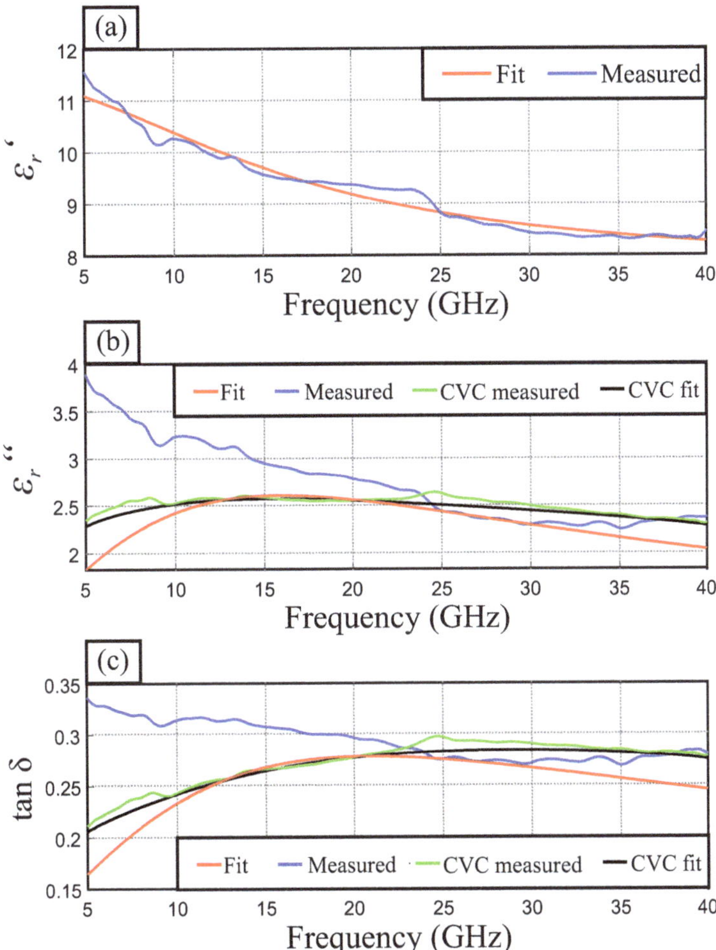

Figure 6. Exemplary results of the mixed material investigation at an carbon black volume fraction of $\zeta_{cb} = 0.07$, indicating the real (**a**) and imaginary part (**b**) permittivity as well as the loss tangent (**c**) for (—) measured values, (—) Debye model fitted values, (—) CVC of measured values and (—) CVC of Debye model fitted values.

Consequently, we used the direct-fitted data for setting up the mixing equations. Figure 7 shows several 3D plots indicating the properties of processable indicator materials such as the complex permittivity (a) + (b), total reflection coefficient (c), and the expected, maximum heating (d) for the scenario described above. In the following investigation, we are using an indicator material foil with a volume fraction of $\zeta_{CB} = 0.05$, which allows for homogeneous mixing and a temperature increase of more than $2\,°\text{C}$ at $10\,\text{GHz}$ when applying $500\,\text{mW}$ microwave power.

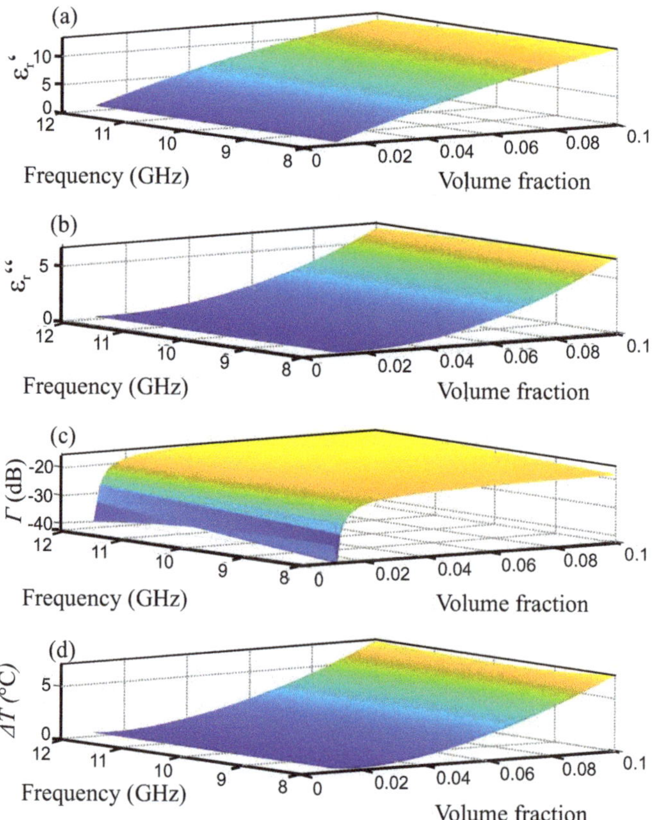

Figure 7. Resulting parameters of the fitted material mixing equation showing (**a**) real part permittivity, (**b**) imaginary permittivity, (**c**) total reflection coefficient, (**d**) maximum heating. Subfigures (**c**,**d**) apply for the above described indicator material and measurement scenario.

5. Results

In order to investigate the applicability of FRMWT and to validate the presented models, several simulations and measurements have been performed on a simple test scenario. The chosen scenario is the fundamental TE10-mode within a WR90 rectangular waveguide utilizing a IEC 60154-2:2016 flange as introduced in Figure 4. The indicator material was selected from the considerations made in Section 4 and provides a complex permittivity of $\underline{\varepsilon}_r = 7.2 + j1.6$; as operating frequency we chose 10 GHz and an excitation power of 500 mW in both, simulations and measurements. As the waveguide will operate in fundamental mode, we expect a *cos*-shaped E-field distribution along side the x-axis of the waveguide. However, since the square of the electrical field strength contributes the thermal heating as shown in (6) we expect a cos^2-shaped heat distribution. It needs to be mentioned that because the waveguide is truncated it will excite evanescent modes as well, which will slightly interfere with the fundamental mode resulting in minimal distortion of the thermal foot print.

5.1. Measurement Setup

From theoretical approximations in Section 2.2 and 3, we can conclude that the observable temperature increase for FRMWT applications will be only several kelvin, when applying reasonable microwave power. Therefore, we need to build a measurement environment, which prevents thermal and environmental disturbances. In this work, the thermal footprint of the device under test will be recorded by means of a high resolution thermal camera. Usually, these kind of cameras are portable so that it can be integrated into a measurement chamber. Further key properties of this chamber are:

- Housing:
  The measurement chamber needs to be closed and air tight for enabling different measurement environments such as gases. Moreover, it must prevent forced convection.
- Utilized Materials:
  The housing should be impervious for infrared radiation for preventing disturbing environmental reflections. The metal content should be kept low or needs a microwave absorber covering in order to avoid disturbing reflections.
- Peripheral Connections:
  Gas in- and outlets allow for FRMWT in different surrounding gases. Microwave connectors are necessary for feeding the structure under test.

The constructed measurement chamber was built of Plexiglas plates as these sufficiently attenuate infrared propagation within the spectrum of interest between 3 and 15 μm. The chamber measures a total size of (height, width, length) 50 cm × 55 cm × 45 cm. Gas in- and outlets can be used for flooding the whole setup with gases, while an SMA-through-hole connector allows for the connection of microwave devices within a frequency range of 0.1–30 GHz. Furthermore, an optical breadboard as lower wall allows for an universal connection of mechanical components. Figure 8 shows a photography of the described measurement chamber. As thermal camera we used a VarioCAM HD research 900 from Infratec. This thermal camera records a spectral range from 7.5 to 14 μm by means of an uncooled microbolometer focal plane array, which results in an image resolution of 2048 × 1536 pixels and a thermal resolution of 20 mK. Because the indicator material was manufactured from epoxy resin as matrix material we used an emissivity of 0.96 and a spectral range of 8 μm to 14 μm for all recordings, which refers to the values recommended by the thermal camera manufacturer [26]. The influence of further parameters is neglected in this work, because we mainly focus on temperature differences and its evolution over time.

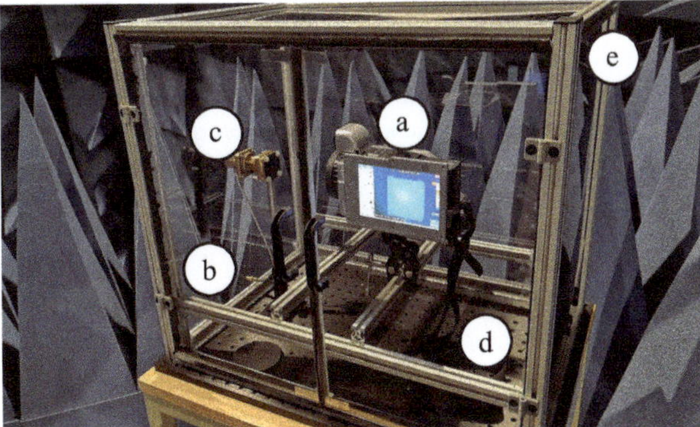

Figure 8. Photography of the realized setup containing: (**a**) Thermal camera, (**b**) through-hole microwave connection, (**c**) microwave device under test, (**d**) universal mechanical mounting board, (**e**) rear absorber wall.

5.2. Simulation and Measurement Results

In order to investigate the applicability of the indicator material the FRMWT-concept, multi-physics simulations as well as measurements with a high definition thermal camera were performed and compared. For all simulations we used the simulation software CST Microwave Studio 2020 , which allows for the coupling of three-dimensional full wave analysis with a thermal simulation of the corresponding loss distributions. In terms of the microwave investigation the time domain solver with a simulation accuracy of -30 dB was utilized. In order to ensure accurate simulation results, a hexahedral mesh type was used with a mesh of approximately 1.2 million meshcells. In a subsequent step, thermal simulations were carried out, basing on the microwave simulation results in order to track the heating behavior over time and to illustrate the temperature distribution caused by the microwave losses.

For this purpose, the convective heat transfer coefficient α_{th} was set to $15.3 \, \mathrm{Wm^{-2} \, K^{-1}}$. Due to the gap between material and waveguide, convection was assumed on both sides. Thermal conduction was not considered further due to the experimental setup. In addition, the material parameters of the indicator foil were included. These contain the previously measured permittivity $\underline{\varepsilon}_r = 7.2 + \mathrm{j}1.6$. Further material parameters were calculated from the raw materials' datasheets in order to agree with the measurement setup. Here, we set the heat capacity to be $c_w = 894 \, \mathrm{J \, (kg \, K)^{-1}}$, the mass density to be $\rho = 1100 \, \mathrm{kg \, m^{-3}}$ and the material thickness to $d = 0.9$ mm. The remaining geometrical dimensions apply as described above. All simulations were carried out with an microwave excitation power of 500 mW. The following thermography images were captured on a truncated WR90 rectangular waveguide surrounded by an IEC 60154-2:2016 flange. The test recording was performed at an operating frequency of 10 GHz and a power of 500 mW, for keeping the measurement results comparable to the simulation. For the verification of the proposed transient model from Equation (21), we recorded the setup's hotspot temperature over time in both simulation and measurement. By subtracting the initial temperature, the temperature increase over time is obtained, which is presented in Figure 9 and compared to the transient FRMWT model from (21) for different thermal retardation values β_{th}. The steady state temperature difference was determined to be 2.5 K and 2.71 K in simulation and measurement, respectively. The discrepancy between simulation and measurement is 0.21 K, which is an acceptable, relative difference of 7.7% and can be explained by a different implementation of the thermal losses within the simulation tool. Because of the slightly different final temperature, the transient behavior also slightly differs between simulation and measurement as presented in Figure 9. However, the measurement fits very good to the transient FRMWT model from (21) when choosing the thermal retardation to $\beta_{th} = 0.35$. After the determination of all relevant parameters, the thermal time constant is calculated to be $\tau_{th} = 12$ s. Consequently, the following thermal images are recorded at $t_{rec} = 35$ s, which perfectly satisfies Equation (22). The recorded heat distributions, resulting from simulation and measurement, are shown in Figure 10a,b, respectively. In both images, the hotspot is located in the waveguide's center position as expected because of the E-field maximum of the fundamental mode. Towards the edges of the material the heating decreases as expected. However, in y-direction the expected rectangular shape is flattened due to a remaining thermal conductivity and fringe fields in the flange-material-interspace.

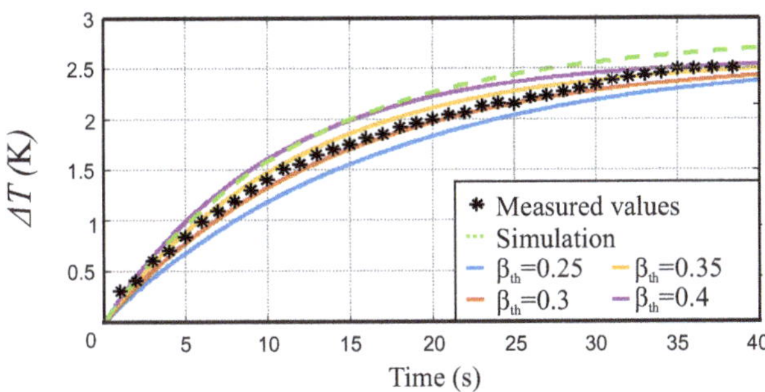

Figure 9. Transient temperature observation of the investigated waveguide's center, indicating the FRMWT model for different model parameters (solids), simulation results (– –) and measurement results (∗).

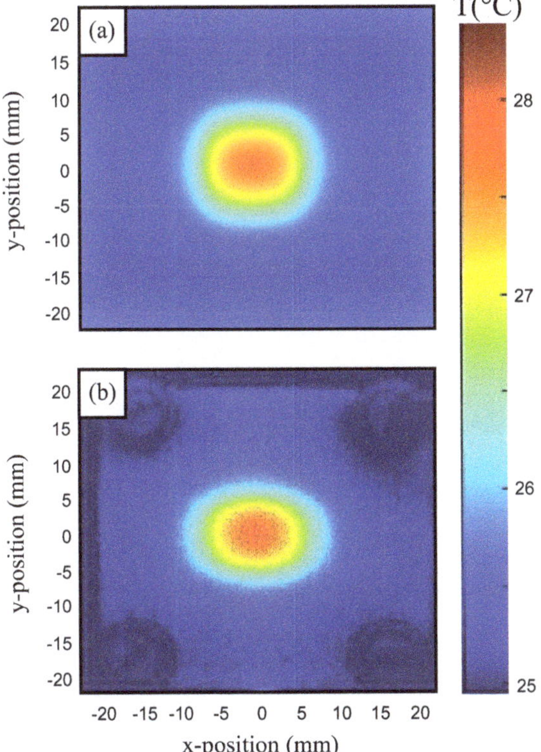

Figure 10. Thermography images of the investigated waveguide from (**a**) simulation and (**b**) thermal camera.

Figure 11 shows the temperature distributions alongside the waveguide's x-axis in comparison to a normalized $\cos^2$ function. Both, simulation and measurement results, clearly reveal the expected field distribution. However, the simulation result fit the $\cos^2$-reference a little better. An explanation for this could be the combination of evanescent fields at the truncated waveguide flange as well as the neglected thermal conduction in simulation.

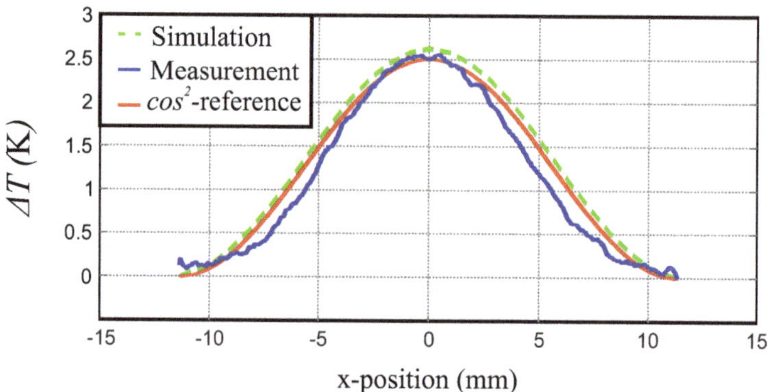

Figure 11. Temperature distribution of the heating in the waveguide's center alongside the x-axis for simulation (– –) and measurements (–) compared to the theoretical cos^2-shape (–).

6. Discussion and Conclusions

In this contribution, we present a method called Field Representation Microwave Thermography (FRMWT) for the fast and accurate recording of electromagnetic fields in the microwave range. Theoretical background from microwave engineering, thermodynamics, and fluid mechanics has been derived in order set up a holistic theory, which allows the prediction of the desired field illustration. Regarding the necessary indicator material, the so called complex valued completion (CVC) has been introduced, which allows for the measurement validation of high loss materials. Investigations on mixed high-loss materials showed the necessity of the CVC for dielectrics, as straight forward measurements delivered erroneous values for the material permittivity's imaginary part. Further, within simulations and practical experiments, we demonstrated the applicability of FRMWT using the example of the fundamental TE10-mode of rectangular waveguides. Within these investigations, we proved both, the theoretical transient FRMWT model as well as the ability of the field illustration by recording the predicted cos^2-shape of the TE10 mode's thermal foot print. Compared to previous work, which already showed the applicability of field representation in low frequency ranges up to 100 MHz and in the optical region, this work accesses the gap-region of micro- and mmWaves from 1 GHz up to 100 GHz. Although the proposed theoretical FRMWT model makes use of several approximations, it provides an excellent and holistic foundation for future measurement setups. The introduced lossy material modeling and complex valued completion combines previous work on dielectric material mixing and gives easy access to future material modeling. Of course, its accuracy is limited to the preciseness of a priori material-knowledge and the accuracy of the necessary broad material characterization. The presented measurement setup for the FRMWT already showed a good functionality. However, as we still make use of an extremely expensive thermal camera, the current setup is only useful for laboratory measurements. In order to establish the FRMWT in practical fields, the thermal camera must be substituted. In future work, further investigations on more complex propagation modes at higher frequencies will be performed and a post-processing concept will be evaluated, which delivers key facts on the investigated mode such as the mode purity.

Author Contributions: Conceptualization, C.B.; methodology, C.B. and T.M.; software, K.O.; validation, B.H. and C.B.; formal analysis, C.B., K.O. and B.H.; investigation, C.B.; resources, T.M.; data curation, C.B.; writing—original draft preparation, C.B.; writing—review and editing, C.B.; visualization, C.B.; supervision, C.B. and T.M.; project administration, C.B. and T.M.; funding acquisition, C.B. All authors have read and agreed to the published version of the manuscript.

Funding: This research was funded by MERCUR Research Center within the TORONTO project (AN-2019-0026).

Institutional Review Board Statement: Not applicable.

Informed Consent Statement: Not applicable.

Data Availability Statement: Not applicable.

Conflicts of Interest: The authors declare no conflict of interest.

Abbreviations

The following abbreviations are used in this manuscript:

FRMWT	Field Representation Microwave Thermography
CVC	Complex Valued Completion

References

1. Thompson, J.E.; Kristiansen, M.; Hagler, M.O. Optical measurement of high electric and magnetic fields. *IEEE Trans. Instrum. Meas.* **1976**, *IM-25*, 1–7. [CrossRef]
2. Cecelja, F.; Balachandran, W. Electrooptic sensor for near-field measurement. *IEEE Trans. Instrum. Meas.* **1999**, *48*, 650–653. [CrossRef]
3. Li, Z.; Yuan, H.; Cui, Y.; Ding, Z.; Zhao, L. Measurement of Distorted Power-Frequency Electric Field With Integrated Optical Sensor. *IEEE Trans. Instrum. Meas.* **2019**, *68*, 1132–1139. [CrossRef]
4. Puranen, L.; Jokela, K. Simultaneous measurements of RF electric and magnetic near fields-theoretical considerations. *IEEE Trans. Instrum. Meas.* **1993**, *42*, 1001–1008. [CrossRef]
5. Yan, Z.; Wang, J.; Zhang, W.; Wang, Y.; Fan, J. A Miniature Ultrawideband Electric Field Probe Based on Coax-Thru-Hole via Array for Near-Field Measurement. *IEEE Trans. Instrum. Meas.* **2017**, *66*, 2762–2770. [CrossRef]
6. Pokovic, K.; Schmid, T.; Kuster, N. Millimeter-resolution E-field probe for isotropic measurement in lossy media between 100 MHz and 20 GHz. *IEEE Trans. Instrum. Meas.* **2000**, *49*, 873–878. [CrossRef]
7. Zhang, H.; Yang, R.; He, Y.; Foudazi, A.; Chengand, L.; Tian, G. A Review of Microwave Thermography Nondestructive Testing and Evaluation. *Sensors* **2017**, *17*, 1123. [CrossRef] [PubMed]
8. Baghdasaryan, Z.; Babajanyan, A.; Pdabashyan, L.; Lee, H.; Firedmann, B.; Lee, K. Visualization of Microwave Heating for mesh-Patterned Indium-tin-Oxide by a Thermo-Elastic Optical Indicator Microscope. *Armen. J. Phys.* **2018**, *11*, 175–179.
9. Lee, H.; ; Arakelyan, S.; Friedman, B.; Lee, K. Temperature and microwave near field imaging by thermo-elastic optical indicator microscopy. *Sci. Rep.* **2016**, *6*, 39696. [CrossRef] [PubMed]
10. Yang, B.; Dong, Y.; Hu, Z.; Liu, G.; Wang, Y.; Du, G. Noninvasive Imaging Method of Microwave Near Field Based on Solid-State Quantum Sensing. *IEEE Trans. Microw. Theory Tech.* **2018**, *66*, 2276–2283. [CrossRef]
11. Faure, S.; Bobo, J.F.; Carrey, J.; Issac, F.; Prost, D. Composant Sensible Pour Dispositif de Mesure de Champ Electromagnétique par Thermofluorescence, Procédés de Mesure et de Fabrication Correspondants. French Patent 1758907, 26 September 2017.
12. Ragazzo, H.; Faure, S.; Carrey, J.; Issac, F.; Prost, D.; Bobo, J.F. Detection and Imaging of Magnetic Field in the Microwave Regime With a Combination of Magnetic Losses Material and Thermofluorescent Molecules. *IEEE Trans. Magn.* **2019**, *55*, 6500104. [CrossRef]
13. Ragazzo, H.; Prost, D.; Bobo, J.-F.; Faure, S. Characterization of electromagnetic fields of radiating systems by thermo-fluorescence. In Proceedings of the 2020 International Symposium on Electromagnetic Compatibility—EMC EUROPE, Rome, Italy, 23–25 September 2020; pp. 1–6. [CrossRef]
14. Birle, M.; Leu, C. Dielectric Heating in Insultaning Materials at High DC and AC Voltages Superimposed by High Frequency High Voltages. In Proceedings of the International Symposium on High Volatage Engineering, Seoul, Korea, 25–30 August 2013.
15. Incropera, F.P.; de Witt, D.P. *Fundamentals of Heat and Mass Transfer*; Wiley: New York, NY, USA, 1996.
16. Herwig, H.; Schäfer, P. A Combined pertubation/finite-difference procedure applied to temperature effects and stability in laminar boundary layer. *Arch. Appl. Mech.* **1996**, *66*, 264–272. [CrossRef]
17. Moran, M.J.; Shapiro, H.N. *Fundamentals of Engineering Thermodynamics*; Wiley: New York, NY, USA, 1995.
18. VDI-GVC. (Ed.) Heat Transfer Modes and Basic Principles of Their Description. In *VDI Heat Atlas*; Springer: Berlin/Heidelberg, Germany, 2010. [CrossRef]
19. VDI-GVC. (Ed.) Heat Conduction and Overall Heat Resistances. In *VDI Heat Atlas*; Springer: Berlin/Heidelberg, Germany, 2010. [CrossRef]
20. Hattenhorst, B.; Mallach, M.; Baer, C.; Musch, T.; Barowski, J.; Rolfes, I. Dielectric phantom materials for broadband biomedical applications. In Proceedings of the 2017 First IEEE MTT-S International Microwave Bio Conference (IMBIOC), Gothenburg, Sweden, 15–17 May 2017; pp. 1–4. [CrossRef]
21. Gutierrez, S.; Just, T.; Sachs, J.; Baer, C.; Vega, F. Field-Deployable System for the Measurement of Complex Permittivity of Improvised Explosives and Lossy Dielectric Materials. *IEEE Sens. J.* **2018**, *18*, 6706–6714. [CrossRef]
22. Sihvola, A. *Electromagnetic Mixing and Formulas and Applications*; IET Electromagnetic Waves Series 47; IET: England, UK, 2008.
23. Chester, G.V.; Thellung, A. The law of Wiedemann and Franz. *Proc. Phys. Soc.* **1961**, *77*, 1005–1013. [CrossRef]

24. Jackson, J.D. *Classical Electrodynamics*, 3rd ed.; John Wiley& Sons: New York, NY, USA, 1999.
25. Weerasundara, R.; Raju, G.G. An efficient algorithm for numerical computation of the complex dielectric permittivity using Hilbert transform and FFT techniques through Kramers Kronig relation. In Proceedings of the 2004 IEEE International Conference on Solid Dielectrics, ICSD 2004, Toulouse, France, 5–9 July 2004; Volume 2, pp. 558–561. [CrossRef]
26. Leistner, C.; Löffelholz, M.; Hartmann, S. Model validation of polymer curing processes using thermography. *Polym. Test.* **2009** *77*, 105893. [CrossRef]

Article

Model-Based Systems Engineering Applied to Trade-Off Analysis of Wireless Power Transfer Technologies for Implanted Biomedical Microdevices

Juan A. Martínez Rojas [1,*], José L. Fernández [2], Rocío Sánchez Montero [1], Pablo Luis López Espí [1] and Efren Diez-Jimenez [1]

[1] Department of Signal Theory and Communications, Escuela Politécnica Superior, Universidad de Alcalá, Campus Universitario, Ctra. de Madrid a Barcelona km 33.600, 28007 Alcalá de Henares, Spain; rocio.sanchez@uah.es (R.S.M.); pablo.lopez@uah.es (P.L.L.E.); efren.diez@uah.es (E.D.-J.)

[2] Independent Consultant, Model-Based Systems Engineering Methodologist, 28807 Madrid, Spain; joseluis.fernandezs@upm.es

* Correspondence: juanan.martinez@uah.es

Abstract: Decision-making is an important part of human life and particularly in any engineering process related to a complex product. New sensors and actuators based on MEMS technologies are increasingly complex and quickly evolving into products. New biomedical implanted devices may benefit from system engineering approaches, previously reserved to very large projects, and it is expected that this need will increase in the future. Here, we propose the application of Model Based Systems Engineering (MBSE) to systematize and optimize the trade-off analysis process. The criteria, their utility functions and the weighting factors are applied in a systematic way for the selection of the best alternative. Combining trade-off with MBSE allow us to identify the more suitable technology to be implemented to transfer energy to an implanted biomedical micro device.

Keywords: trade-off analysis; medical MEMS; wireless power transfer

Citation: Rojas, J.A.M.; Fernández, J.L.; Sánchez Montero, R.; Espí, P.L.L.; Diez-Jimenez, E. Model-Based Systems Engineering Applied to Trade-Off Analysis of Wireless Power Transfer Technologies for Implanted Biomedical Microdevices. *Sensors* **2021**, *21*, 3201. https://doi.org/10.3390/s21093201

Academic Editors: Andrea Randazzo, Cristina Ponti and Alessandro Fedeli

Received: 25 March 2021
Accepted: 1 May 2021
Published: 5 May 2021

Publisher's Note: MDPI stays neutral with regard to jurisdictional claims in published maps and institutional affiliations.

1. Introduction

New sensors and actuators based on MEMS technologies are increasingly complex and quickly applied to new products. MEMS-based devices demand system engineering approaches, which were previously limited to very large projects, and it is expected that this need will increase in the future, particularly in biomedical applications.

During product research and development, one important step is the powering of the implantable microdevices [1]. For this choice, there are many different technologies that should be analyzed in order to properly select the best option for each product. This trade-off is normally carried out based on previous experience and/or using a classical weighted approach. However, when the trade-off is started from current literature reviews, the amount of data is so large that it becomes necessary to develop more complex and systematic tools for reaching the optimum selection for microdevice applications.

In this work, we propose the application of Model Based Systems Engineering (MBSE) to systematize and optimize the trade-off analysis process. The criteria, their utility functions and the weighting factors are applied in a systematic way for the selection of the best alternative. Combining trade-off studies with MBSE allows us to identify the more suitable technology to be implemented to transfer energy to an implanted biomedical MEMS device. At present, a fully automated algorithm able to perform the complete decision process in the design of a new device is unfeasible. MBSE and trade-off analysis involve a wise combination of science, engineering and art. However, this cognitive endeavor can be made more rigorous by applying certain rules and mathematical techniques. In this work, we show how such techniques are applied to a very difficult challenge, which tries to push the limits of the MEMS engineering to design an implantable medical device with size in the order of 1 mm^3. Logically, the selected criteria will be severely constrained by this demand.

In Section 2, the systems engineering methodology used in this study is briefly explained. It is a very general, comprehensive and powerful approach, but in this case, we show its application to the trade-off analysis of the best present Wireless Power Transfer (WPT) alternatives for the intended biomedical implantable microdevice.

In Section 3, the state-of-the-art WPT for medical applications is reviewed. In Section 4, we detail the different steps which permit a complete and accurate assessment of different engineering alternatives. This trade-off analysis is described in the context of a larger systems engineering design effort, which is explained in Section 5.

In Section 6, we comment on the results of the trade-off analysis and its implications. The information obtained in this study permits a logical selection of the best WPT system for our application and provides valuable clues to other kind of important studies, such as sensitivity analysis.

2. Trade-Off and Heuristics in the MBSE Methodology ISE&PPOOA

At present, there are several MBSE methodologies, more or less complete or general in scope [2]. The ISE&PPOOA methodology is a logically consistent approach to MBSE which combines the best features of the traditional and modern tools for optimal design; for example, N^2 charts with SysML diagrams. ISE&PPOOA proposes two views of the architecture of the modeled system [3]. These views are the functional architecture and the physical architecture. Below, we briefly describe each of them.

The functional architecture uses diverse SysML diagrams [4] and tables. We understand a function as a transformation to be performed by the system that consumes mass energy or data and generates new ones or transforms them.

Combining diagrams and tables, the functional architecture represents the functional hierarchy using a SysML block definition diagram. This diagram is complemented with activity diagrams for the main system functional flows to represent the system behavior. The N^2 chart is a table used for an interface description where the main functional interfaces are identified. A textual description of the system functions is provided, as well.

The physical architecture, or architecture of the solution, is developed in two main steps, the results of which are the so-called modular architecture and refined architecture. The modular architecture represents a first version of the solution architecture representing its main logical blocks. These logical block or modules are blocks allocating functions based on the principles of maximum cohesion and minimum coupling between them.

The transition from the modular architecture to the refined architecture is where we apply a combination of trade-off analysis and design heuristics. Although the scope of systems engineering application of trade-off studies is wider, trade-off analysis is prescribed here for choosing and ranking alternative solutions to be applicable to the system component level. Instead of trade-off analysis we recommend the use of design heuristics [3] to implement those non-functional requirements that apply to the architecture design so that they are implemented as design patterns or at the level of system connectors.

Solution architecture is represented by the system decomposition into subsystems and parts using a SysML block definition diagram. This diagram is complemented with SysML internal block diagrams representing the system physical blocks with either logical or physical connectors for each identified subsystem, and activity and state diagrams for behavioral description as needed. A tabular description of the system parts may be provided as well. Functional allocation may be represented either in tabular form or at the system blocks, allocation by definition, or as partitions in the activity diagrams, allocation by usage, represented using SysML notation.

We get benefit of this systematic approach for system engineering to develop an optimal trade-off selection method. In this case, we apply this methodology to WPT technologies for the specific case of an implantable MEMS device. The benefits of integrating the design process for a suitable medical WPT in the range of a 1 mm^3 into a complete MBSE design framework are manifold, due to its complexity, which demands the exploration of multiple alternatives that have not been already tested for such a small size.

3. State-of-the-Art Wireless Power Transfer Technology for Biomedical Applications

Many patients can take advantage of implanted biomedical devices. New micro-electromechanical systems (MEMS) technologies allow the development of increasingly complex and miniaturized devices. This reduces many risks associated with the treatment, from surgery to tissue compatibility, due to reduced invasiveness. Another very important factor for implanted devices is energy consumption. Permanent wires inside the body are an unacceptable solution in most cases, while batteries can be dangerous and must be replaced using surgery.

The potential of MEMS for low energy consumption opens the possibility of remote powering without external contacts. The technologies that permit remote powering are collectively known as Wireless Power Transfer (WPT). These technologies are relatively recent and are under development. Although there are many publications exploring detailed aspects of some technologies, there are relatively few reviews comparing different alternative approaches in a systematic and weighted way.

A very recent and complete review was written by Khan et al. in 2020 [5], who compared several WPT technologies using the following parameters: implant type, implant WPT system size, distance from power source to device, type of radiation and frequency, input power, efficiency, test model, SAR (specific absorption rate), safety considerations and technological maturity. The different WPT approaches studied in this article were: NRCC (Non-Radiative Capacitive Coupling), NRIC (Non-Radiative Inductive Coupling), NRMRC (Non-Radiative Magnetic Resonance Coupling), NRRMF (Non-Radiative and Radiative Mid-Field), RFF (Radiative Far-Field), APT (Acoustic Power Transfer) and OPT (Optical Power Transfer). The general conclusion of this review, after a qualitative study of the performance of these technologies, using a Low–Medium–High scale, was that NRIC and NRMRC were better than other WPT techniques due to their moderate size, range and higher PTE performance. Additionally, more complete studies on tissue safety exist for NRIC and NRMRC WPT. APT was comparable to NRIC and NRMRC WPT in terms of performance; the other technologies (NRCC, NRRMF, RFF and OPT) were still too immature from a technological point of view.

Another recent review was provided by Zhou et al. [6], who compared several WPT systems, investigating their key performances such as power transfer capability, power level, efficiency, safety requirements and some others. They organized their review by medical applications, instead of energy types or technologies, but their study was restricted to electromagnetic near-field WPT devices.

Moore et al., in 2019, reviewed the state-of-the-art electromagnetic WPT, especially magnetic resonance, in medicine [7]. They found 17 relevant journal papers and/or conference papers and separated them into defined categories: Implants, Pumps, Ultrasound Imaging and Gastrointestinal (GI) Endoscopy. They found no strong correlation between the system parameters and biomedical applications.

Mahmood et al., in 2019, proposed a new ultrasound sensor-based WPT for low-power medical devices [8]. A 40 kHz ultrasound transducer was used to supply power to a wearable heart rate sensor for medical application. The system consisted of a power unit and a heart rate measurement unit. The power unit included an ultrasonic transmitter and receiver, rectifier, boost converter and super-capacitors. At 4 F, the system achieved 69.4% transfer efficiency and 0.318 mW power at 4 cm. They also compared their work with previous ultrasound WPT systems. They remarked that power and efficiency decreased as the air gap increased to more than 4 cm.

Kakkar in 2018 designed an ultra-low power system architecture for implantable medical devices based on an embedded processor platform chip [9]. This work explored the partitioning of a chip, as well as the trade-offs associated with design choices, especially intelligent power management. He described several design requirements implying ultra-small volume, improved resolution for both sensing and stimulation, integrated electronics design, autonomous operation without batteries and intelligent power management. Concrete values for power operation were provided in figures, being in the order of 250 microwatts.

Shadid and Noghanian, in 2018, wrote a literature survey on WPT for biomedical devices based on inductive coupling, concentrating on the applications using near-field power transfer methods [10]. They compared different systems with the following parameters: frequency, output power, transmitter dimensions, receiver dimensions, gap and efficiency. They plotted efficiency versus delivered power and efficiency versus frequency for the reviewed systems.

Taalla et al., in 2018, presented a comparison of inductive and ultrasonic WPT techniques used to power implantable devices [11]. The inductive and ultrasonic techniques were analyzed studying their sizes, operating distance, power transfer efficiency, output power and overall system efficiency standpoints. They concluded that the inductive coupling approach can deliver more power with higher efficiency compared to the ultrasonic technique, but the ultrasonic technique can transmit power to longer distances.

Agarwal et al., in 2017, presented a comparison of various power transfer methods based on their power budgets and WPT range [12]. Power requirements of specific implants such as cochlear, retinal, cortical and peripheral were also considered. Patient's safety concerns with respect to electrical, biological, physical, electromagnetic interference and cyber security were also explored. Their conclusion was that EM, NRIC and NCC were better for high-power devices, but for low-power in the order of fewer milliwatts, ultrasonic, mid-field, or far-field technologies are promising.

Dinis et al., in 2017, presented a review of the state-of-the-art implantable electronic devices with wireless power capabilities, ranging from inductive coupling to ultrasounds [13]. They compared the different power transmission mechanisms and showed that the power that current technologies can safely transmit to an implant is reaching its limit. In order to overcome these difficulties, they proposed a new approach, capable of multiplying the available power inside a brain phantom for the same specific absorption rate (SAR) value. They compared previous devices using WPT link distance, antenna/transducer size, received power at the implant, link efficiency and the calculated power density (obtained by dividing the received power by the antenna/transducer size), and ultrasound seemed to be the best WPT solution. Moreover, biological energy harvesters for implantable devices using biologically renewable energy sources, such as muscle movement, vibrations or glucose, were briefly discussed.

The review of Kim et al., published in 2017, focused on Near-Field Wireless Power and Communication for biomedical applications [14]. They proposed that near-field magnetic wireless systems had advantages in water-rich environments, such as biological tissues, due to lower power absorption. However, various issues in near-field magnetic systems remained, such as transmission range, misalignment and limited channel capacity for communications. They suggested that mid-field coupling based wireless powering was convenient for smaller-sized implants using the sub-GHz range. Finally, they indicated that the Q-factor of the coils and their cross coupling are the primary factors that need to be taken into account for system performance optimization.

Lu and Ma (2016) made a review of the best architectures for efficient WPT, allowing further device miniaturization and higher power loss reduction [15]. The main contribution of this article was its best design guidelines for WPT systems based on near-field inductive coupling wireless power transfer. They remarked that operating at a higher WPT frequency can lead to significant size reduction of passive components and better Q values with smaller inductance. However, higher frequencies produced higher tissue absorption inside the human body. They explained that the selection of a correct WPT system architecture for portable or implantable biomedical applications was a trade-off between device volume, efficiency, regulation accuracy, speed and functionality.

Altawy and Youssef, in 2016, studied the trade-off between security, safety and availability in implantable medical devices [16]. They discussed the challenges and constraints associated with securing such systems and focused on the tradeoff between security measures required for blocking unauthorized access to the device and the safety of the patient in emergency situations where such measures must be dropped to allow access. They analyzed the up-to-date proposed solutions and discussed their strengths and limitations.

Safety and thermal aspects of implanted medical devices were described in [17] by Campi et al. in 2016. They studied a WPT system based on magnetic resonant coupling applied to a pacemaker for recharging its battery using a low operational frequency (20 kHz). Other safety considerations for an implantable rectenna for far-field WPT can be found in [18]. Other very informative specific monographs dedicated to WPT for biomedical devices can be found in [19–22].

All these reviews and monographs articles give exhaustive and detailed information about current trends in WPT. However, even with this information, so many alternatives make it difficult to properly select the most adequate solution for a specific application. Here, we propose a rigorous engineering approach for a trade-off study based on systems engineering and particularly using the models of the system developed by the ISE&PPOOA MBSE methodology. This new comprehensive approach permits conclusion with the most convenient choice and it also allows the ranking of the rest of the solutions for an eventual selection.

4. Steps of Trade-Off Studies in the ISE&PPOOA MBSE Context

Based on diverse processes for trade-off analysis found in the literature, we use here a trade-off analysis process that can be integrated with the ISE&PPOOA architecting design using its outputs and producing inputs to the ISE&PPOOA MBSE process presented in Chapter 4 of the ISE&PPOOA book [3]. Traditional approaches such as those found in the NASA report [23] in 1994 do not use SysML system models. However, recent approaches such as IBM use SysML notation and diagrams [24]. The steps of the proposed trade-off subprocess of ISE&PPOOA are presented in Figure 1.

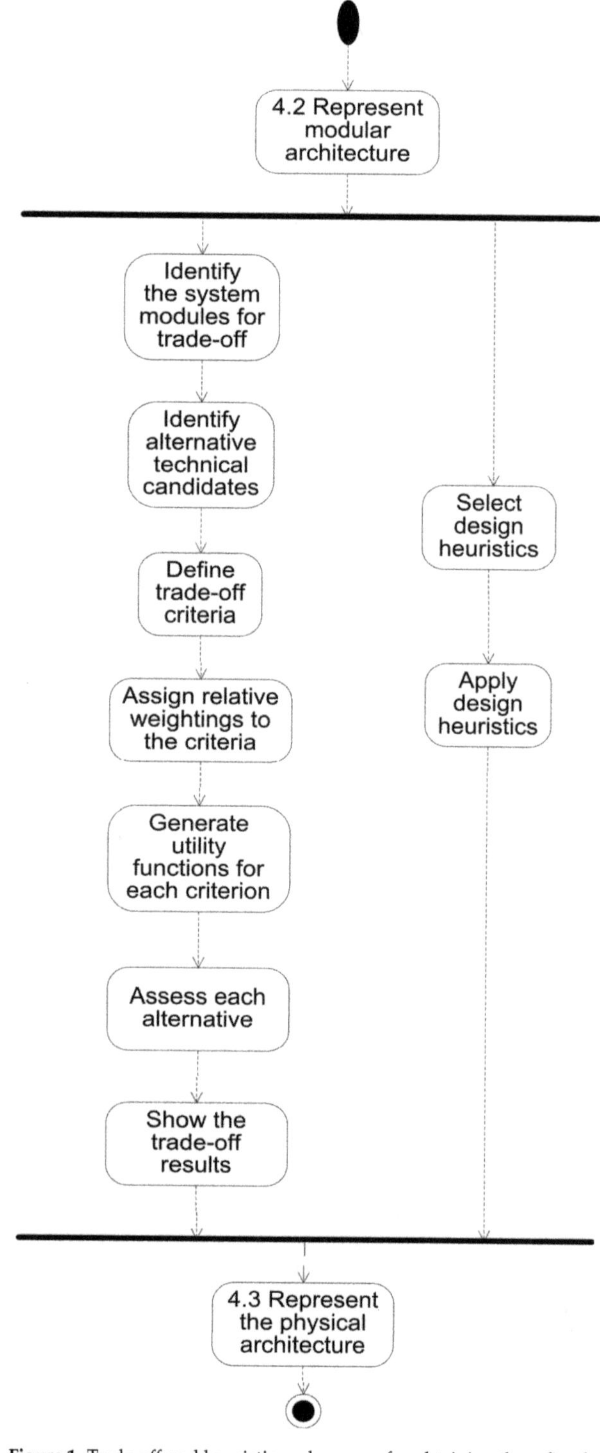

Figure 1. Trade-off and heuristics subprocess for obtaining the refined physical architecture.

4.1. Identify the System Modules for the Trade-Off Study

From the modular architecture obtained in step 4.2 of the ISE&PPOOA process described elsewhere [3], modules (blocks) are selected, which are the logical building elements clustering cohesive functionality. These may be implemented using alternative technical solutions that are identified in the next step.

4.2. Identify Credible Alternative Technical Candidates for Implementing the System Logical Building Blocks or Modules under Consideration

The list of technical alternatives selected during brainstorming sessions may be reduced if the system requirements are considered to have been met [25]. Some alternatives may be discarded based either on cost, technology readiness or other criteria used to eliminate alternatives. The remaining alternatives that need to be assessed should be described in detail.

4.3. Define Trade-Off Criteria

Objectives related to stakeholder needs and system requirements are transformed into a set of performance, cost and other criteria to be used for the trade-off study.

4.4. Assign Relative Weightings to the Criteria

There are diverse methods for deriving numerical values to the weights to be assigned to the criteria. If a pair-wise comparison is used, we recommend the Analytical Hierarchical Process (AHP) [23] to establish the relative weights to the criteria at the same level, so that all weights sum to 1.0. Another method we recommend and use in the present WPT case is the "swing weight matrix" [26]. Swing weights are assigned to the criteria based not only on importance but on the variation of their scales as well. The reason is that it does not make sense to consider one criterion more important than another without considering the degree of variation among the consequences for the alternatives under trade-off analysis.

The swing matrix is a matrix where the top row defines the value measure importance and the left column represents the range of value measure variation. As recommended by Parnell, weights should descend in magnitude as we move in the diagonal from top left cell to the bottom right cell of the swing weight matrix. Multiple criteria can be placed in the same cell [27].

4.5. Generate the Utility Function for Each Criterion

For the trade study, it is necessary to represent, as part of the system model, the selected assessment criterion and the utility or value functions associated to each criterion. Utility functions can be discrete or continuous. Utility functions follow three basic shapes: linear, curve and S shape curve.

For example, when an increasing utility function is created for a particular criterion, the systems engineer ascertains whether the project stakeholders consider it as the minimum value of the measure to be accepted, mapping it to the 0 value on the score scale (y-axis). The measure beyond which an alternative provides no additional value is mapped to the highest score scale (y-axis). It is important to pick the appropriate inflection points for drawing the curve, which may be either convex or concave.

4.6. Assess Each Alternative

Every alternative should be estimated for a given criterion in terms of its score, based on the applied utility function. Then, the relative weights assigned to each criterion are used to compute the objective function that combines the weights and scores. The sum combining function is frequently used. The assessment is based on the created utility functions, using criteria values from each alternative, obtained from test data, vendor provided data, simulations, prototypes, engineering practice or literature.

4.7. Show the Trade-Off Results

Generally, a summary table of criteria versus alternatives is presented to summarize the results from the preceding steps. Based on these results, a decision is made. A sensitivity analysis is recommended to determine the robustness of the alternatives selected based on their highest rank in the trade-off analysis [27].

Concurrently, design heuristics are selected and used to refine the architecture. The use of heuristics is recommended to implement non-functional requirements that cannot be allocated to some building blocks but are to be implemented as design patterns or layouts of connectors between the building elements.

5. Application of Trade-Off Analysis within the ISE&PPOOA MBSE Approach to Select the Best WPT Alternative for an Implanted Biomedical Device

In this work, we are interested in the design of a challenging WPT system able to power a newly created micromotor for medical intravascular surgery. This demands the previous study of the best technological alternatives for its implementation. Many severe constraints are associated with this problem, including lack of data in the intended size range of 1 mm^3. Due to this, a comprehensive design framework such as the ISE&PPOOA methodology is needed. We shall study the most promising WPT alternatives to power a 1 mm^3 MEMS device inside a human artery, using the information reviewed in the Section 3.

Below, we describe the main outcomes produced by performing the trade-off steps described in the previous section.

5.1. Identify the System Modules for the Trade-Off Study

The modular architecture previously obtained using ISE&PPOOA can be seen as an internal block diagram represented using SysML notation in Figure 2. This architecture was designed for a complex autonomous implanted system with sensors, actuators and communication capabilities. In order to simplify the trade-off study, we limit ourselves to the system parts related to WPT, functionality that can be reduced to the blocks "Power Source" and "Internal Electrical Power Generator".

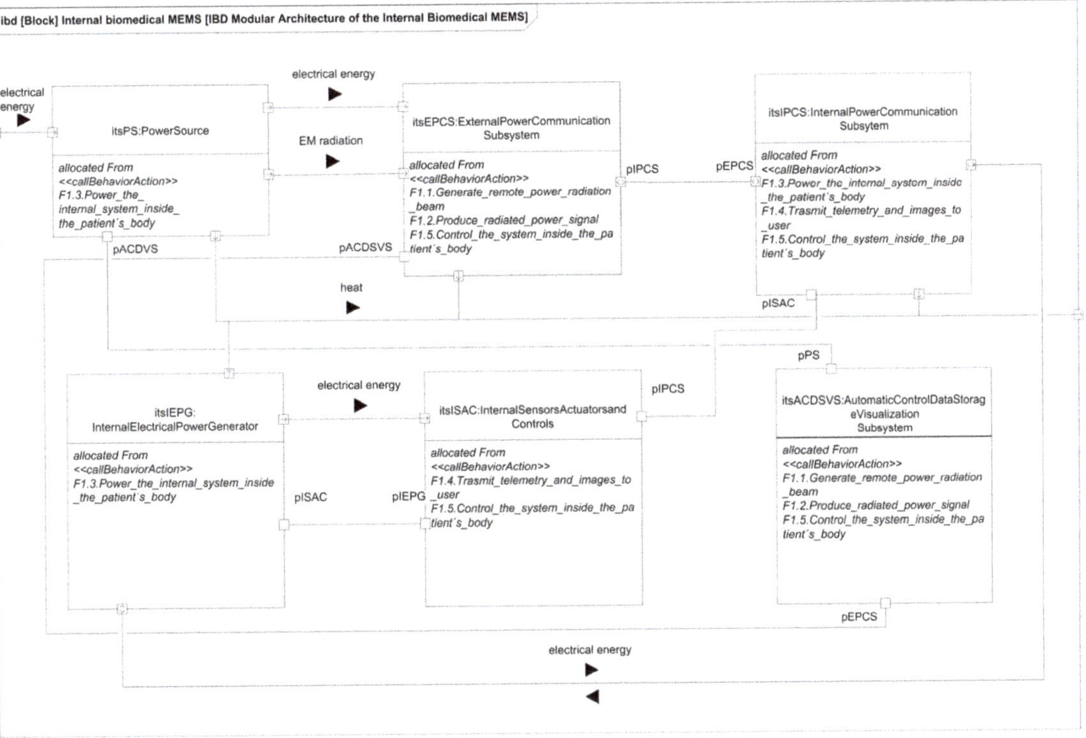

Figure 2. Logical blocks of a biomedical implanted system with WPT.

5.2. Identify Credible Alternative Technical Candidates for Implementing the System Logical Building Blocks or Modules under Consideration

After a comprehensive literature review of the state-of-the-art WPT, summarized in the Introduction, possible candidates to power our MEMS-based device in the one cubic millimeter range are:

- Non-Radiative Inductive Coupling;
- Non-Radiative Magnetic Resonance Coupling;
- Non-Radiative Mid-Field;
- Radiative Far-Field;
- Acoustic Power Transfer.

5.3. Define Trade-Off Criteria

Based on the system requirements [25] related to energy transfer and energy harvesting by the device inside the patient body, we select the following trade-off criteria:

- Input power (W), the initial delivered power to the implanted medical device outside the body;
- Power transfer effectiveness (%), the ratio of power produced by the implanted device and the input power;
- Implant WPTRx size (mm), the largest dimension of the implanted medical device;
- Effective operation distance (mm), the maximum distance between the external power source and the implanted device in air (or water for ultrasound) for successful performance;
- Specific absorption rate, SAR (W/kg), the power absorbed per mass of tissue;
- Mechanical complexity (low–medium–high), which depends on the number, size, shape and materials of the system parts. It is also directly related to manufacturing

costs. The higher the number of parts, the smaller their size, the more complex their shapes and the costlier their materials, the higher will be the mechanical complexity;
- Technical maturity (low–medium–high), an abstract measure of the degree of consolidation and performance of a technological solution.

Other safety criteria besides SAR and resilience impact the architecture at the connectors level, as well, so we recommend the use of heuristics instead of trade-off studies to implement these safety and resilience requirements.

Both the criteria selection process and the application of heuristics cannot be fully automated without human intervention, so a certain degree of subjectiveness, but also creativity, is unavoidable. Additionally, many unknown variables are present in the microdevice design, due to the exploratory and frontier research nature of this project. Thus, the trade-off criteria have been selected trying to constrain the space of alternatives (trade-space) as much as possible in order to produce a reasonable solution for a physical architecture, avoiding the introduction of spurious information that cannot be confirmed at present in the intended range of sizes. Another important consideration in the selection process was that the criteria values obtained from the literature revision could be reasonably extrapolated to our design objectives to build the utility curves. The final result of this process is the refined physical architecture of the best WPT alternative (APT) after the trade-off study, which can be seen in Figure 3.

It is not easy or even possible to know a priori which set of trade-off variables will be optimal for every problem, because it is the result of many constraints, especially the stakeholders' needs [25], which are usually presented in an ambiguous or less than desirable rigorous way. A very useful piece of advice to confirm that the criteria selection process is correct is based on the exploration of the resulting utility curves. This is a powerful approach to refine the criteria selection process itself in an iterative way, which is inherent to the MBSE design approach. From a mathematical point of view, the curves should be smooth (continuous and differentiable) and restricted to a few reasonable types, such as polynomial, logarithmic or exponential functions. The extremal and inflection points should indicate critical values of the corresponding trade-off criterion. Any lack of regularity, bijectivity (with suitable domain restrictions if necessary, like in exponential or even degree polynomials, for example, in order to be invertible) or oscillations indicate some serious problem with the utility curve and the associated trade-off variable, which must be corrected either by changing the criterion completely or revising the prescribed values. In our case, it can be observed that all utility curves in Figures 4–10 fulfill these mathematical conditions. These mathematical-based "metacriteria" for trade-off analysis verification are not explicitly described in the systems engineering literature, as far as we know, and can be considered a valuable contribution.

5.4. Assign Relative Weightings to the Criteria

We apply the swing weight matrix for our trade-study because it considers variation in the measured range as well as importance. Thus, the swing weight matrix is more complete than other approaches that only consider importance. The swing matrix obtained for the trade-off criteria selected in the previous step is shown in Table 1. Weights are assigned based on the technical literature review, previous experience with MEMS projects and stakeholders' needs. One common criticism of trade-off analysis based on weighting criteria outside the systems engineering community is that it involves some degree of subjectiveness. However, this human assessment is unavoidable, because there are many competing stakeholder interests, generally formulated in an incomplete and ambiguous manner, in addition to very complex physical and technical constraints. At present, no algorithm or artificial intelligence approach is able to complete this step automatically without human assistance. Similar observations can be applied to other widely used decision-making approaches such as Analytical Hierarchy Process.

Table 1. Swing matrix of the studied criteria. Weights are assigned based on technical literature values and stakeholders needs.

		Level of Importance of the Value Measure		
		Very Important	**Important**	**Less Important**
Variation in Measure Range	*High*	Input power: 100 Power transfer effectiveness: 100		
	Medium	SAR: 75		
	Low	Implant WPTRx size: 90	Mechanical complexity: 50	Effective operation distance: 25 Technical maturity: 25

5.5. Generate the Utility Function for Each Criterion

Utility functions for the selected criteria can be seen in Figures 4–10. These utility functions are built and represented using the recommendations described in ISE&PPOOA book [3] admitting the mathematical conditions described in more detail in Section 5.3.

5.6. Assess Each Alternative and Show the Trade-Off Results

The results of this trade-off analysis can be seen in Table 2. Weights have been normalized with respect to the total weight sum of the swing matrix values (weight sum = 4.65), converting the swing weights into measure weights (or global weights), so that their sum is now 1. The final weighted sum for a given technological alternative is calculated as the sum of the products of the measure weights of each criterion by the values obtained from the utility curves for the same criterion. In other words, the values of the weights column are multiplied by the corresponding utility function values of a given alternative column and then summed to produce the final score. Comparing the obtained scores, we can select the preferred alternative. In this case, the Acoustic Power Transfer solution is the best, followed by the Radiating Far Field technology.

Table 2. Summary of results of the trade-off analysis. Weights are normalized with respect to the total weight sum of the swing matrix values (weight sum = 465).

Swing Matrix Values		Utility Curve Values (for the Best Case)				
Criteria		**NRIC**	**NRMRC**	**NRMF**	**RFF**	**APT**
Input power	0.215	9	5	6	9	6
Power transfer effectiveness	0.215	3	7	4	9	9
Implant WPTRx size	0.193	9	9	8	5	9
Effective operation distance	0.054	1	3	5	7	7
SAR	0.161	1	3	5	3	7
Mechanical complexity	0.107	9	7	7	5	5
Technical maturity	0.054	9	9	7	7	5
	Weighted sum	5.981	6.197	5.896	6.609	7.272

6. Results and Discussion

The results indicate that the most appropriate technology is APT (Acoustic Power Transfer) due to its transfer effectiveness, size, effective operation distance and SAR. After we identified the technology, we were able to decompose in more detail the system blocks that have to implement it, and we used a BDD diagram made with standard SysML notation (Figure 3).

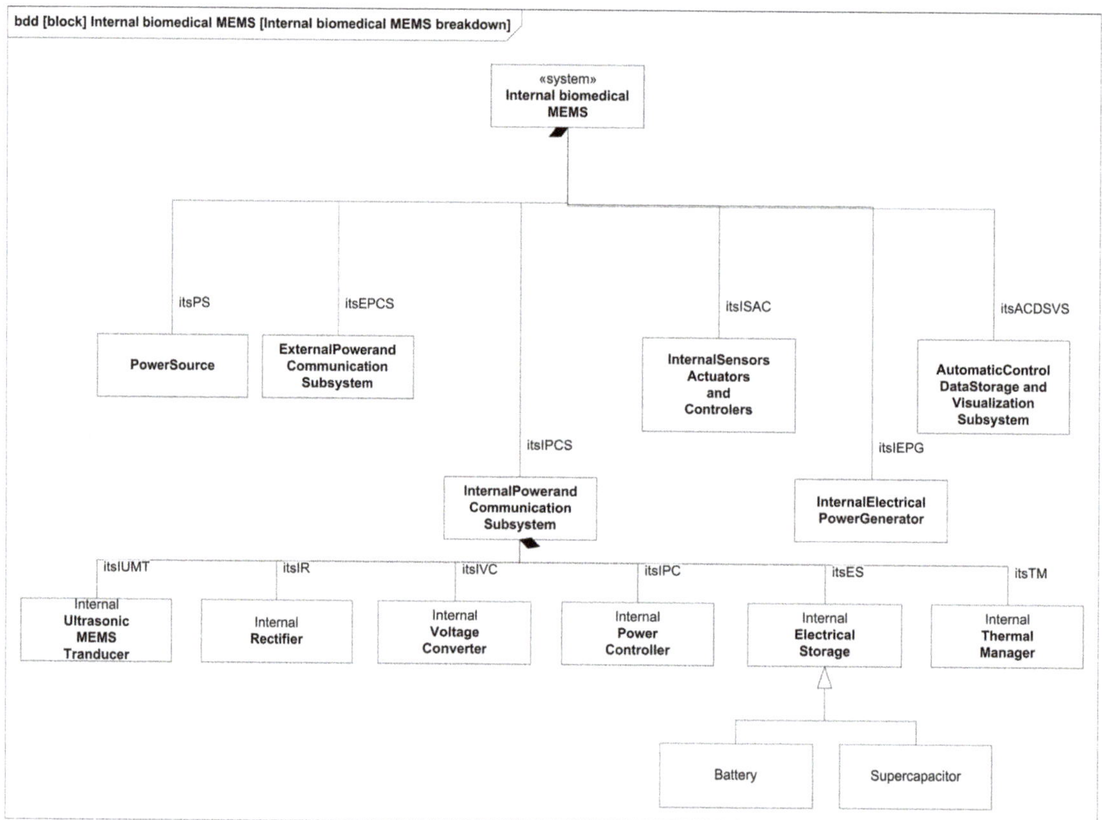

Figure 3. Refined physical architecture of the best WPT alternative (APT) after the trade-off study.

From the detailed decomposition of these blocks, it is possible to build, by using design heuristics and design patterns, the IBD diagrams that would define the refined architecture of the solution which, for brevity reasons, are not shown here.

Utility curves were obtained compiling relevant data from the revised scientific publications, as shown in Section 3, and verifying its mathematical correctness following the conditions explained in Section 5.3.

Specifically, data for the utility curves were collected from:

Table 2 (page 46) of reference [5];

Table 1 (page 33–34) of reference [7];

Table 1 (page 4) of reference [8];

Table 2 (page 4–5), Table 3 (page 6–7) and Table 4 (page 8) of reference [10];

Table 1 (page 2101) and Table 2 (page 2102) of reference [11];

Table 1 (page 9) of reference [13].

The values correspond to the best results of the state-of-the-art technologies combined. We used a quantitative safety criterion (SAR), but safety heuristics could be applied to refine the architecture as well.

A sensitivity analysis is recommended to determine the robustness of the alternatives. A complete sensitivity analysis is not performed here, but a very useful qualitative analysis can be performed from the information derived from the utility curves, which are far richer in content than their mere use as trade-off tools could suggest.

Input power (Figure 4) is a decreasing linear function in a semilogarithmic scale, which implies that its dynamic range is large. Thus, a small variation in other variables can produce a large variation in the required input power, making this variable very sensitive to small changes in the design for all the studied alternatives. This variable is also one of the most important, so that it can be deduced that the overall performance of the WPT system will be very sensitive. Thus, the input power values can vary for the same alternative with different values of the other variables. Moreover, input power is very relevant from a safety point of view, because it is limited to levels which cannot produce any damage, pain or discomfort on the skin or inside human tissues. Other safety considerations related to input power would be possible electromagnetic interferences and damage to other devices or to human operators.

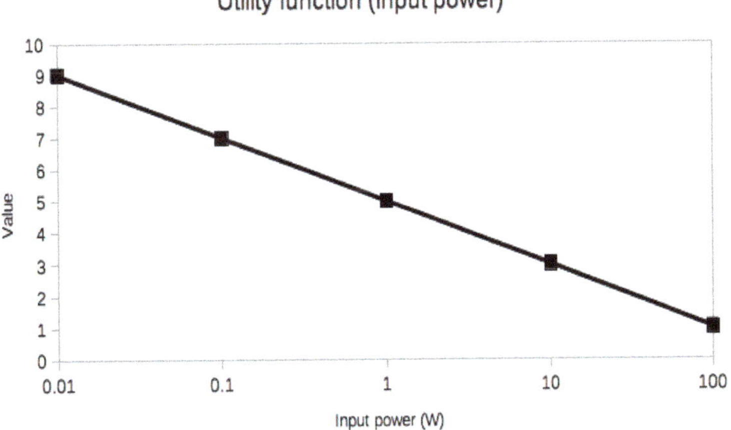

Figure 4. Utility function for Input Power.

Power Transmission Effectiveness (PTE) (Figure 5) is a key variable with the largest range of variation of all studied criteria, even larger than the input power; it is the most sensitive value and the latest technological advances are critical to fix it. Its mathematical behavior is an increasing linear function in a semilog scale. For example, ultrasound-based energy harvesters only a few years ago could have a PTE as low as 0.001%, while the most recent ones can achieve 40%. Obviously, such a large increment has changed this alternative from a feasible WPT technology for medical implants to the most promising one in our trade-off analysis for MEMS. Thus, the two most sensitive variables, input power and PTE, have opposite contributions to the performance of the WPT system, because we want the largest PTE with the lowest input power.

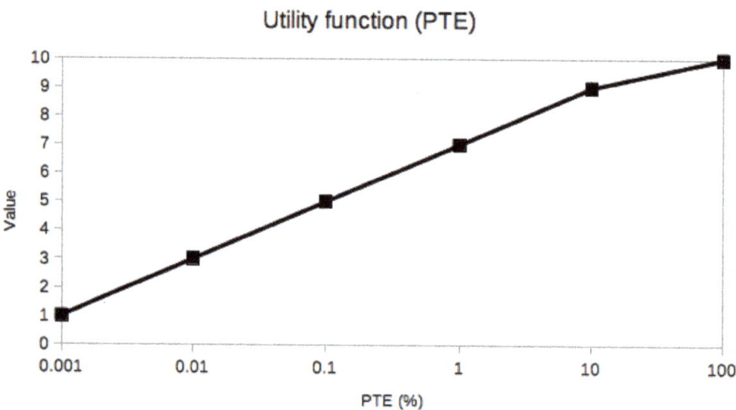

Figure 5. Utility function for Power Transmission Effectiveness.

Implant size (Figure 6) is one of the most important constraints in our example due to an extreme constraint: it must be fit in a volume of the order of 1 mm^3. However, PTE decreases exponentially and power input increases exponentially when the device size is decreased, even by small amounts, due to the small range of variation of this variable with a decreasing linear utility function. A sensitivity analysis would reveal that even small building tolerances in the manufacturing process with current technologies could radically change the performance of the final implant, which is unacceptable for a medical device.

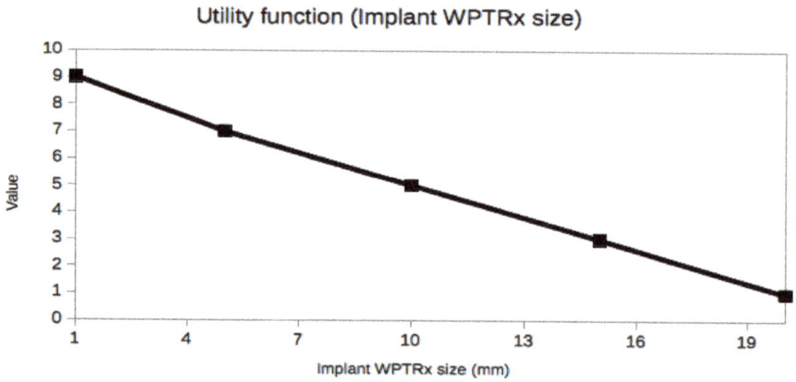

Figure 6. Utility function for Implant WPTRx Size.

The effective operation distance (Figure 7) has a nonlinear response curve. This variable determines the depth of the implant. It can be observed that present technologies do not allow very deep implants in practice if millimeter size devices are desired. The range of variation is small, so that small variations in depth or distance from the external power source or both can produce very large changes in input power and PTE. However, in this case, the sensitivity is even worse than in the case of the device size, because the utility function increase is not linear and the steepest variation can be seen in the smallest distance values, in the range of a few mm. This implies that in order to have a stable power supply, very strict positioning mechanisms should be used, which could be impractical or even unfeasible if the implant can or must be moved.

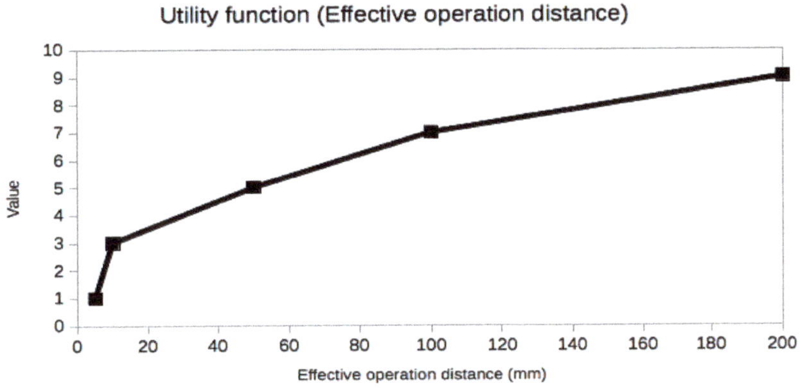

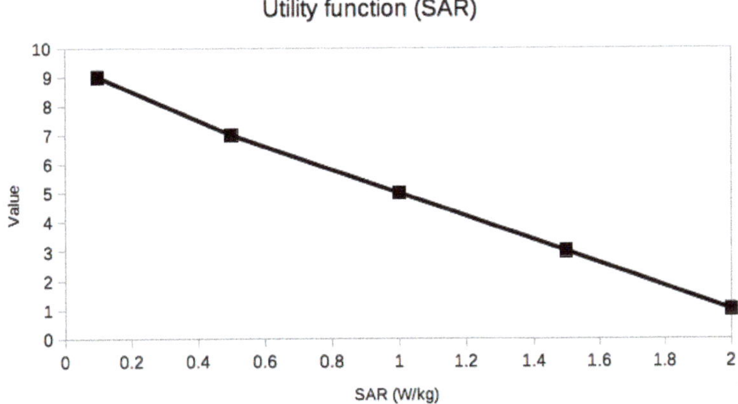

Figure 7. Utility function for Effective Operation Distance.

SAR (Figure 8) is one of the most important safety related metrics, although the input power has to be limited as well. We can see that its range is severely constrained and most of the state-of-the-art devices are dangerously near the limit for the best values of the other variables. Thus, a sensitivity analysis would indicate that even a modest decrease in the SAR values in order to comply with legal restrictions would enormously impact the values of input power and PTE, surely forcing the implant size to be considerably larger, frustrating the goal of 1 mm^3 total volume if radically new design ideas are not found.

Figure 8. Utility function for Specific Absorption Rate.

The last two variables, mechanical complexity (Figure 9) and technical maturity (Figure 10), are very different from the previous ones. They are abstract metrics that try to describe the effort and cost of building the WPT system without using economic or monetary values, presently unknown, because we are dealing with a present research project pushing the limits of near-future MEMS technology. In fact, their values are more categorical than numerical, although an easy conversion can be done in order to draw the utility curves, which are simply linear in these cases. In this study, these two variables are not as critical, but within a constrained project budget, much more detailed utility curves should be obtained in this regard. Surely, with accurate costs, their influence would severely constrain the feasible technological options and the conclusions of the trade-off analysis could be very different.

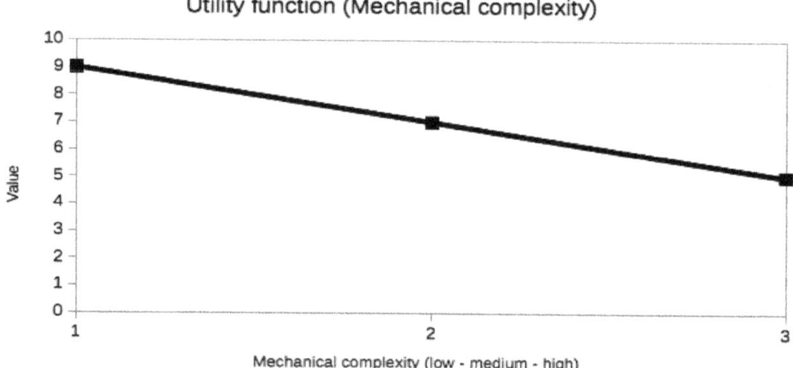

Figure 9. Utility function for Mechanical complexity.

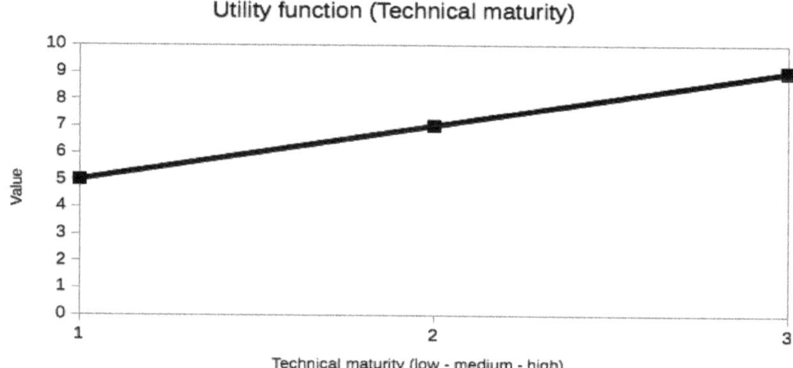

Figure 10. Utility function for Technical maturity.

7. Conclusions

The combination of MBSE and trade-off analysis is a very useful engineering best practice for the selection of the most suitable technology to be applied to solve a particular problem. Here, MBSE allows us to identify the system level where the application of trade-off should be accomplished. In this way, we identified the system building blocks and which functions should be implemented to solve the problem of a wireless energy transfer and energy harvesting system for a new type of micromotor inside the patient's body.

The results obtained by this trade-off study are consistent with the results obtained by other researchers and published in the literature, but expand and detail them within the context of a real research systems engineering effort with many unknowns because it pushes the limits of present MEMS technologies to sizes of the order of 1 mm^3. The methodological approach we propose here helps to inform better design decisions in the early phases of the project, saving costly redesign efforts in later phases. Moreover, we have detailed the mathematical conditions that the utility functions must have in order to be useful for trade-off studies. Finally, we have shown how the information from the trade-off analysis can be used to make a valuable qualitative sensitivity study of the system.

Author Contributions: Conceptualization, J.A.M.R. and J.L.F.; methodology, J.L.F.; formal analysis, J.A.M.R. and J.L.F.; investigation, J.A.M.R. and P.L.L.E.; resources, J.A.M.R. and R.S.M.; data curation, R.S.M. and P.L.L.E.; writing—original draft preparation, E.D.-J. and J.A.M.R.; visualization, P.L.L.E.; supervision, E.D.-J. and J.L.F.; project administration, E.D.-J.; funding acquisition, E.D.-J. All authors have read and agreed to the published version of the manuscript.

Funding: This work has been conducted for the UWIPOM2 project, which received funding from the European Union's Horizon 2020 research and innovation programme under grant agreement No 857654.

Institutional Review Board Statement: Not applicable.

Informed Consent Statement: Not applicable.

Data Availability Statement: Not applicable.

Acknowledgments: The authors want to recognize the work of Alba Martínez Pérez during preparation of the figures and text edition.

Conflicts of Interest: The authors declare no conflict of interest.

References

1. Kim, W.S.; Jeong, M.; Hong, S.; Lim, B.; Park, S. Il Fully implantable low-power high frequency range optoelectronic devices for dual-channel modulation in the brain. *Sensors* **2020**, *20*, 3639. [CrossRef] [PubMed]
2. Estefan, J. *Survey of Model-Based Systems Engineering (MBSE) Methodologies*; INCOSE-TD-2007-003-01 Version/Revision: B, 10 June 2008; INCOSE: Seattle, WA, USA, 2008.
3. Fernandez, J.L.; Hernandez, C. (Eds.) *Practical Model-Based Systems Engineering*; Artech House: Norwood, MA, USA, 2019; ISBN 978-1-63081-579-0.
4. OMG. Systems Modeling Language. Available online: http://www.omgsysml.org/ (accessed on 30 April 2021).
5. Khan, S.R.; Pavuluri, S.K.; Cummins, G.; Desmulliez, M.P.Y. Wireless power transfer techniques for implantable medical devices: A review. *Sensors* **2020**, *20*, 3489. [CrossRef] [PubMed]
6. Zhou, Y.; Liu, C.; Huang, Y. Wireless Power Transfer for Implanted Medical Application: A Review. *Energies* **2020**, *13*, 2837. [CrossRef]
7. Moore, J.; Castellanos, S.; Xu, S.; Wood, B.; Ren, H.; Tse, Z.T.H. Applications of Wireless Power Transfer in Medicine: State-of-the-Art Reviews. *Ann. Biomed. Eng.* **2019**, *47*, 22–38. [CrossRef] [PubMed]
8. Mahmood, M.F.; Mohammed, S.L.; Gharghan, S.K. Ultrasound sensor-based wireless power transfer for low-power medical devices. *J. Low Power Electron. Appl.* **2019**, *9*, 20. [CrossRef]
9. Kakkar, V. An Ultra Low Power System Architecture for Implantable Medical Devices. *IEEE Access* **2019**, *7*, 111160–111167. [CrossRef]
10. Shadid, R.; Noghanian, S. A Literature Survey on Wireless Power Transfer for Biomedical Devices. *Int. J. Antennas Propag.* **2018**, *2018*, 4382841. [CrossRef]
11. Taalla, R.V.; Arefin, M.S.; Kaynak, A.; Kouzani, A.Z. A review on miniaturized ultrasonic wireless power transfer to implantable medical devices. *IEEE Access* **2019**, *7*, 2092–2106. [CrossRef]
12. Agarwal, K.; Jegadeesan, R.; Guo, Y.X.; Thakor, N.V. Wireless Power Transfer Strategies for Implantable Bioelectronics. *IEEE Rev. Biomed. Eng.* **2017**, *10*, 136–161. [CrossRef] [PubMed]
13. Dinis, H.; Colmiais, I.; Mendes, P.M. Extending the limits of wireless power transfer to miniaturized implantable electronic devices. *Micromachines* **2017**, *8*, 359. [CrossRef] [PubMed]
14. Kim, H.; Member, S.; Hirayama, H.; Zhang, R.U.I.; Choi, J.; Member, S. Review of near-field wireless power and communication for biomedical applications. *IEEE Access* **2017**, *5*, 21264–21285. [CrossRef]
15. Lu, Y.; Ma, D.B. Wireless power transfer system architectures for portable or implantable applications. *Energies* **2016**, *9*, 1087. [CrossRef]
16. Altawy, R.; Youssef, A.M. Security Tradeoffs in Cyber Physical Systems: A Case Study Survey on Implantable Medical Devices. *IEEE Access* **2016**, *4*, 959–979. [CrossRef]
17. Campi, T.; Cruciani, S.; De Santis, V.; Feliziani, M. EMF Safety and Thermal Aspects in a Pacemaker Equipped with a Wireless Power Transfer System Working at Low Frequency. *IEEE Trans. Microw. Theory. Tech.* **2016**, *64*, 375–382. [CrossRef]
18. Liu, C.; Guo, Y.X.; Sun, H.; Xiao, S. Design and safety considerations of an implantable rectenna for far-field wireless power transfer. *IEEE Trans. Antennas Propag.* **2014**, *62*, 5798–5806. [CrossRef]
19. Sun, T.; Xie, X.; Wang, Z. *Wireless Power Transfer for Medical Microsystems*; Springer: Berlin/Heidelberg, Germany, 2013; Volume 9781461477, ISBN 9781461477020.
20. Van Schuylenbergh, K.; Puers, R. *Inductive Powering: Basic Theory and Application to Biomedical Systems*; Springer: Dordrecht, The Netherlands, 2009; ISBN 978-90-481-2412-1.

21. Yilmaz, G.; Dehollain, C. *Wireless Power Transfer and Data Communication for Neural Implants*; Springer International Publishing: Berlin, Germany, 2017; ISBN 978-3-319-49336-7.
22. Zhong, W.; Xu, D.; Hui, R.S.Y. *Wireless Power transfer—Between Distance and Efficiency*; Springer Nature: Cham, Switzerland, 2020; ISBN 9789811524400.
23. Goldberg, B.E.; Everhart, K.; Stevens, R.; Babbitt, N.; Clemens, P.; Stout, L. *System Engineering Toolbox for Design-Oriented Engineers*; NASA Reference Publication 1358; NASA: Huntsville, AL, USA, December 1994.
24. Bleakley, G.; Lapping, A.; Whitfield, A. Determining the right solution using SysML and model based systems engineering (MBSE) for trade studies. In Proceedings of the 21st Annual International Symposium of the International Council on Systems Engineering, INCOSE 2011, Denver, CO, USA, 23 June 2011; Volume 21, pp. 783–795.
25. Fernández, J.L.; Martínez Rojas, J.A.; Díez-Jiménez, E. Modeling the Mission Dimension: The case of an intravascular medical device. In Proceedings of the International Council on Systems Engineering Workshop 2021, INCOSE; 2021. Available online: https://www.omgwiki.org/MBSE/doku.php?id=mbse:incose_mbse_iw_2021 (accessed on 30 April 2021).
26. Parnell, G.S.; Trainor, T.E. Using the swing weight matrix to weight multiple objectives. In Proceedings of the 19th Annual International Symposium of the International Council on Systems Engineering, INCOSE 2009, Singapore, 20 July 2009; Volume 19, pp. 283–298.
27. Smith, E.D.; Young, J.S.; Plattelli-Palmarinl, M.; Bahill, A.T. Ameliorating mental mistakes in tradeoff studies. *Syst. Eng.* **2007**, *10*, 222–240. [CrossRef]

Article

Detecting Axial Ratio of Microwave Field with High Resolution Using NV Centers in Diamond

Cui-Hong Li [1,2,*], **Deng-Feng Li** [3], **Yu Zheng** [3], **Fang-Wen Sun** [3], **A. M. Du** [1,2] and **Ya-Song Ge** [1,2]

[1] Key Laboratory of Earth and Planetary Physics, Institute of Geology and Geophysics, Chinese Academy of Sciences, Beijing 100029, China; amdu@mail.iggcas.ac.cn (A.M.D.); ysge@mail.iggcas.ac.cn (Y.-S.G.)

[2] University of Chinese Academy of Sciences, Beijing 100049, China

[3] CAS Key Lab of Quantum Information, University of Science and Technology of China, Hefei 230026, China; ldf2015@mail.ustc.edu.cn (D.-F.L.); bigz@mail.ustc.edu.cn (Y.Z.); fwsun@ustc.edu.cn (F.-W.S.)

* Correspondence: licuihong14@mails.ucas.ac.cn

Received: 7 April 2019; Accepted: 14 May 2019; Published: 21 May 2019

Abstract: Polarization property characterization of the microwave (MW) field with high speed and resolution is vitally beneficial as the circularly-polarized MW field plays an important role in the development of quantum technologies and satellite communication technologies. In this work, we propose a scheme to detect the axial ratio of the MW field with optical diffraction limit resolution with a nitrogen vacancy (NV) center in diamond. Firstly, the idea of polarization selective detection of the MW magnetic field is carried out using a single NV center implanted in a type-IIa CVD diamond with a confocal microscope system achieving a sensitivity of 1.7 $\mu T/\sqrt{Hz}$. Then, high speed wide-field characterization of the MW magnetic field at the submillimeter scale is realized by combining wide-field microscopy and ensemble NV centers inherent in a general CVD diamond. The precision axial ratio can be detected by measuring the magnitudes of two counter-rotating circularly-polarized MW magnetic fields. The wide-field detection of the axial ratio and strength parameters of microwave fields enables high speed testing of small-scale microwave devices.

Keywords: MW magnetic field; axial ratio; polarization; NV center

---

1. Introduction

Building and measuring circularly-polarized microwave (MW) fields are of great importance in many fields like magnetic resonance [1,2] and mobile satellite communications [3–5]. Conventionally, precision measurement of MW field strength is realized by converting the field strength to easily-measurable electrical signals like using the calorimetric method [6] and the technique of peak demodulation [7]. The antenna axial ratio measurement is conducted with the help of a linearly-polarized antenna [4]. However, the calorimetric method is susceptible to environment temperature, and the technique of peak demodulation does not perform well at weak field strength. The assistant antenna perturbs the properties of the antenna under test (AUT) and usually has a large volume. Quantum-based MW field detection technologies, such as super quantum interference device [8] and cold atoms [9], which offer high sensitivity, require extreme measurement conditions. Vapor cell devices [10,11] that are capable of high sensitivity MW field sensing at near room temperature have a spatial resolution confined to sub-100 μm.

Recently, quantum sensing based on a point defect, the nitrogen vacancy (NV) center, in diamond has attracted wide attention for its long coherence time at room temperature [12]. It has already been deeply researched and applied to magnetic field, electric field, and temperature sensing [13–15] in many subjects. As its electron spin state can be coherently manipulated by the MW field, it can also be applied to MW field sensing. In previous works, the NV-based MW magnetic field sensing schemes mainly focused on detecting MW magnetic field strength [16,17].

In this paper, we propose a scheme to detect the axial ratio of an MW device with high resolution using NV centers in diamond. The axial ratio detection is based on MW magnetic field strength detection of the two counter-rotating circularly-polarized components of an MW field. An NV center is a negatively charged point defect in diamond consisting of a substitutional nitrogen atom and an adjacent carbon vacancy. Its ground electron-spin triplet state can be selectively and coherently manipulated by a specified circularly-polarized MW field. The MW magnetic field strength of specific polarization can be derived from Rabi frequency or magnetic resonance spectra contrast [16,18]. Firstly, the MW axial ratio detection is demonstrated with a single NV center in diamond with a conventional confocal microscope. Then, wide-field detecting of the MW magnetic field, which greatly promotes sensing speed, is conducted with a wide-field microscopy system. The MW axial ratio detection technique facilitates the test and design of MW devices aimed at quantum spin manipulation [19–21] and satellite communication.

2. Principle

The axial ratio is an important parameter to describe a polarized MW field. Broadly, an arbitrarily-polarized MW field can be treated as an elliptically-polarized wave. The arbitrarily- polarized MW field spread along the z direction can be described as a combination of two circularly-polarized microwave fields, $\mathbf{B}_+$ and $\mathbf{B}_-$, spread along the z direction with the same frequency as the source MW field that rotate in different directions in the xoy plane (Figure 1).

$$\mathbf{B} = \mathbf{B}_+ + \mathbf{B}_-, \tag{1}$$

where B_+ (B_-) indicates the magnetic field strength of the σ^+ (σ^-) circularly-polarized field. The axial ratio can be described as the ratio of the magnitudes of the major and minor axes defined by the magnetic field vector. It can be figured out that the length of the major (minor) axis of the ellipse equals $2|B_+ + B_-|$ ($2|B_+ - B_-|$).

$$AR = |\frac{B_+ + B_-}{B_+ - B_-}|. \tag{2}$$

Obviously, $AR = 1$ for a circularly-polarized MW field, and $AR = \infty$ for a linearly-polarized MW field.

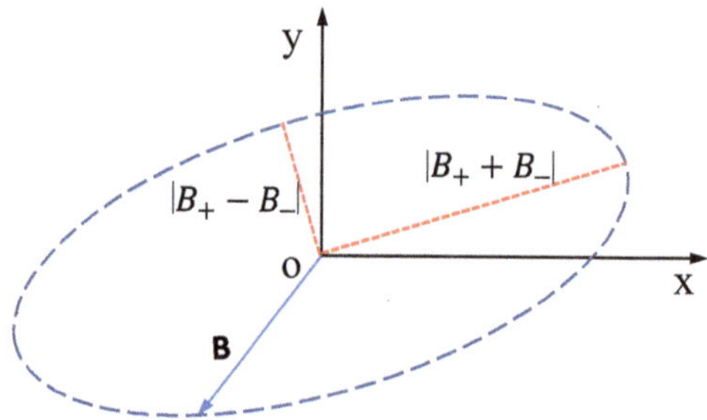

Figure 1. At a fixed point in space (or for fixed z), the magnetic vector **B** of a polarized microwave (MW) field traces out an ellipse in the xoy plane.

Axial ratio detection of MW fields with NV centers in diamond is possible as the strength of two counter-rotating MW magnetic fields can be selectively acquired by a spin-1 system. NV electron spin in its ground state can be optically pumped to the excited state and emit red fluorescence while decaying to the ground spin state.The spin state selective intersystem crossing (ISC) process [22] in the NV center enables the state read out and state initialization of NV electron spin. The NV electron in the $m_s = 0$ state emits more photons then the NV center in the $m_s = \pm 1$ state. Successive illumination initializes the NV center to the $m_s = 0$ state. As shown in Figure 2a, the electron spin ground triplet state of NV centers has an energy splitting of $D = 2.87$ GHz between the $m_s = \pm 1$ state and $m_s = 0$ state under zero magnetic field. The spin transition $m_s = 0 \leftrightarrow m_s = -1(+1)$, which can be simply treated as a two-level system (TLS) and can be selectively driven by $\sigma^-(\sigma^+)$ circularly-polarized microwave field perpendicular to the NV axis [20]. According to the semi-classical theory of light–matter interaction, the transition frequency, i.e., Rabi frequency Ω_R, and the optically-detected signal contrast C_R that corresponds to the electron spin state population of the TLS under a driving MW magnetic field can be written as:

$$\Omega_R = \sqrt{\Delta^2 + \Omega_0^2}, \quad C_R = C_{R0}\frac{\Omega_0^2}{\Omega_0^2 + \Delta^2}, \tag{3}$$

where $\Delta = \omega_0 - \omega_m$ is the frequency difference between the frequency corresponding to the energy separation $\hbar\omega_0$ of the TLS and the frequency of the driving microwave field ω_m. $\Omega_0 = \gamma_e B_{mw}$ is the Rabi frequency at resonance, i.e., $\omega_0 = \omega_m$; $\gamma_e = 2.8$ MHz/G is the gyromagnetic ratio of NV electron spin. B_{mw} is the amplitude of the resonant component of the driving microwave magnetic field. C_{R0} is the maximum signal contrast between the $m_s = 0$ state and $m_s = \pm 1$ state. Thus, the MW magnetic field strength of particular polarization can be deduced from Rabi frequency.

Furthermore, due to the Zeeman effect, the resonance frequency of transition $m_s = 0 \leftrightarrow m_s = \pm 1$ moves to $\omega_\pm = D_{gs} \pm \gamma_e B_z$ under a bias magnetic field B_z along the NV axis (Figure 2a). This enables wide-band detecting of the microwave magnetic field. Simply by turning over the permanent magnet to apply bias magnetic fields in reversed direction, the strength of both σ^- and σ^+ circularly-polarized components of the MW magnetic field at a certain frequency perpendicular to the NV axis that rotate in different directions can be extracted by detecting Rabi frequencies. Therefore, the axial ratio of the MW magnetic field at the location of the NV center perpendicular to the NV axis can be deduced.

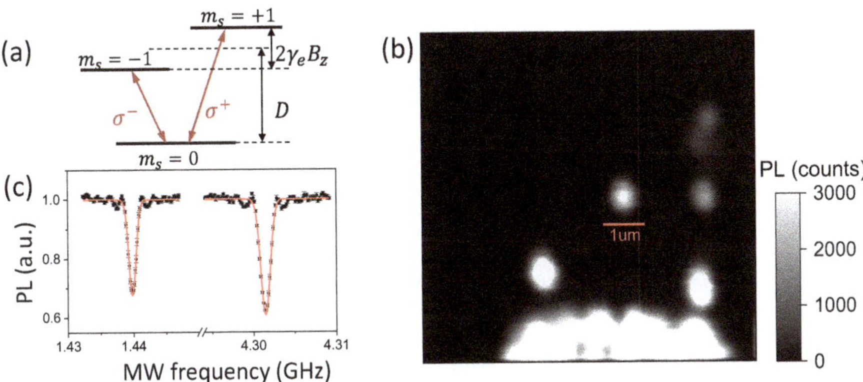

Figure 2. (**a**) The ground electron spin triplet state of the NV center with zero field splitting of $D = 2.87$ GHz between the $m_s = 0$ state and $m_s = \pm 1$ state. A magnetic field of B_z along the direction of the NV axis leads to an energy level splitting of $2\gamma_e B_z$, where $\gamma_e = 2.8$ MHz/G is the gyro-magnetic ratio of NV electron spin. The spin transition $m_s = 0 \leftrightarrow m_s = -1(+1)$ can be selectively addressed by the $\sigma^-(\sigma^+)$ polarized MW field. (**b**) Scanning picture of a diamond sample containing a single NV center. The marked bright spot is the single NV center used for MW field detection in this experiment. The other bright signals below are from ensemble NV centers. (**c**) Optically-detected magnetic resonance spectra of nitrogen vacancy (NV) center under a bias magnetic field of about 500 G. The data are fitted with the Gauss function. The first (second) dip at $f_1 = 1.44$ GHz ($f_2 = 4.30$ GHz) corresponds to the transition $m_s = 0 \leftrightarrow m_s = -1 (+1)$ state.

3. Experiments and Results

3.1. MW Magnetic Field Detection Using a Single NV Center in Diamond

The detection of the MW magnetic field is firstly demonstrated with a single NV center in diamond. The experiment setup was based on a home-built confocal microscope [23]. A 532-nm green laser was used to initialize and detect the NV center. A microscope objective (NA = 0.9, Olympus, Kyoto, Japan) was used to focus the laser and collect fluorescence photons. The diamond sample was a single crystal synthetic (100)-oriented electronic-grade type-IIa diamond with dimension of $2 \times 2 \times 0.5$ mm^3 from the Element Six company. The diamond sample typically had less than a 0.03 ppb NV concentration before implantation. The investigated individual NV center was generated by ^{14}N ion implantation. The implantation dosage was 10^{11} cm^{-2}, and the estimated average depth of NV was about 20 nm. The MW field (R&S SMB 100A signal generator, Munich, Germany) was delivered to the sample via a coplanar waveguide (CPW) fabricated on the top surface of the diamond substrate. The diamond sample was mounted on a NanoCube Piezo System (PI 611.3S, Karlsruhe, Germany) enabling nano-scale scanning. A permanent magnet mounted on a three-axis translation stage applied a bias magnetic field of between zero and roughly 600 G, aligning with the NV axis (z axis). Figure 2b is a fluorescence picture obtained by scanning the diamond sample. The marked bright spot in the picture is a single NV center chosen for MW magnetic field sensing. Second order photon correlation measurement of the bright spot with strong antibunching observed at zero delay $g_2(0) \sim 0.11 < 0.5$ is proof that a single NV center was investigated [24] (data not shown).

To conduct polarization selective MW magnetic field sensing, a bias magnetic field was applied to split the resonant spectra lines corresponding to $m_s = 0 \leftrightarrow m_s = -1$ and $m_s = 0 \leftrightarrow m_s = +1$ transitions. In our experiment, a bias magnetic field of about 500 G was applied. As shown in Figure 2c, the optically-detected magnetic resonance (ODMR) spectra revealed an energy level splitting of 2.86 GHz between the $m_s = -1$ state and $m_s = +1$ state. Although the MW field from the signal generator was linearly polarized, the transition $m_s = 0 \leftrightarrow m_s = -1$ was only sensitive to the σ_- MW field when the MW magnetic field strength was not too strong according to the rotating wave approximation in quantum optics [19]. We implemented the measurement of the strength of the σ^- circularly-polarized component of the MW magnetic field with the NV center by detecting Rabi frequency. The MW field frequency was adjusted in resonance to the $m_s = 0 \leftrightarrow m_s = -1$ transition. The Rabi frequency detection sequence is shown in Figure 3a. A laser pulse was firstly applied to initialize the NV electron to the ground $m_s = 0$ state. Then, an MW manipulation pulse was inserted. Finally, the population of the electron spin state was detected by another laser pulse. To obtain the manipulation frequency of the MW field, the MW pulse varied in time duration. Moreover, a photon count read pulse-2was applied after the NV center was polarized to the $m_s = 0$ state again to provide a reference for drifts of the NV center. Figure 3b shows the Rabi oscillation detected with the MW field source power of about 7.8 mW applied. The oscillation signal was fitted by $1 - C_0 e^{-t/t_0} sin(wt + \phi)$. The manipulation frequency of 0.63 MHz indicated the σ^- polarized MW magnetic field strength of $B_- = 0.22$ G at the NV center spot perpendicular to the NV axis. According to the fitting error of $\delta B_- = 0.0024$ G and detecting time of $t_{meas} = 1000$ s of the Rabi nutation curve, the sensitivity for MW field detection was $\eta = \delta B \sqrt{t_{meas}} = 1.7$ μT/$\sqrt{Hz}$. The Rabi oscillation frequency can also be derived by Fourier transformation (Figure 3c).

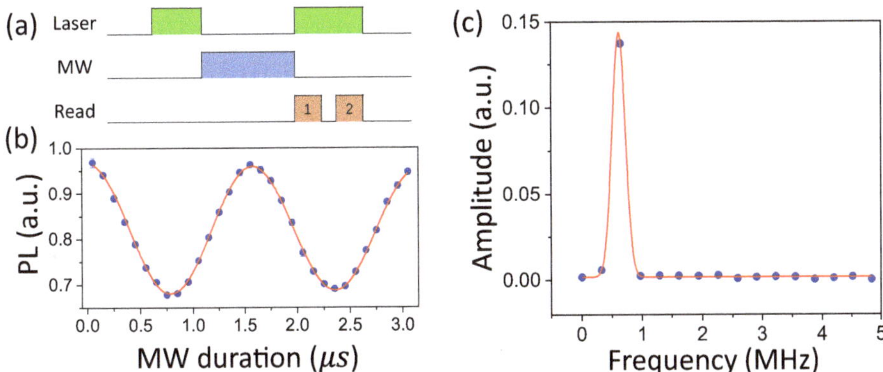

Figure 3. (**a**) Rabi frequency measurement sequence. A first green pulse of 3 µs was applied to initialize the NV electron spin to the $m_s = 0$ state; then, the MW field was applied to interact with the NV electron spin; finally, the electron spin state was determined by simultaneously applying the laser pulse and photon counts read pulse (0.3 µs). To avoid the drift effect of the NV center on determining the NV state, the spin state photon counts were normalized to another photon counts read pulse that was applied in succession to the first one after the NV center was polarized again. (**b**) Rabi oscillation signal driven by a resonant MW field of 1.44 GHz. Each data point on the diagram was obtained by repeating the measurement sequences one million times. The oscillation frequency of 0.63 MHz indicates an MW magnetic field strength of $B_{\sigma^-} = 0.22$ G at the NV spot. The fitting error yields a sensitivity of $1.7\ \mu T/\sqrt{Hz}$. (**c**) Fourier transformation of (**b**).

As $B_{MW} \propto \sqrt{P}$, $\omega_{MW} = \gamma_e B_{MW}$, the Rabi frequency increased linearly with the square root of the resonant MW field power. The linear dependence was verified by detecting Rabi oscillation signals with different MW field powers applied. Figure 4 shows the linear dependence of Rabi frequency on microwave magnetic field strength. This is consistent with the measurement principle.

Turning over the permanent magnet to apply a bias magnetic field in the reverse direction, the σ^+ component of the MW magnetic field perpendicular to the NV axis can be detected similarly. Therefore, the axial ratio of the MW magnetic field at the NV point perpendicular to the NV axis can be derived.

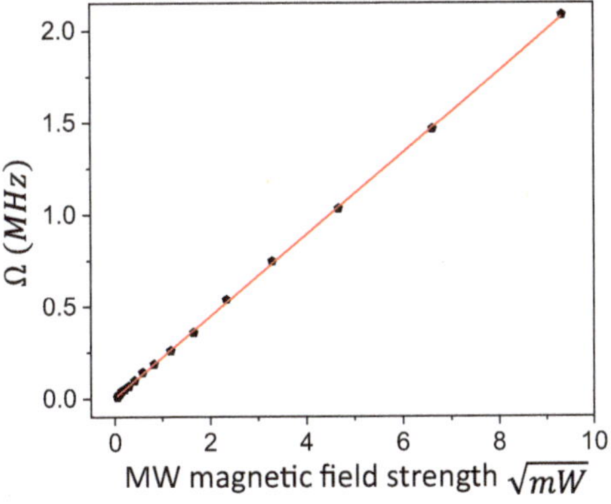

Figure 4. Rabi frequency versus MW source power applied. The data are fitted by a linear function.

3.2. Wide-Field MW Magnetic Field Imaging with Ensemble NV Centers

A single NV center is capable of detecting the strength of the particular circularly-polarized component of the MW magnetic field with high spatial resolution and high sensitivity. Combining the wide-field microscopy system and ensemble NV centers, the detection speed can be crucially improved, and the spatial resolution is only confined to the optical diffraction limit, which is generally at the micron or sub-micron degree. For many practical applications, wide-field microscopy that offers MW field strength imaging at a short time scale and requires simpler experimental technologies could be highly beneficial.

For wide-field detection of the MW magnetic field, the experiment was based on a home-built wide-field back focal imaging microscope system. As shown in Figure 5a, the green laser was focused at the back focal area of the object ($NA = 0.7$) to illuminate the sample with collimated light. The red fluorescence emitted by ensemble NV centers at the focal plane was collected by the same objective and was separated from the source light by a dichroic mirror. Then, the light was filtered and focused on an EMCCD (Andor Ultra iXon 897, Oxford, UK) for imaging. The spatial resolution of the microscope system was about 700 nm, as the wavelength of fluorescence emitted from the NV center was roughly between 600 nm and 750 nm. The sample used in this experiment was a $2.6 \times 2.6 \times 0.3$ mm^3 general single crystal (100)-oriented CVD diamond from Element Six company with inherent ensemble NV centers. The inherent NV concentration in the diamond was estimated to be about 3×10^{14} cm^{-3} (2 ppb) by comparing the fluorescence photon counts of the NV centers in the diamond to that of a single NV center in type-IIa diamond under the confocal microscope system. The fluorescence emitted from the diamond was ascribed to the NV$^-$ center according to its zero photon line at 637 nm detected by the spectrometer FHR 640 (data not shown). A metal micro-strip line with a width of 8 μm was fabricated on the top surface of the diamond to deliver the microwave field.

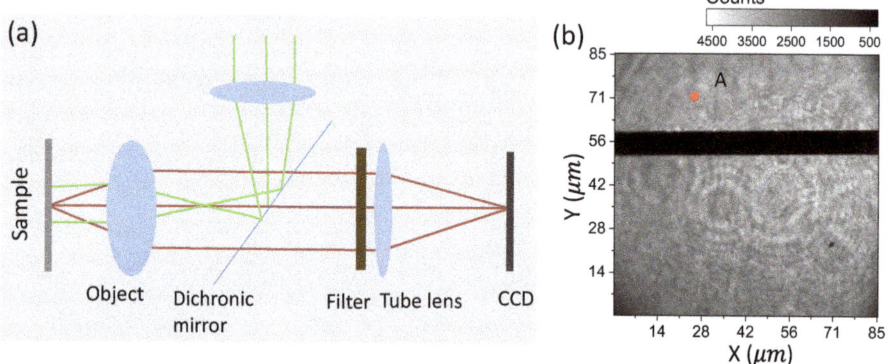

Figure 5. (a) Schematic diagram of wide-field fluorescence microscopy. The green light was focused at the back focal are of the object to excite NV centers with collimated light. The red fluorescence emitted by the NV centers was collected by the same object. The dichroic mirror was used to separate the laser and fluorescence. At last, the fluorescence was further filtered and focused on the EMCCD. (b) Wide-field fluorescence imaging of the diamond sample without the MW field fed in. The dark strip indicates the position of the metal MW strip line.

Figure 5b is a fluorescence picture taken by EMCCD with an exposure time of 2 *s*. The metal strip line on the top surface blocked the fluorescence from the NV center, as shown in Figure 5b. Rough detection of the MW magnetic field was carried out by comparing the signals with and without the resonant MW magnetic field applied [18]. At a zero magnetic bias field, the NV ground electron spin energy splitting was $D_{gs} \sim 2.87$ GHz. Taking pictures of the same region as in Figure 5b with an MW field of $f = 2.87$ GHz applied, the NV fluorescence brightness decreased in the vicinity of the metal strip line (data not shown). The signal contrast decreased with the distance between NV centers and the MW strip line.

Taking pictures of the same region while scanning the MW frequency applied to the strip line, the ODMRsignal of the region can be extracted. To verify the wide-field MW magnetic field sensing method, ODMR signals of Spot A were extracted with the MW magnetic fields of different source powers P transmitted to the MW strip line (Figure 6a). Obviously, the magnetic resonance signal contrast increased with the power of the MW field applied, as the red square in Figure 6b indicates [18]. Thus, simply by tuning the NV electron spin transition frequency in resonance with the MW field, the qualitative MW magnetic field strength of specific polarization and spread direction [19] can be determined from the NV fluorescence signal contrast. Besides, the blue dot in Figure 6b shows that the full width at half maximum (FWHM) of the NV magnetic resonance signal decreased with reduced MW source power till a threshold value. The minimum FWHM of NV ODMR spectra reflects the NV electron spin decoherence property [18]. Moreover, the NV axial directions in the diamond were the same as the four tetrahedral diamond axes [25]. The MW magnetic field spread along each NV axis can be detected. Thus, the axial ratio of an MW field spread along an arbitrary direction can be extracted. Further applying the Rabi measurement sequence in the wide-field microscopy system, precise detection of MW magnetic field strength and the axial ratio is attainable [26,27].

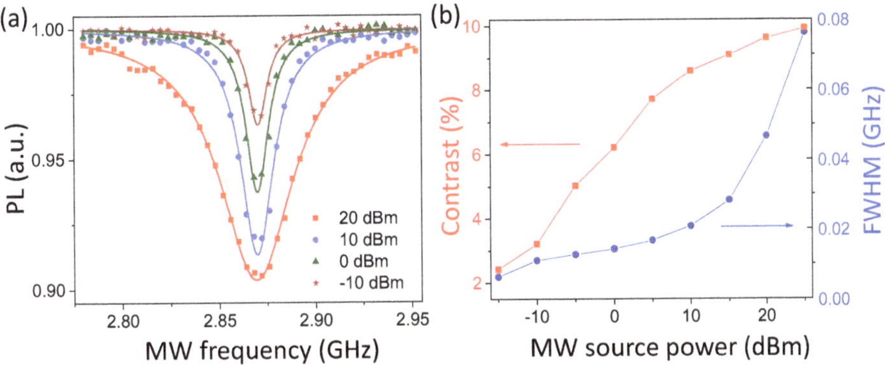

Figure 6. (a) ODMRspectra of Spot A in Figure 5b extracted with four different MW sources' power fed in. The line width and dip contrast increase with the MW source power. (b) ODMR signal contrast (red square) and FWHM (blue dot) versus MW source power extracted from ODMR spectra of Spot A with different MW source powers fed in.

4. Conclusions

In summary, we proposed a new method that can detect the axial ratio of microwave fields with high resolution for wide-field and a short time scale. The measurement method relied on detection of two circularly-polarized MW magnetic fields that rotate in different directions in the plane perpendicular to the NV axis. The axial ratio detecting of the MW magnetic field was firstly demonstrated using a single NV center in diamond, achieving a sensitivity of 1.7 $\mu T/\sqrt{Hz}$. Then, wide-field detection of the MW magnetic field with an imaging field size of $85 \times 85~\mu m^2$ was displayed. The single NV center in the type-IIa CVD diamond investigated was generated by ion implantation, and the ensemble NV centers in the general CVD diamond were inherent. In a future study, we will apply this technique to test the polarization-controllable MW devices that we designed for selective manipulation of NV sublevels.The method we proposed can effectively benefit the design of MW fields with special polarization properties at the sub-millimeter to millimeter scale and thus benefit the development of techniques relying on quantum spin manipulation [28–30] and a small-sized antenna.

Author Contributions: Conceptualization, C.-H.L.; methodology, C.-H.L.; software, Y.Z.; validation, D.-F.L., A.M.D.; formal analysis, C.-H.L.; investigation, C.-H.L.; resources, F.-W.S.; data curation, D.-F.L.; writing—original draft preparation, C.-H.L.; writing—review and editing, C.-H.L., A.M.D.; visualization, C.-H.L.; supervision, F.-W.S., A.M.D.; project administration, Y.-S.G.; funding acquisition, Y.-S.G., A.M.D.

Funding: This research was funded by "Development of High Sensitivity Diamond NV Magnetic Sensor" in Chinese Academy of Sciences Research Equipment Development Project, the Strategic Priority Program in Space

Sensors **2019**, *19*, 2347

Science of CAS grant number XDA1502030105, the Strategic Priority Research Program of the Chinese Academy of Sciences grant number XDA14040204, the National Key Research and Development Program of China grant numbers 2016YFB0501300 and 2016YFB0501304 and the APC was funded by "Development of High Sensitivity Diamond NV Magnetic Sensor" in Chinese Academy of Sciences Research Equipment Development Project.

Conflicts of Interest: The authors declare no conflict of interest.

References

1. Chen, M.; Meng, C.; Zhang, Q.; Duan, C.; Shi, F.; Du, J. Quantum metrology with single spins in diamond under ambient conditions. *Natl. Sci. Rev.* **2017**, *5*, 346–355. [CrossRef]
2. Yasukawa, T.; Sigillito, A.; Rose, B.; Tyryshkin, A.; Lyon, S. Addressing spin transitions on Bi 209 donors in silicon using circularly polarized microwaves. *Phys. Rev. B* **2016**, *93*, 121306. [CrossRef]
3. Iwasaki, H. A circularly polarized small-size microstrip antenna with a cross slot. *IEEE Trans. Antennas Propag.* **1996**, *44*, 1399–1401. [CrossRef]
4. Wang, P.; Wen, G.; Li, J.; Huang, Y.; Yang, L.; Zhang, Q. Wideband circularly polarized UHF RFID reader antenna with high gain and wide axial ratio beamwidths. *Prog. Electromagn. Res.* **2012**, *129*, 365–385. [CrossRef]
5. Li, C.; Zhang, F.S.; Zhang, F.; Yang, K. A Wideband Circularly Polarized Antenna with Wide Beamwidth for GNSS Applications. *Prog. Electromagn. Res.* **2018**, *84*, 189–200. [CrossRef]
6. Anikeev, V. Calculation of calorimetric effect of thermo-field electron emission. *IEEE Trans. Dielec. Electr. Insul.* **1999**, *6*, 426–429. [CrossRef]
7. Vremera, E.T.; Brunetti, L.; Oberto, L.; Sellone, M. Power sensor calibration by implementing true-twin microcalorimeter. *IEEE Trans. Instrum. Meas.* **2011**, *60*, 2335–2340. [CrossRef]
8. Black, R.; Wellstood, F.; Dantsker, E.; Miklich, A.; Koelle, D.; Ludwig, F.; Clarke, J. Imaging radio-frequency fields using a scanning SQUID microscope. *Appl. Phys. Lett.* **1995**, *66*, 1267–1269. [CrossRef]
9. Böhi, P.; Riedel, M.F.; Hänsch, T.W.; Treutlein, P. Imaging of microwave fields using ultracold atoms. *Appl. Phys. Lett.* **2010**, *97*, 051101. [CrossRef]
10. Shi, H.; Ma, J.; Li, X.; Liu, J.; Li, C.; Zhang, S. A Quantum-Based Microwave Magnetic Field Sensor. *Sensors* **2018**, *18*, 3288. [CrossRef]
11. Horsley, A.; Du, G.X.; Treutlein, P. Widefield microwave imaging in alkali vapor cells with sub-100 μm resolution. *New J. Phys.* **2015**, *17*, 112002. [CrossRef]
12. Rondin, L.; Tetienne, J.; Hingant, T.; Roch, J.; Maletinsky, P.; Jacques, V. Magnetometry with nitrogen-vacancy defects in diamond. *Rep. Prog. Phys.* **2014**, *77*, 056503. [CrossRef]
13. Wang, N.; Liu, G.Q.; Leong, W.H.; Zeng, H.; Feng, X.; Li, S.H.; Dolde, F.; Fedder, H.; Wrachtrup, J.; Cui, X.D.; et al. Magnetic Criticality Enhanced Hybrid Nanodiamond Thermometer under Ambient Conditions. *Phys. Rev. X* **2018**, *8*, 011042. [CrossRef]
14. Gross, I.; Akhtar, W.; Garcia, V.; Martínez, L.; Chouaieb, S.; Garcia, K.; Carrétéro, C.; Barthélémy, A.; Appel, P.; Maletinsky, P.; et al. Real-space imaging of non-collinear antiferromagnetic order with a single-spin magnetometer. *Nature* **2017**, *549*, 252–256. [CrossRef]
15. Kost, M.; Cai, J.; Plenio, M.B. Resolving single molecule structures with nitrogen-vacancy centers in diamond. *Sci. Rep.* **2015**, *5*, 11007. [CrossRef]
16. Wang, P.; Yuan, Z.; Huang, P.; Rong, X.; Wang, M.; Xu, X.; Duan, C.; Ju, C.; Shi, F.; Du, J. High-resolution vector microwave magnetometry based on solid-state spins in diamond. *Nat. Commun.* **2015**, *6*, 6631. [CrossRef]
17. Shao, L.; Liu, R.; Zhang, M.; Shneidman, A.V.; Audier, X.; Markham, M.; Dhillon, H.; Twitchen, D.J.; Xiao, Y.F.; Lončar, M. Wide-Field Optical Microscopy of Microwave Fields Using Nitrogen-Vacancy Centers in Diamonds. *Adv. Opt. Mater.* **2016**, *4*, 1075–1080. [CrossRef]
18. Dréau, A.; Lesik, M.; Rondin, L.; Spinicelli, P.; Arcizet, O.; Roch, J.F.; Jacques, V. Avoiding power broadening in optically detected magnetic resonance of single NV defects for enhanced dc magnetic field sensitivity. *Phys. Rev. B* **2011**, *84*, 195204. [CrossRef]
19. London, P.; Balasubramanian, P.; Naydenov, B.; McGuinness, L.; Jelezko, F. Strong driving of a single spin using arbitrarily polarized fields. *Phys. Rev. A* **2014**, *90*, 012302. [CrossRef]

20. Alegre, T.P.M.; Santori, C.; Medeiros-Ribeiro, G.; Beausoleil, R.G. Polarization-selective excitation of nitrogen vacancy centers in diamond. *Phys. Rev. B* **2007**, *76*, 165205. [CrossRef]

21. Herrmann, J.; Appleton, M.A.; Sasaki, K.; Monnai, Y.; Teraji, T.; Itoh, K.M.; Abe, E. Polarization-and frequency-tunable microwave circuit for selective excitation of nitrogen-vacancy spins in diamond. *Appl. Phys. Lett.* **2016**, *109*, 183111. [CrossRef]

22. Goldman, M.L.; Doherty, M.; Sipahigil, A.; Yao, N.Y.; Bennett, S.; Manson, N.; Kubanek, A.; Lukin, M.D. State-selective intersystem crossing in nitrogen-vacancy centers. *Phys. Rev. B* **2015**, *91*, 165201. [CrossRef]

23. Li, C.H.; Dong, Y.; Xu, J.Y.; Li, D.F.; Chen, X.D.; Du, A.; Ge, Y.S.; Guo, G.C.; Sun, F.W. Enhancing the sensitivity of a single electron spin sensor by multi-frequency control. *Appl. Phys. Lett.* **2018**, *113*, 072401. [CrossRef]

24. Sipahigil, A.; Goldman, M.L.; Togan, E.; Chu, Y.; Markham, M.; Twitchen, D.J.; Zibrov, A.S.; Kubanek, A.; Lukin, M.D. Quantum interference of single photons from remote nitrogen-vacancy centers in diamond. *Phys. Rev. Lett.* **2012**, *108*, 143601. [CrossRef]

25. Yahata, K.; Matsuzaki, Y.; Saito, S.; Watanabe, H.; Ishi-Hayase, J. Demonstration of vector magnetic field sensing by simultaneous control of nitrogen-vacancy centers in diamond using multi-frequency microwave pulses. *Appl. Phys. Lett.* **2019**, *114*, 022404. [CrossRef]

26. Horsley, A.; Appel, P.; Wolters, J.; Achard, J.; Tallaire, A.; Maletinsky, P.; Treutlein, P. Microwave Device Characterization Using a Widefield Diamond Microscope. *Phys. Rev. Appl.* **2018**, *10*, 044039. [CrossRef]

27. Horsley, A.; Du, G.X.; Pellaton, M.; Affolderbach, C.; Mileti, G.; Treutlein, P. Imaging of relaxation times and microwave field strength in a microfabricated vapor cell. *Phys. Rev. A* **2013**, *88*, 063407. [CrossRef]

28. Dong, Y.; Zheng, Y.; Li, S.; Li, C.C.; Chen, X.D.; Guo, G.C.; Sun, F.W. Non-Markovianity-assisted high-fidelity Deutsch–Jozsa algorithm in diamond. *NPJ Quantum Inf.* **2018**, *4*, 3. [CrossRef]

29. Jensen, K.; Leefer, N.; Jarmola, A.; Dumeige, Y.; Acosta, V.M.; Kehayias, P.; Patton, B.; Budker, D. Cavity-enhanced room-temperature magnetometry using absorption by nitrogen-vacancy centers in diamond. *Phys. Rev. Lett.* **2014**, *112*, 160802. [CrossRef]

30. Wang, J.; Feng, F.; Zhang, J.; Chen, J.; Zheng, Z.; Guo, L.; Zhang, W.; Song, X.; Guo, G.; Fan, L.; et al. High-sensitivity temperature sensing using an implanted single nitrogen-vacancy center array in diamond. *Phys. Rev. B* **2015**, *91*, 155404. [CrossRef]

Letter

Terahertz Frequency-Scaled Differential Imaging for Sub-6 GHz Vehicular Antenna Signature Analysis

Jose Antonio Solano-Perez [1,*], María-Teresa Martínez-Inglés [2], Jose-Maria Molina-Garcia-Pardo [1], Jordi Romeu [3], Lluis Jofre-Roca [3], Christian Ballesteros-Sánchez [3], José-Víctor Rodríguez [1] and Antonio Mateo-Aroca [4]

[1] Departamento Tecnologías de la Información y las Comunicaciones, Universidad Politécnica de Cartagena, Cartagena, 30202 Murcia, Spain; josemaria.molina@upct.es (J.-M.M.-G.-P.); jvictor.rodriguez@upct.es (J.-V.R.)
[2] Centro Universitario de la Defensa, Universidad Politécnica de Cartagena, Base Aérea de San Javier. Academia General del Aire, 30720 Murcia, Spain; mteresa.martinez@cud.upct.es
[3] CommSenslab, Department of Signal Theory and Communications, School of Telecommunications Engineering Technical University of Catalonia (Universitat Politecnica de Catalunya, UPC) Campus Nord UPC, Edif. D-3 Jordi Girona, 1-3, 08034 Barcelona, Spain; romeu@tsc.upc.edu (J.R.); jofre@tsc.upc.edu (L.J.-R.); christian.ballesteros@tsc.upc.edu (C.B.-S.)
[4] Departamento Automática, Ingeniería Eléctrica y Tecnología Electrónica, Universidad Politécnica de Cartagena, Cartagena, 30202 Murcia, Spain; Antonio.Mateo@upct.es
* Correspondence: ja.solano@upct.es

Received: 25 August 2020; Accepted: 28 September 2020; Published: 2 October 2020

Abstract: The next generation of connected and autonomous vehicles will be equipped with high numbers of antennas operating in a wide frequency range for communications and environment sensing. The study of 3D spatial angular responses and the radiation patterns modified by vehicular structure will allow for better integration of the associated communication and sensing antennas. The use of near-field monostatic focusing, applied with frequency-dimension scale translation and differential imaging, offers a novel imaging application. The objective of this paper is to theoretically and experimentally study the method of obtaining currents produced by an antenna radiating on top of a vehicular platform using differential imaging. The experimental part of the study focuses on measuring a scaled target using an imaging system operating in a terahertz band—from 220 to 330 GHz—that matches a 5G frequency band according to frequency-dimension scale translation. The results show that the induced currents are properly estimated using this methodology, and that the influence of the bandwidth is assessed.

Keywords: frequency-dimension scale; terahertz; measurements; differential imaging

---

1. Introduction

The future generation of connected vehicle, along with the autonomous vehicle it evolves into, will require significantly increasing the number of antennas on its surface operating at different frequency bands from sub-6 GHz to millimeter-wave (mmWave), as effective communication and environment sensing are ensured in this way in respect of other vehicles and different base-stations [1].

The operation of these antennas and a focus on a 3D spatial angular response (3D radiation pattern) may be significantly influenced (perturbed) by the nearby vehicular structure. The task of studying these effects may require ascertaining the distribution of the currents induced by the antenna on the vehicular surface by means of numerical or experimental methods. The process of obtaining these currents for realistic vehicle geometries, especially when using experimental techniques, may require complex and bulky setups.

The new imaging systems operating in terahertz frequencies provide a very interesting tool that is applicable for solving the stated problem in relation to the estimation of the induced currents on the vehicular structure by the antenna. This imaging system provides the reconstructed image and shape obtained via the scattered field of metallic or non-metallic targets [2].

The concept of using terahertz frequencies and imaging techniques has been reviewed in order to confirm its applicability to the present problem. In [3], the ultrawide-band (UWB) imaging radar system used in this paper is tested at mmWave and terahertz bands, validating the spatial resolution on the order of millimeters and its imaging capabilities. The work presented in [4] is an imaging system that inspects a polymer radome in the frequency range of 70–170 GHz, reporting good performance in resolution and proposing to increase the frequency up to 300 GHz for improved resolution. Continuing with frequencies around 300 GHz, the three-dimensional (3D) imaging system presented in [5] is a synthetic aperture radar (SAR) operating at 340 GHz, utilizing the Fourier transform in two dimensions for the reconstruction process. In [6], the operating frequency is increased to 522 GHz. Finally, to conclude the background review, a novel reconstruction process in time domain operating in the microwave and mmWave frequency bands is presented in [7].

Extensive discussion has focused on the structural scattering and the scattering of an antenna's radiation mode [8]. It is intended to reconstruct the currents associated with scattering of the radiation mode of an antenna.

The main objective of this paper is to investigate the distribution of currents induced by the antenna on the vehicular surface by means of differential image reconstruction processes, and using a previously developed multi-frequency system [3] operating in the terahertz frequency band and extending this concept to explore an area. This experimental method is a non-invasive way of obtaining the induced currents without placing the antenna in the supporting structure. A frequency-dimension scale translation is applied to reduce complexity from an experimental point of view. As the defining aspect, a differential imaging concept is exploited to process the images captured during the experiment. In this specific case of measurements, it could be applied to the study and optimization of the co-location of several antenna systems in vehicles. To the authors' best knowledge, the uniqueness of this paper concerns dealing with a differential imaging method to obtain the current distribution without influencing the measure.

As discussed previously, the terahertz imaging radar system makes it possible to obtain the spatial distribution of the reflectivity of an object for the analytical and experimental assessment of induced currents in the vehicle structure. This system uses a double focal system with an independent transmitter antenna and receiver antenna, being connected to both ports of a vector network analyzer (VNA) to obtain the scattering parameters over a defined frequency range [3]. Using frequency-dimension scale translation, and as a proof-of-concept, the behavior of a car with a length of 4.5 m operating at the new 6 GHz (3.5 GHz) 5G frequency band was studied through a 1:43 scale metallic car prototype of 10.5 cm length, measured at the R-band (220 GHz–330 GHz). The 1:43 scale metallic car corresponds with a frequency range from 2.5 to 3.8 GHz. The results are differential measurements calculated from the difference between the scattering image reconstructed by the imaging system with the antenna (ON state or short-circuited monopole antenna) and without the antenna (OFF state or open-circuited monopole antenna). Then, an analysis of the differential image provides key information about the current car surface distribution.

This paper is structured as follows: Section 2 jointly develops the theoretical formulation—related to the total image and differential image reconstruction process with the measurement description. Section 3 includes the total image reconstruction for the different measurement cases. Section 4 shows the results related to differential imaging. Finally, Section 5 presents the conclusions.

2. The Imaging Process

In this section, the two-step process of imaging is formulated, first in terms of total image, and then in a novel differential form, to apply to the reconstruction of the currents created by a radiating antenna on top of a vehicle.

2.1. Formulation of the Total Image Reconstruction Process

The ultrawide-band (UWB) near-field imaging radar system used for the measurement is a flat SAR system composed of a transmitter antenna (*Tx*) and a receiver antenna (*Rx*) with fixed spatial separation, named as *X* offset distance, being a monostatic configuration but with a bi-static angle. The *Tx–Rx* system is fixed and explores the environment by moving the object in an *XY* plane using a positioning table. The system points toward the target located at a distance r_t in the *Z*-axis, as shown in Figure 1.

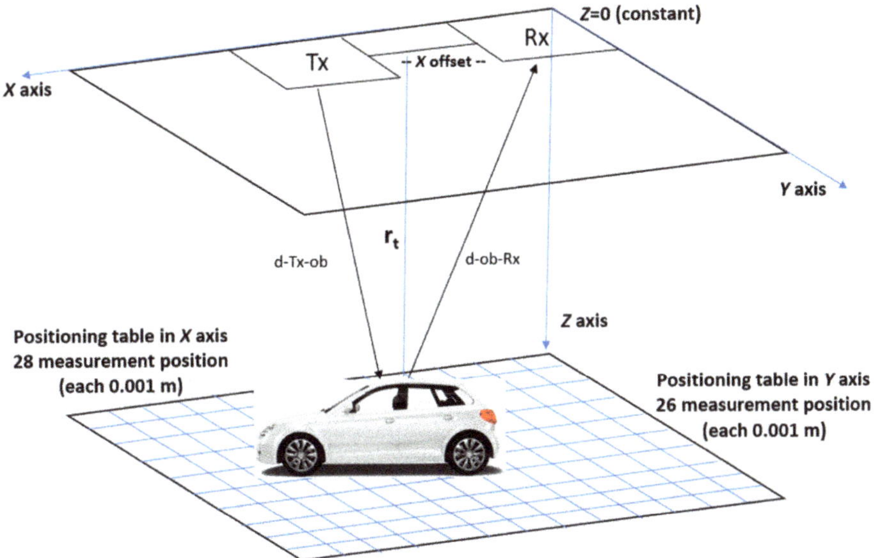

Figure 1. Measurement arrangements to explore the object in an *XY* plane using a positioning table.

Figure 1 shows the measurement arrangements, depicting the *Tx–Rx* and the object.

The objective is to get the spatial distribution image of the electric contrast associated with the object under test. The electric contrast is written as $c(\vec{r}) = \frac{(\varepsilon_t(\vec{r}) - \varepsilon_b)}{\varepsilon_b}$, with $\varepsilon_t(\vec{r})$ being the object's complex dielectric permittivity relative to its environment (background) constant value $\varepsilon_b(\vec{r})$.

The imaging system generates an illuminating field with polarization in the *y*-axis (transversal) that is incident on the object under investigation, represented by a metallic car in Figure 1. $\vec{E}(\vec{r}, f; \vec{r}_{t_i})$ denotes the electric field generated in point $\vec{r}_{t_i}$ by the *Tx* antenna placed at $\vec{r}_t$ and operating at frequency f.

$$\vec{J}_{eq}(\vec{r}, f; \vec{r}_{t_i}) = j\omega\varepsilon_b\, c(\vec{r})\vec{E}(\vec{r}, f; \vec{r}_{t_i}) \tag{1}$$

The equivalent current $\vec{J}_{eq}$ is the scattering field source generated by the object. Then, $\vec{J}_{eq}$ is understood as a "trace" of the original object (image).

Figure 1 shows the *Tx* and *Rx* antennas fixed in position, and the object is moved using *XY* plane geometry located at a distance of $\vec{r}_t$, on a set of positions (N_x, N_y) across *x*-axis and *y*-axis, to gather a 2D array that includes monostatic measurements.

For targets with highly conductive behavior ($\sigma \gg \omega\varepsilon_0\varepsilon_r'$), like the metallic car used in this paper, the equivalent currents tend to produce the conduction current $\vec{J}_{eq} = \sigma\vec{E}$ [9].

The equivalent current $\vec{J}_{eq}\left(\vec{r}_t, f; \vec{r}_{t_i}\right)$ generates a scattered field $(\vec{E}_s)$ in the point $\vec{r}_{t_i}$ of the object, which is measured by the *Rx* antenna at each point of measurement. $\vec{E}_s$ has the following equation:

$$\vec{E}_s\left(\vec{r}_{R_j}, f; \vec{r}_{t_i}\right) = -j\omega\mu_0 \int_{V_0} \vec{J}_{eq}\left(\vec{r}_t, f; \vec{r}_{t_i}\right) G\left(\left|\vec{r}_{R_i} - \vec{r}_t\right|, f\right) dV \tag{2}$$

where ω is the angular frequency, μ_0 is the magnetic permeability in vacuum, $\vec{J}_{eq}$ is the equivalent current, and $G\left(\left|\vec{r}_{R_i} - \vec{r}_t\right|, f\right)$ is the Green's function that matches the geometry of the problem. For a 3D measurement geometry, Green's function is written as $G(r) = \frac{e^{-jk_b r}}{r}$, where $k_b = \omega\sqrt{\mu_0\varepsilon_0\varepsilon_b}$.

For the case where the object is uniformly illuminated using high-directivity antennas, $\vec{E}_s$ is written as:

$$\vec{E}_s\left(\vec{r}_{R_j}, f; \vec{r}_{t_i}\right) = -k_b^2(f)A_d \int_{V_0} c(\vec{r}, f) G\left(\left|\vec{r}_t - \vec{r}_{t_i}\right|, f\right) G\left(\left|\vec{r}_{R_i} - \vec{r}_t\right|, f\right) dV \tag{3}$$

where the parameters of the transmitter and receiver antennas are included in the complex constant named A_d.

The total image reconstruction process introduced by the previous works [3,9–13] is applied, characterized as a bi-focusing technique using multi-frequency. The contrast factor $c(\vec{r})$ is averaged through the complete terahertz band and to the entire reconstruction space, which can be expressed as

$$c(\vec{r}) = A_o \sum_{f_{min}}^{f_{max}} \sum_{i=1}^{N_{T-R}} \frac{\vec{E}_s\left(\vec{r}_{R_j}, f; \vec{r}_{t_i}\right)}{k_b^2(f)} e^{jk_b(f)\left(\left|\vec{r}_t - \vec{r}_{t_i}\right| + \left|\vec{r}_{R_i} - \vec{r}_t\right|\right)} \tag{4}$$

where the parameters of the transmitter and receiver antennas are included in the complex constant called A_o. In order to reduce the computation times, fast Fourier transform (FFT) is used to solve Equation (4).

2.2. Formulation of Differential Image Reconstruction Process

The two-step process, described in the previous paragraph, is composed of a first step—the direct problem of obtaining the scattering fields produced by the external illumination responsible for the creation of the currents into the object—and a second step—an inverse problem focusing back on the previously obtained scattered fields to obtain an image of the currents created by the nearby illuminating geometry.

Here, the goal is to obtain the currents created not by the external illuminating geometry but by the vehicle antenna close to the object, but still using the same external illuminating geometry. These vehicular antenna currents, as described below, should be considered as a differential contribution for two different states: an ON state, or short-circuited monopole antenna, installing monopole in contact with vehicle surface, and an OFF state, or open-circuited monopole antenna, removing the monopole antenna.

To extract the differential impact of in-vehicle antenna radiation, the scattering fields will be collected and the corresponding image of the currents on the surface of the vehicle will be obtained for two successive states of the vehicle antenna: short-circuited and open-circuited [14].

In the first case (the vehicle antenna in the ON, short-circuited state), the total scattered fields $\vec{E}_{ant-on}^{total}$ are produced by the combination of the scattered fields due to the currents created on the surface of the object by the external illuminating geometry $\vec{E}_{ant-on}^{ext-illun}$, the scattered fields due to the structural currents flowing into the vehicle antenna structure without interacting with its port $\vec{E}_{ant-on}^{ant-str}$, and the scattered fields due to the antenna's radiating mode, $\vec{E}_{ant-on}^{ant-rad}$ [15]:

$$\vec{E}_{ant-on}^{total} = \vec{E}_{ant-on}^{ext-illun} + \vec{E}_{ant-on}^{ant-str} + \vec{E}_{ant-on}^{ant-rad} \tag{5}$$

In the second case (vehicle antenna in the OFF, open-circuited state), the total scattered fields $\vec{E}_{ant-off}^{total}$ are produced by the combination of the scattered fields due to the currents created on the surface of the object by the external illuminating geometry $\vec{E}_{ant-off}^{ext-illun}$ and the scattered fields due to the structural currents flowing into the vehicle antenna structure without interacting with its port $\vec{E}_{ant-off}^{ant-str}$.

$$\vec{E}_{ant-off}^{total} = \vec{E}_{ant-off}^{ext-illun} + \vec{E}_{ant-off}^{ant-str} \tag{6}$$

For the compact vehicle antenna, resonant antennas are utilized in most vehicles. Then, it may be considered: (i) that interactions of a high order, such as reflections among external illuminating geometry and vehicle antenna or vehicular platform, may be neglected since they are much smaller than the rest; and (ii) that the scattered fields produced by the unchanged parts, like the vehicular platform and the structure of the antenna, are approximately the same for both states of the vehicle antenna: $\vec{E}_{ant-on}^{ext-illun} \cong \vec{E}_{ant-off}^{ext-illun}$ and $\vec{E}_{ant-on}^{ant-str} \cong \vec{E}_{ant-off}^{ant-str}$.

Based on the previous assumption, the reconstructed currents above the vehicle surface reconstructed from the differential scattered fields $\vec{E}_{ant}^{dif}$ can be established as:

$$\vec{E}_{ant}^{dif} = \vec{E}_{ant-on}^{total} - \vec{E}_{ant-off}^{total} \cong \vec{E}_{ant-on}^{ant-rad} \tag{7}$$

being related to the currents produced by the radiating mode of the vehicle antenna, and Equation (4) becomes:

$$c_{ant}^{dif}(\vec{r}) = f\left(\vec{E}_{ant}^{dif}\right) \cong c_{ant-on}^{total}(\vec{r}) - c_{ant-off}^{total}(\vec{r}) \tag{8}$$

2.3. Measurement Description and Configuration

The measurement device was a vector network analyzer (VNA) (Rohde Schwarz model ZVA 67). One VNA port was connected to the frequency converter and the Tx antenna in which the converter was mounted. In the same way, the other VNA port was connected to the frequency converter and the Rx antenna was mounted in this converter as well. The frequency converters allow us to extend the VNA operating frequency band up to 220–330 GHz [16]. The setup configuration developed in [3] measured the S_{21} parameter (scattering parameter) in the terahertz band to reconstruct the image using a UWB near-field multi-frequency bi-focusing algorithm.

The geometry for the measurement depicted in Figure 1 is used to estimate the distribution of the induced currents on a metal car of 10 cm length—using a terahertz band for imaging—through frequency-dimension scale translation to apply the novel concept of differential imaging.

The XY planar measurement geometry is used for illuminating the metallic vehicle using the ultrawide-band (UWB) imaging radar system [3]. This illumination provides optimization of the induced currents on the vehicular surface, although the vertical monopole antenna is installed perpendicular but connected to the vehicle surface. One of the reasons for using a monopole antenna oriented in the direction of the Tx/Rx antennas is because, even when the excitation of the monopole mode may be reduced, its mutual coupling with the Tx/Rx antennas is also reduced and so the multiple reflections between them may certainly be neglected and, at the same time, the surface currents on top of the vehicle due to the antenna should be visible—especially when the *Tx/Rx* antennas move out of the perpendicular direction.

The object is placed on a radiofrequency-absorbing material located 0.5 m from the *Tx/Rx* antennas. The separation between the *Tx* antenna with a frequency converter and the *Rx* antenna with a frequency converter is 0.1 m, installed in a fixed position in order to avoid any perturbation in the measurement associated with coaxial cables and converter movements. The object under measurement is placed on a positioning table that is capable of moving in two axes (*x*-axis and *y*-axis) with steps of 1 mm [16]. Figure 2 displays the frequency converter with the horn antenna mounted and the positioning table with the metallic car.

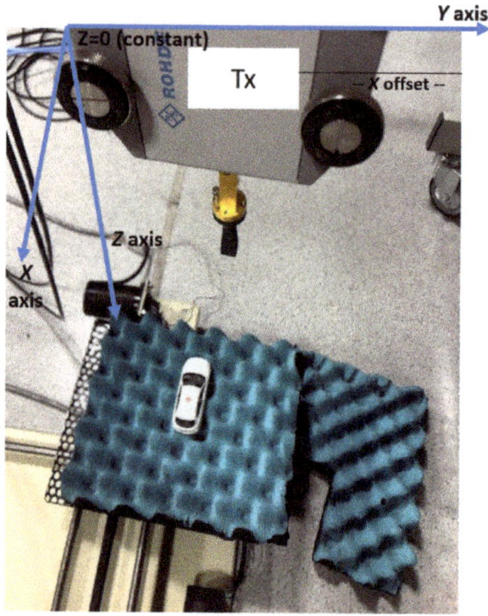

Figure 2. Measurement setup.

The S21 parameter measurements were carried out by sweeping the frequency band from 220 GHz to 330 GHz. The two-frequency converter equipment was manufactured by Rohde & Schwarz (Munich, Germany)®, with the model being the ZVA-Z325 Frequency Converter, J-Band WR-03 [17]. Table 1 includes the VNA configuration parameters.

Table 1. Vector network analyzer (VNA) configuration parameters.

Parameter	Value
Start frequency	220 GHz
End frequency	330 GHz
Sampling points	8192
Intermediate frequency (IF) bandwidth	1 kHz
Emitting power	0 dBm

The *Tx* and *Rx* antennas were horns produced by Flann Microwave LTD (Cornwall, United Kingdom) [18]. The selected horn model was the 32240-25, with a bandwidth (BW) from 217 GHz to 330 GHz. The *E* and *H* plane gain was 23.70 dBi at 217 GHz and 26.99 dBi at 330 GHz. The E and H plane beamwidth was 11.9° at 217 GHz and 7.8° at 330 GHz.

The selected object is a metallic car with a 1:43 scale and 10 cm length. This is a good model for a real car, in terms of both shape and materials. The scale of the car corresponds with the frequency range between 2.5 and 3.8 GHz. A short 15 mm metal wire was placed on top of the car to act as an antenna, as shown in Figure 3. The measurement was performed with (ON state) and without the antenna (OFF state) mounted in the metallic car with the aim of estimating the induced current on the car after using the imaging system to illuminate.

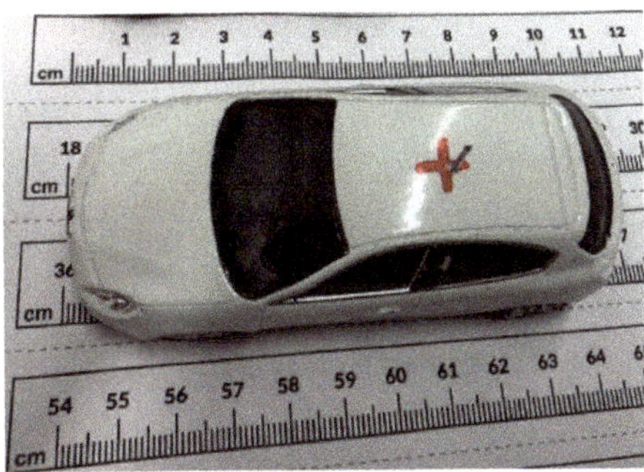

Figure 3. A metallic car with an antenna for differential imaging.

3. Results

In the first step, the total images were obtained based on the two-step process by measuring the scattered fields first and then obtaining the focused field by application of Equation (4) for different frequency ranges and focusing the car with (ON state, short-circuited monopole antenna) and without the antenna (OFF state, open-circuited monopole antenna), as depicted in Figures 4–7.

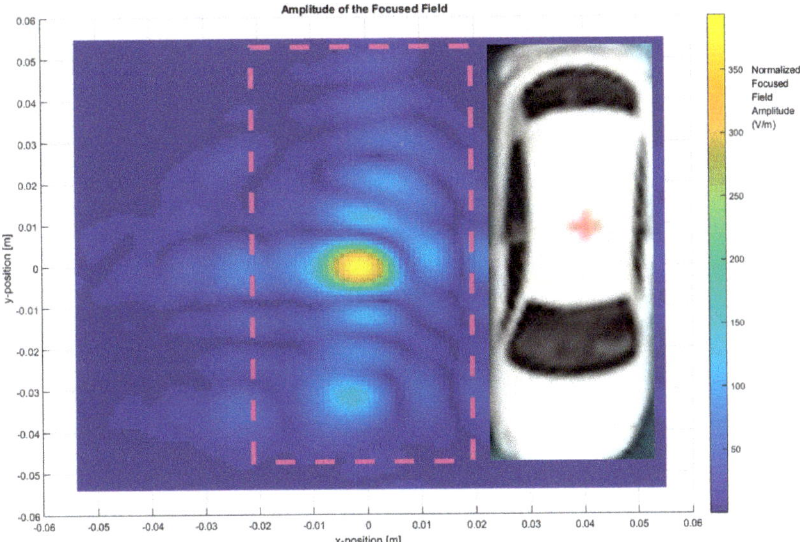

Figure 4. Focused field amplitude of the metallic car without an antenna. Frequency range: 295 GHz–305 GHz (bandwidth (BW) = 10 GHz).

The bandwidth used to reconstruct the image is an important imaging parameter. The first image is shown in Figure 4, which depicts the total image reconstruction without an antenna on top of the metallic car with a frequency of 295 GHz–305 GHz. The top of the metallic car has a curvature that generates a specific focused field distribution.

Then, Figure 5 presents the total image reconstructed without an antenna using a bandwidth of 110 GHz, sweeping from 220 GHz to 330 GHz. This bandwidth provides better resolution performances, meaning the top of the metallic car can be identified in the image as a very clear reconstruction. Additionally, Figure 5b highlights the focused field amplitude value and the distribution due to the reflection in the top of the car, including some minor contributions around the top of the car.

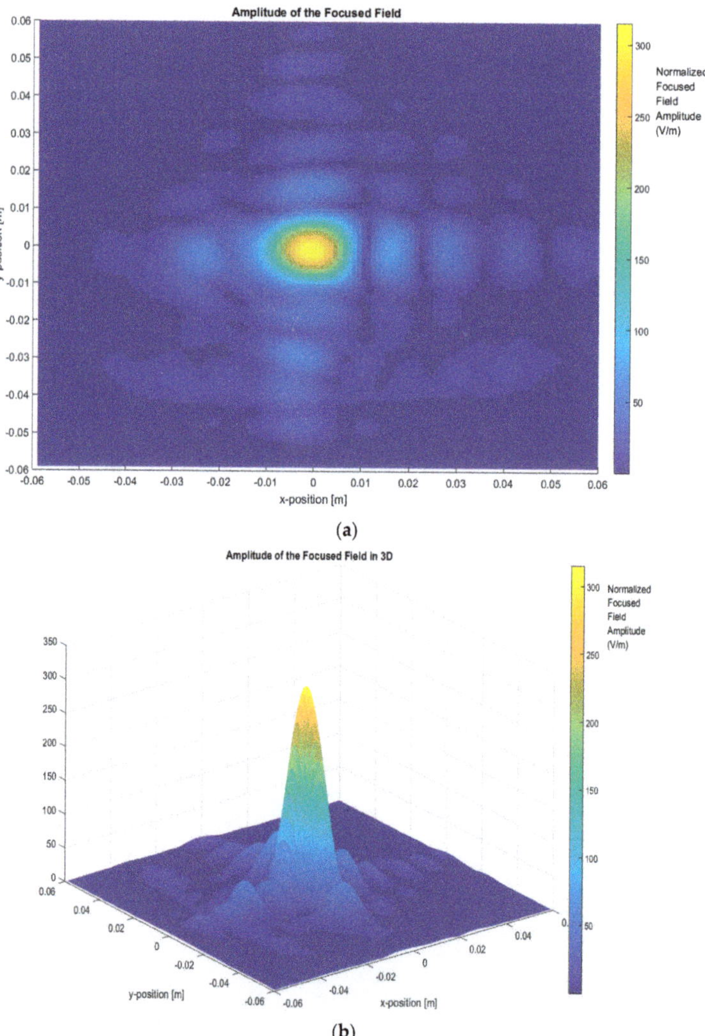

Figure 5. The focused field amplitude of a metallic car without an antenna. Frequency range: 220 GHz– 330 GHz (BW = 110 GHz): (**a**) *XY* plane representation; (**b**) 3D representation of the focused field.

Figure 6 shows the total image reconstructed with the antenna installed on the top and using a bandwidth of 110 GHz, from 220 GHz to 330 GHz. The maximum of the focused fields is identified around the center of the top of the metallic car, where the antenna is installed. The antenna provides reflectivity, but it is masked with the reflectivity of the metallic car top in the graph.

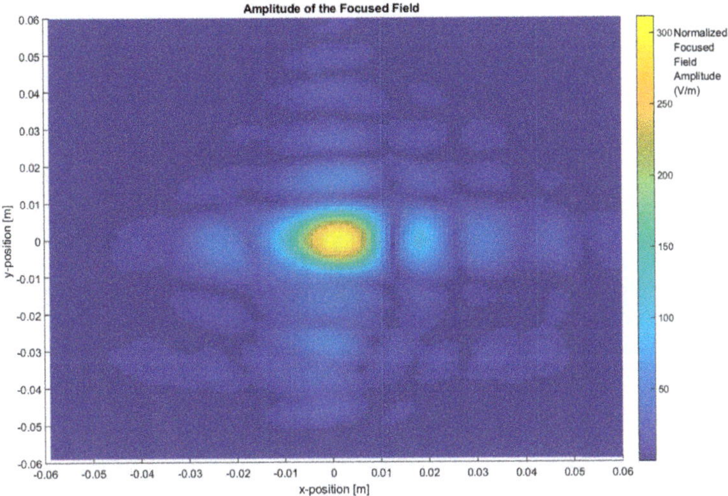

Figure 6. The focused field amplitude of a metallic car with an antenna. Frequency range: 220 GHz–330 GHz (BW = 110 GHz).

Finally, Figure 7 depicts the total image with the antenna installed on the top, using a bandwidth of 10 GHz around 300 GHz. The shape of the focused field is like that in Figure 4 due to the antenna contribution being added to the reflectivity of the metallic car top.

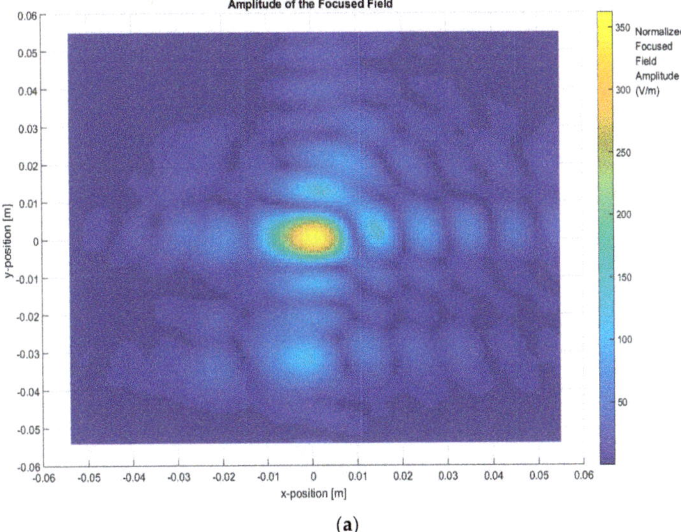

(a)

Figure 7. *Cont.*

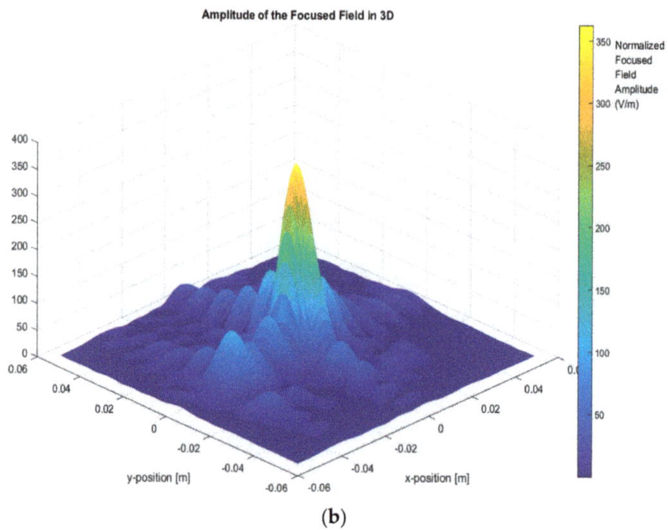

(**b**)

Figure 7. The focused field amplitude of a metallic car with an antenna. Frequency range: 295 GHz–305 GHz (BW = 10 GHz). (**a**) *XY* plane representation; (**b**) 3D representation of the focused field.

4. Discussion

The measured fields are focused on different image planes along the z-axis (z = 0.4787 m) where the capability of the system for focusing on different longitudinal distances (along the z-axis) and different frequency ranges and bandwidths, over different transversal planes (*XY* planes), may be observed. The experimental results show that using a wide bandwidth (110 GHz) is not required in order to obtain a good performance. The performance was of sufficient quality using 10 GHz bandwidth around 300 GHz. The resolutions would be:

$$\Delta z \sim 1/\Delta B \tag{9}$$

$$\Delta x, \Delta y \sim \lambda_0/(2 \sin \theta_0) \tag{10}$$

$$\tan \theta_0 \sim X_0/Z_0 \ or \ Y_0/Z_0 \tag{11}$$

where ΔB is the frequency bandwidth of the measurement, λ_0 is the central frequency wavelength, θ_0 is the target angle visualization seen from the measurement distance, and Z_0 and X_0, are the scanning distances along the x-axis and y-axis, respectively.

Differential imaging is obtained by means of calculating the difference of the experimental data (total image obtained from focused field amplitude) measured using the car, including with a monopole antenna short-circuited (ON state) and without the antenna (OFF state or open-circuited antenna). As stated in Section 2, the short-circuited antenna represents the antenna placed on the top of the car, which is in contact with the car's metal surface.

The objective is to obtain the currents created by the antenna over the surface of the vehicle without getting dazzled ("flash-out") by the higher intensity of the currents on the antenna itself. A differential image may be produced due to the difference in module between the focused field amplitude images with the antenna (see Figures 6 and 7) and without the antenna (see Figures 4 and 5), resulting in the current distribution of the antenna that is shown in Figures 8 and 9.

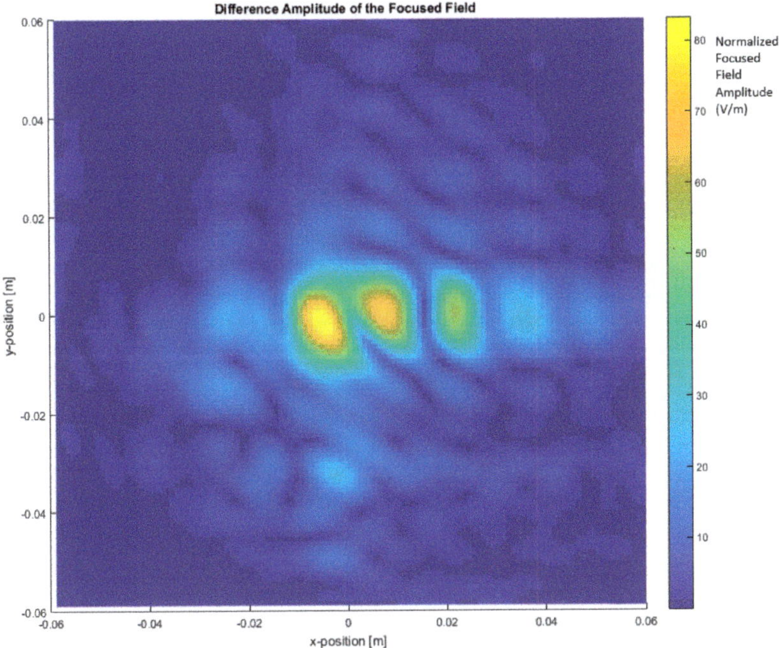

Figure 8. Differential focused field amplitude between the metallic car with and without an antenna. Frequency range: 220 GHz to 330 GHz (BW = 110 GHz).

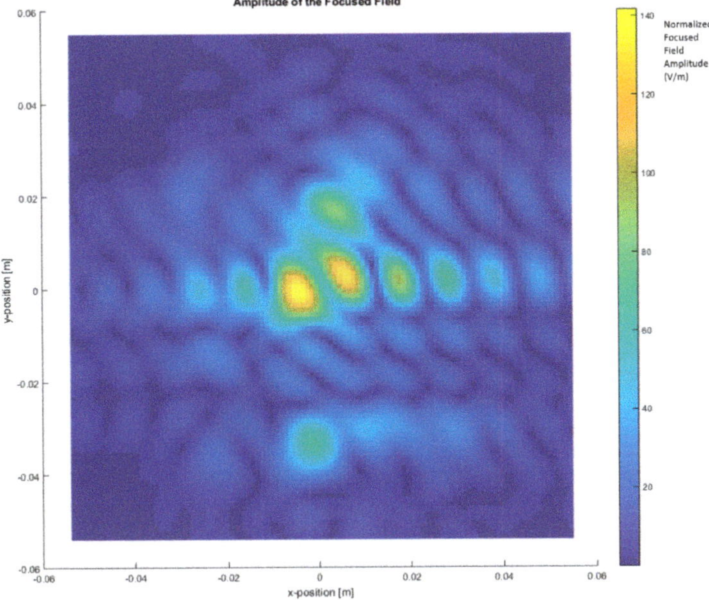

Figure 9. Differential focused field amplitude between the metallic car with and without an antenna. Frequency range: 295 GHz to 305 GHz (BW = 10 GHz).

In Figure 8, it can be identified that the strong central spot corresponding to the antenna has disappeared and, instead, the nearby currents to the antenna are visualized. The maximum can be clearly identified.

Figure 9 depicts the current distribution calculated as a differential focused field amplitude using a frequency range from 295 GHz to 305 GHz (BW = 10 GHz) to assess influence of bandwidth in differential imaging.

The differential focused field amplitude is higher with a bandwidth of 10 GHz around 300 GHz compared with a BW of 110 GHz from 220 GHz to 330 GHz. This effect is due to the gain variation of the horns across the frequency range. The imaging system performs an averaging through all the bandwidths, and the horn antenna gain variation generates this change in the amplitude, with higher amplitude when the BW is lower. However, the image resolution is better when the BW is greater [3]. If the application requires high resolution, then the BW parameter should be maximized.

5. Conclusions

In this work, an application of the approach in the near-field exploration of metallic objects explained in [3] is used to perform the focusing of a metallic car with and without a monopole antenna in order to estimate the induced currents on the vehicle surface using a differential methodology. This method makes it possible to estimate the distribution of the currents induced by the antenna on the vehicle's surface using a very simple method, which is also combined with the frequency-dimension scale translation.

The technique presented in [3] has been extended to a different measurement scenario in a 3D scan in the terahertz band and by using the differential current concept as both a novel form and a major contribution.

The findings verify that, while distance discrimination increased with the bandwidth, the transverse resolution (XY plane) did not change significantly.

The results show promising performance from this technique using differential currents, as depicted in the 300 GHz band. Gaining knowledge of these vehicle-antenna-created currents may offer hints regarding the impact produced by the shapes (borders and corners), discontinuities (slots and windows), or materials (composition or roughness) introduced into the vehicular platform.

Author Contributions: Conceptualization, J.A.S.-P., J.-M.M.-G.-P., J.R., and L.J.-R.; Methodology, J.A.S.-P. and J.-M.M.-G.-P.; Software, J.A.S.-P. and J.-M.M.-G.-P.; Validation, J.A.S.-P., J.-M.M.-G.-P., J.R., and L.J.-R.; Resources, J.-M.M.-G.-P. and M.-T.M.-I.; Data curation, M.-T.M.-I.; Writing—Original Draft Preparation, J.A.S.-P.; Writing—Review and Editing, J.A.S.-P., J.-M.M.-G.-P., J.R., L.J.-R., C.B.-S., and J.-V.R.; Supervision, J.-M.M.-G.-P., J.R., and L.J.-R., M.-T.M.-I., J.-V.R., and A.M.-A.; Project Administration, J.-M.M.-G.-P. and J.-V.R. All authors have read and agreed to the published version of the manuscript.

Funding: This research was funded by the Ministry of Education and Science, Spain (PID2019-107885GB-C33/ AEI/10.13039/501100011033, TEC2016-78028-C3-1-P, TEC2016-78028-C3-2-P and TEC2016-78028-C3-3-P, MDM2016-O6OO), Catalan Research Group 2017 SGR 219, and the European FEDER funds.

Conflicts of Interest: The authors declare no conflict of interest. The funders had no role in the design of the study; in the collection, analyses, or interpretation of data; in the writing of the manuscript; or in the decision to publish the results.

References

1. Chattopadhyay, X.G. Technology, capabilities, and performance of low power terahertz sources. *IEEE Trans. Terahertz Sci. Technol.* **2011**, *1*, 33–53. [CrossRef]
2. Siegel, P. Terahertz technology. *IEEE Trans. Microw. Theory Technol.* **2002**, *50*, 910–928. [CrossRef]
3. Solano-Perez, J.A.; Martínez-Inglés, M.-T.; Molina-Garcia-Pardo, J.-M.; Romeu, J.; Jofre, L.; Rodríguez, J.-V.; Mateo-Aroca, A. Linear and Circular UWB Millimeter-Wave and Terahertz Monostatic Near-Field Synthetic Aperture Imaging. *Sensors* **2020**, *20*, 1544. [CrossRef] [PubMed]
4. Friederich, F.; May, K.H.; Baccouche, B.; Matheis, C.; Bauer, M.; Jonuscheit, J.; Moor, M.; Denman, D.; Bramble, J.; Savage, N. Terahertz Radome Inspection. *Photonics* **2018**, *5*, 1. [CrossRef]

5.	Hao, J.; Li, J.; Pi, Y. Three-Dimensional Imaging of Terahertz Circular SAR with Sparse Linear Array. *Sensors* **2018**, *18*, 2477. [CrossRef] [PubMed]

6.	Dobroiu, A.; Wakasugi, R.; Shirakawa, Y.; Suzuki, S.; Asada, M. Absolute and Precise Terahertz-Wave Radar Based on an Amplitude-Modulated Resonant-Tunneling-Diode Oscillator. *Photonics* **2018**, *5*, 52. [CrossRef]

7.	Fromenteze, T.; Decroze, C.; Abid, S.; Yurduseven, O. Sparsity-Driven Reconstruction Technique for Microwave/Millimeter-Wave Computational Imaging. *Sensors* **2018**, *18*, 1536. [CrossRef] [PubMed]

8.	King, D.D. Measurement and interpretation of antenna scattering. *IEEE Proc. IRE* **1949**, *37*, 770–777. [CrossRef]

9.	Lin, D.B.; Chu, T.H. Bistatic frequency-swept microwave imaging: Principle, methodology, and experimental results. *IEEE Trans. Microw. Theory Technol.* **1993**, *41*, 855–861. [CrossRef]

10.	Jofre, L.; Broquetas, A.; Romeu, J. UWB Tomographic Radar Imaging of Penetrable and Impenetrable Objects. *Proc. IEEE* **2009**, *97*, 451–464. [CrossRef]

11.	Alvarez, Y.; Rodriguez-Vaqueiro, Y.; Gonzalez-Valdes, B.; Mantzavinos, S.; Rappaport, C.M.; Las-Heras, F.; Martínez-Lorenzo, J.Á. Fourier-based Imaging for Multi-static Radar System. *IEEE Trans. Microw. Theory Technol.* **2014**, *62*, 1798–1810. [CrossRef]

12.	Kim, Y.J.; Jofre, L.; de Flaviis, F.; Feng, M.Q. Microwave Reflection Tomographic Array for Damage Detection of Civi Structures. *IEEE Trans. Antennas Propag.* **2003**, *51*, 3022–3032.

13.	Broquetas, A.; Palau, J.; Jofre, L.; Cardama, A. Spherical Wave Near-Field Imaging and Radar Cross-Section Measurement. *IEEE Trans. Antennas Propag.* **1998**, *46*, 730–735. [CrossRef]

14.	Bouazza, B. Millimeter-Wave MIMO Close Range Imaging. Master's Thesis, Universitat Politecnica de Catalunya, Barcelona Spain, July 2020.

15.	Capdevila, S.; Jofre, L.; Romeu, J.; Bolomey, J.C. Multi-Loaded Modulated Scatter Technique for Sensing Applications. *IEEE Trans. Instrum. Meas.* **2013**, *62*, 794–805. [CrossRef]

16.	Arrick Robotics—Stepper Motor, Positioning, Automation, Mobile Robots, Resources. Available online: http://www.arrickrobotics.com/ (accessed on 1 December 2019).

17.	Rohde & Schwarz GmbH & Co. KG. Available online: https://www.rohde-schwarz.com (accessed on 10 June 2020).

18.	Homepage Flann—Flann Microwave. Available online: http://flann.com/ (accessed on 10 June 2020).

MDPI

St. Alban-Anlage 66

4052 Basel

Switzerland

Tel. +41 61 683 77 34

Fax +41 61 302 89 18

www.mdpi.com

Sensors Editorial Office

E-mail: sensors@mdpi.com

www.mdpi.com/journal/sensors